Interfacial Instability

Springer
New York
Berlin
Heidelberg
Hong Kong
London
Milan
Paris
Tokyo

L.E. Johns   R. Narayanan

Interfacial Instability

With 61 Figures

 Springer

L.E. Johns
Department of Chemical Engineering
University of Florida
Gainesville, FL 32611
USA
johns@che.ufl.edu

R. Narayanan
Department of Chemical Engineering
University of Florida
Gainesville, FL 32611
USA
ranga@che.ufl.edu

Library of Congress Cataloging-in-Publication Data
Johns, L.E. (Lewis E.)
    Interfacial instability / L.E. Johns, R. Narayanan.
      p. cm.
    Includes bibliographical references and index.
    ISBN 0-387-95480-5 (alk. paper)
      1. Surface tension.   2. Jets—Fluid dynamics.   3. Stability.   4. Interfaces (Physical
sciences)   I. Narayanan, Ranga.   II. Title.
    QC183 .J595 2002
    541.3'3—dc21   2002020935

ISBN 0-387-95480-5   Printed on acid-free paper.

Printed in the United States of America.

9 8 7 6 5 4 3 2 1   SPIN 10874281

Typesetting: Pages created by the authors using a Springer TEX macro package.

www.springer-ny.com

Springer-Verlag   New York Berlin Heidelberg
A member of BertelsmannSpringer Science+Business Media GmbH

To Ruth and Vasudha

Preface: A Guide for the Reader

This book presents a set of essays on interfacial instability, drawing attention to the surface tension acting at an interface between two phases. The stability of a liquid jet is the model problem. In this problem, surface tension works on the pressures of the fluids, and thence on their velocities. In other problems, it works on gradients of temperature and concentration. It affects the freezing point in solidification, it affects the solubility in precipitation and it affects the rate of electron transfer in electrodeposition.

The essays are nine in number and there are three appendices. A set of derivations is collected at the end of many of the essays to serve as a basis for discussion. These are called discussion notes to set them apart from endnotes. The endnotes ordinarily explain background information which the reader may need in order to enjoy an essay, however, on one or two occasions an endnote presents the derivation of a formula needed in the essay.

The subject matter is tied together by two or three common themes.

Each essay presents a problem where the domain on which the problem is defined is not fixed in advance. Determining the domain is then part of solving the problem. This is one theme and it calls for some explanation. This is given in the second essay, which presents a way in which a domain perturbation can be carried out. As our plan is to project the nonlinear equations onto a reference domain, it is necessary to explain how this can be done. The equations that hold at a deflecting surface present the main problem and, although intuition might lead to the correct perturbation equations in simple cases, in other cases a definite method might be welcome. Appendix A presents several short, but interesting, illustrations of the ideas worked out in the second essay.

Most problems start with a base state upon which a perturbation is imposed. The base state will be a solution to the nonlinear equations and it will depend on the values to which its input or control variables are set. The base state may be stable or it may be unstable to the perturbation. Determining the critical values of the input variables at which the base state loses stability is a second theme.

This brings us to the aims of the essays and to another important theme. The essays were written to fulfill two needs at one and the same time. They can be used by teachers who would like to add lectures on interfacial instabilities to their courses and they can be used by graduate students trying to bring themselves to the point of doing their research on interfacial problems. To achieve both aims, we have selected problems that can be solved by hand, using only pencil and paper, or chalk and blackboard. This is the third main theme. Little numerical work is required and it follows at once that no nonlinear calculations are going to be presented. A lot of physics can be learned by doing the linear problem first, physics that can help direct a nonlinear calculation.

The simpler the base solution, the further we can go using only a pencil and a piece of paper. This guides the selection of the topics presented in this book. But, by limiting ourselves to pencil-and-paper work, we are not limited to the same problems that have been presented to students for many years. The fifth, seventh, eighth and ninth essays are largely unfamiliar and so, of course, is the second essay. The first, third, fourth and sixth essays may be familiar, but that is not to say that no new results can be found therein. It is surprising just how much turns up in the course of reworking the classical problems.

Now, let us indicate the way in which the essays fit together.

The first four essays draw attention to the breakup of a cylindrical thread of liquid, called a jet. This is addressed first by way of Rayleigh's work principle and then by way of a perturbation calculation. Rayleigh's work principle is a useful rule of thumb, and we try to use it as much as possible throughout the essays. But forces cause instabilities, and forces are what the perturbation equations are all about. As a deflecting surface calls into question the domain on which the fluid lies, writing the perturbation equations draws on the material presented in the second essay.

The first and third essays explain the problem in the case of an infinitely long jet. Then, in the fourth essay, the jet is made of finite length in order to introduce a control variable. This brings in the various ways in which a jet can meet its end walls and introduces their influence on the question of its stability. The finite jet calculation retains the flavor of the infinite jet in only one case. Save this, the fundamental modes of displacement do not have discernible wavelengths.

Two end conditions, free and fixed, are of interest. By free conditions, contact at right angles is intended. The other case is called fixed. This end condition, which can be realized quite easily in an experiment, precludes symmetric eigenfunctions. At the same time it eliminates contact line motion, which presents something of a technical difficulty if the no-slip condition must be taken into account.

At first, the fluids are take to be inviscid. This simplifies our calculations and it determines the critical point, the point of zero growth, entirely correctly. This is just as one might guess. Critical points are determined by a balance of forces. If viscosity does not enter into this balance, then critical points can be determined as though the fluids were inviscid. Still, the viscosity of the fluids is not entirely neglected. Its role is explained, first by way of a picture, and then by way of an energy calculation.

Now, something a little odd turns up in the three essays on jets and bridges and asserts itself as a theme well into the sixth essay. It is this: Maintaining the volume of a jet fixed, as its surface is displaced, turns out to be crucial in using Rayleigh's work principle, however, on carrying out the perturbation calculation, it is, at first, of little consequence. It is not until the jet is made finite that holding its volume fixed leads to an important perturbation equation, but even then, it is not important for all end conditions.

In the fifth essay, our attention turns to the effect of rotation in an effort to see if Rayleigh's criterion for detecting an adverse stratification of angular momentum carries the same force, when a liquid cylinder is confined by its own surface tension, as it does when it is confined by a rigid wall.

The sixth essay, then, brings to a close the sequence of essays wherein surface tension plays its role through pressure differences across surfaces and the effect of pressure on fluid motions. This essay deals with the stability of a heavy fluid lying over a light fluid, and it adds the effect of gravity to the effect of surface tension. Due to this, the base solution turns out to be a little more interesting, yet the problem is very much like the jet problem, and all the questions raised and answered there are again raised and answered in this essay.

In the seventh essay, the stability of freezing and melting fronts is taken up; in the eighth essay, attention is directed to the stability of the front dividing a solid and the solution into which the solid is dissolving or out of which it is precipitating.

These two problems are very much alike, but one interesting difference is discovered and it is tied up in the fact that in both cases surface tension, although it is ordinarily stabilizing, can make a destabilizing contribution to the shape of the front. These two essays continue the theme of displaced surfaces, and the role of surface tension in determining their stability. They lead into the ninth essay, where two such surfaces must be taken into account in order to determine the stability of the surface of a metal being plated out of an electrolytic solution. Surface tension continues to play its stabilizing role, but now by way of its influence on the electron transfer reaction that must take place at both surfaces.

Some of the background information for this essay, mainly information on the rates of electron transfer reactions, may not be widely known, and so it is presented, for it is essential to a physical explanation of the instability.

It is in this essay that our explanations for the causes of the instabilities by way of pictures achieve their greatest precision. It is here that the signature of the base solution on its stability can be most confidently decoded.

As a guide to the reader, we list references to books and papers that got us started on these essays or that helped us over some of the rough points:

- S. Chandrasekhar. *Hydrodynamic and Hydromagnetic Stability.* Dover, New York, 1981.
- P.G. Drazin and W.H. Reid. *Hydrodynamic stability.* Cambridge University Press, Cambridge, 1981.
- J.W.S. Rayleigh. On the Instability of Jets. *Scientific Papers*, i:361, 1899.
- J.W.S. Rayleigh. On the Instability of a Cylinder of Viscous Liquid Under Capillary Force. *Scientific Papers*, iii:585, 1899.
- J. Plateau. *Statique Expérimentale et Théorique Des Liquides Soumis Aux Seules Forces Moléculaires, vol. 2.* Gauthier-Villars, Paris, 1873.
- D.D. Joseph. Domain Perturbations: The Higher Order Theory of Infinitesimal Water Waves. *Arch. Rat. Mech. An.*, 51:295, 1973.
- J. Ponstein. Instability of Rotating Cylindrical Jets. *Appl. Sci. Res. Sec. A*, 8:425, 1959.
- P.G. Saffman and G.I. Taylor. The Penetration of a Fluid into a Porous Medium or Hele-Shaw Cell Containing a More Viscous Liquid. *Proc. Roy. Soc. Lond. A*, 245:155, 1958.
- J.S. Langer. Instabilities and Pattern Formation in Crystal Growth. *Rev. Mod. Phys.*, 52(1):1, 1980.
- J.ÓM. Bockris and A.K.N. Reddy. *Modern Electrochemistry.* Plenum, New York, 1973.

Gainesville, Florida, USA   *L.E. Johns*
November 2001   *R. Narayanan*

Contents

1

The Stability of Liquid Jets by Rayleigh's Work Principle

1.1   The Stability of Liquid Jets

To begin a series of essays on hydrodynamic stability we can do no better than to go back more than 100 years and explain Rayleigh's first calculations on the stability of a cylindrical thread of liquid, a liquid jet.

To explain how liquid jets break up, Rayleigh introduced an intuitive principle which he used to predict this and several other instabilities.

The principle is this:

> Let a fluid, in a state of motion or rest, whose stability is to be determined, be displaced to a nearby state. If it requires work to do this, then the original state of the fluid will be stable to this displacement, otherwise it will be unstable.

This first essay presents the application of this work principle to the breakup of an infinitely long thread of liquid in the form of a right circular cylinder.

To determine whether or not a cylindrical jet is stable to a certain displacement, the work required to carry out the displacement must be obtained. This work is proportional to the increase in the surface area of the jet, the surface tension being the factor of proportionality. An increase in the surface area corresponds to an increase in the potential energy and this is a requirement that work must be done to carry out the displacement.

A jet is stable to a displacement that increases its surface area, unstable to one that decreases it.[1]

1.1.1   The Physics of the Instability

A surface in the shape of a cylinder is far more interesting than a surface in the shape of a plane. Small periodic displacements introduced onto a plane increase its area and are stabilizing. This might be called the direct effect and that is all there is in the case of a plane, other than to notice that long wavelength displacements do not increase the area as strongly as do short wavelength displacements. However, in the case of a cylinder, there is a little more. Again there is the direct effect, again this increases the surface area of the cylinder and again this is stabilizing. Yet, a small periodic displacement of a cylinder produces a second effect and it has to do with longitudinal variation, not with transverse variation. Longitudinal variation increases the diameter of the cylinder at crests and decreases it at troughs. As Rayleigh must have seen, this is the important distinction between plane and cylindrical surfaces. This comes into play when we take into account the fact that the volume of a jet must remain fixed as its surface is displaced. Indeed, the volume of a cylindrical jet increases under a crest by more than it decreases under a trough unless the diameter of the jet is made to decrease, and decreasing the diameter of the jet decreases its area. This is illustrated in Figure 1.1.

As a diameter decrease attends only longitudinal variation, longitudinal variation is dangerous, whereas transverse variation is not. And, as the diameter decrease turns out to be independent of the wavelength of the displacement, long wavelength displacements, which do not increase the area strongly by their direct effect on area, ought to be dangerous, whereas short wavelength displacements ought to be safe. Due to this, a displacement may increase the area or it may decrease the area and which of these obtains depends on the wavelength of the displacement. Nothing of this sort happens in the case of a liquid bounded by a plane surface. There, the constant-volume condition is satisfied by all periodic displacements, the short wavelength displacements increasing the surface area by more than the long wavelength displacements, but all displacements increase the surface area.

---

[1] Forces cause fluid motions and a mechanism advanced to explain an instability must refer to the forces acting and the way in which a displacement brings forces into play which cause it to grow. Rayleigh's work principle does no such thing. Still, it sheds light on the physical origins of the instability and it is predictive. An explanation based on forces can be found at the end of the third essay (cf. Figure 3.2).

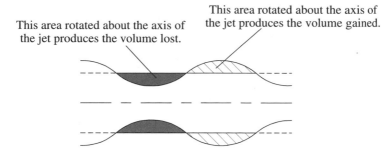

This area rotated about the axis of
the jet produces the volume lost.

This area rotated about the axis of
the jet produces the volume gained.

As the volume gained would exceed the volume lost, the diameter
of the jet must be decreased to hold the volume fixed. This decreases
the surface area of the jet.

Figure 1.1. *Why a Periodic Displacement Would Increase the Volume of a Jet
Unless its Diameter Decreases*

1.1.2 Rayleigh's Work Principle

To produce the calculation[2] required to establish the stability of a jet in
the way that Rayleigh carried it out, let the axis Oz lie along the axis of
a liquid thread in its cylindrical configuration before it is displaced. Its
cylindrical surface is denoted by $r = R_0$, where r denotes the usual radial
distance measured from the axis of the jet to a point on its surface.

Then, introduce a displacement of the surface of the jet and, as longitudinal variation must be more dangerous than transverse variation, take it
to be

$$r = R + \epsilon \cos kz \qquad (1.1)$$

where ϵ denotes the amplitude of the displacement, taken to be small, k,
$k = 2\pi/\lambda$, denotes the wave number of the longitudinal variation, λ denotes
the corresponding wavelength and $2R$ denotes the diameter of the jet. The
value of R depends on the value of ϵ and it is set to maintain the volume of
the jet fixed on displacement.[3] It turns out, of course, to be less than R_0.

The potential energy increase due to the displacement of a surface is a
multiple of its area increase. To determine the area of a displaced cylinder,

---

[2]Starting on page 291 of his second volume, Plateau presents a derivation that does
not differ by much from Rayleigh's derivation. Although its conclusions are open to
several interpretations, we are certain Plateau thought, among other things, that it
establishes the critical wavelength of an infinite jet at the value $2\pi R_0$. His jet was not
infinite, bridging the space between two solid disks while adhering to their edges.

[3]The possibility $k = 0$ must be ruled out. It would require $R + \epsilon$ to be R_0 in order
to maintain the volume of the jet fixed on displacement. But this is no displacement.

when it is infinitely long, the calculation is carried out over one wavelength. Let A denote the surface area of a displaced jet and let V denote the corresponding volume, then A and V can be determined by the formulas

$$A = \int_0^\lambda 2\pi r \, \frac{ds}{dz} \, dz \tag{1.2}$$

and

$$V = \int_0^\lambda \pi r^2 \, dz \tag{1.3}$$

where s denotes the arc length along the curve in which a plane through the axis intersects the surface.

Turning first to the area, using

$$ds = \left[\left[\frac{dr}{dz} \right]^2 + 1 \right]^{\frac{1}{2}} dz$$

or, when ϵ is small,

$$ds = \left[1 + \frac{1}{2} \left[\frac{dr}{dz} \right]^2 \right] dz$$

and substituting equation (1.1) into equation (1.2), we get

$$A = \int_0^\lambda 2\pi [R + \epsilon \cos kz] \left[1 + \frac{1}{2} \epsilon^2 k^2 \sin^2 kz \right] dz$$

Then, carrying out the integration, the term to first order in ϵ vanishes, and so, to second order, there obtains

$$A = 2\pi R\lambda + \frac{1}{2} \pi R \epsilon^2 k^2 \lambda$$

and hence the area per unit length, to sufficient accuracy, turns out to be

$$\frac{A}{\lambda} = 2\pi R + \frac{1}{2} \pi R \epsilon^2 k^2 \tag{1.4}$$

The second term, being positive, makes it look like the area per unit length of a cylinder increases on displacement so that, if this were all there were to it, a liquid jet in the form of a right cylindrical thread would be stable to small displacements. However, the value of R must be determined to meet the condition that the volume of the jet remain fixed and, as R must be less than R_0, the first term predicts a possibly off-setting decrease in area.

To determine R, substitute equation (1.1) into equation (1.3), to obtain

$$V = \int_0^\lambda \pi [R + \epsilon \cos kz]^2 \, dz$$

and carry out the integration to find that

$$\frac{V}{\lambda} = \pi R^2 + \frac{1}{2}\pi\epsilon^2$$

Then, as $\frac{V}{\lambda}$ must be πR_0^2, this requires R to be

$$R = R_0\left[1 - \frac{1}{2}\frac{\epsilon^2}{R_0^2}\right]^{\frac{1}{2}}$$

which, again to sufficient accuracy, is

$$R = R_0 - \frac{1}{4}\frac{\epsilon^2}{R_0} \tag{1.5}$$

The increase in the surface area of the jet, per unit length, due to the displacement $r = R + \epsilon\cos kz$, is then

$$\frac{1}{2}\pi\frac{\epsilon^2}{R_0}[R_0^2 k^2 - 1]$$

and in terms of the wavelength of the displacement, this is

$$\frac{1}{2}\pi\frac{\epsilon^2}{R_0\lambda^2}\left[[2\pi R_0]^2 - \lambda^2\right]$$

For small wavelength displacements this is positive and the jet is stable. For large wavelength displacements, it is the reverse. An infinitely long jet always breaks up. Its initial circumference, $2\pi R_0$, determines the wavelength ranges of the stable and the unstable displacements. The critical wavelength is the circumference of the undisplaced jet.

Rayleigh made this calculation to determine the critical wavelength at which displacements just begin to grow. The jet is unstable to displacements at all longer wavelengths and, among these unstable displacements, Rayleigh thought that one may grow faster than the others and that the wavelength of this fastest growing displacement ought to determine what is seen at jet breakup (viz., it ought to determine the diameters of the droplets so formed and their spacing). A photograph, taken by us and presented as Figure 1.2, illustrates the instability.[4]

To predict these growth rates, the kinetic energy of the jet must be taken into account. Although Rayleigh determined the growth rates of unstable displacements by doing this, we do not repeat his calculation, as it does not add much to our understanding of the cause of the instability. In fact, Rayleigh also obtained this result by a perturbation calculation and that is going to be the subject of our third essay.

Now all of this might seem extraordinary due to the fact that it might be difficult to form a liquid thread into a very long right circular cylinder

---

[4]The liquid thread is honey lying on a flat surface.

Figure 1.2. *Photographs Illustrating the Instability*

when that configuration is certainly unstable. Yet, jet breakup at the critical wavelength might not seem unreasonable if a control variable could be introduced and its critical value determined. Then, on one side of the critical value, all displacements would be stable, whereas if the control variable were moved to its critical value, and just beyond, at least one displacement would become unstable.

To stabilize the liquid jet, then, is our next objective.

1.2   The Stability of Liquid Bridges

The foregoing tells us that a simple way to exercise control over the stability of a liquid thread would be to filter out the long wavelength disturbances. It is enough to let the thread be in the form of a right circular cylinder of radius R_0 and length L_0 spanning the space between two solid plane walls, a distance L_0 apart. This is called a liquid bridge. We require the surface of the liquid bridge to meet the solid walls at right angles.

Then, to work out the stability of the displacement

$$r = R + \epsilon \cos kz, \quad 0 < z < L_0$$

the calculation is much as it was before, except that the requirement that the contact angle be always $\frac{1}{2}\pi$ limits the values of k to those wavelengths that fit between the solid walls. Now, an integer number of half-wavelengths must fit and this can be so if and only if

$$\frac{1}{2}\lambda = \frac{L_0}{n}$$

whence k must be restricted to the values

$$n\frac{\pi}{L_0}, \quad n = 1, \ 2, \ \cdots$$

The displacement corresponding to $n = 0$ cannot satisfy the constant-volume requirement unless $\epsilon = 0$ and so it is omitted.

Denoting the surface area of the displaced bridge by A and its volume by V, we find, as above, that

$$A = 2\pi R L_0 + \frac{1}{2}\pi R \epsilon^2 k^2 L_0$$

and

$$V = \pi R^2 L_0 + \frac{1}{2}\pi \epsilon^2 L_0$$

where the calculation of A is carried to second order in ϵ.

Again, V must remain fixed, now at the value $\pi R_0^2 L_0$, as the cylindrical liquid bridge undergoes its displacement and this determines R, as it depends on R_0 and ϵ. The result, to order ϵ^2, is again

$$R = R_0\left[1 - \frac{\epsilon^2}{4R_0^2}\right]$$

and using this, the surface area, to sufficient accuracy, is

$$2\pi R_0 L_0 + \frac{1}{2}\pi\frac{L_0}{R_0}\epsilon^2[R_0^2 k^2 - 1]$$

whence the increase in the area due to displacement is

$$\frac{1}{2}\pi\frac{L_0}{R_0}\epsilon^2[R_0^2 k^2 - 1]$$

This is just as before, only now, using $k = n\pi/L_0$, $n = 1, 2, \cdots$, the increase in area is

$$\frac{1}{2}\pi\frac{L_0}{R_0}\epsilon^2\left[\frac{n^2\pi^2 R_0^2}{L_0^2} - 1\right]$$

The bridge is stable to displacements corresponding to large values of n; only small values of n might be dangerous and the most dangerous displacement corresponds to $n = 1$. If the bridge is stable to this displacement, it is stable to all displacements, otherwise it is unstable. The stability criterion is then

$$\frac{\pi^2 R_0^2}{L_0^2} > 1 \quad \text{or} \quad L_0 < \pi R_0$$

The critical value of L_0 is then πR_0. If L_0 is less than this, the bridge is stable to small displacements. As L_0 increases just beyond πR_0, the bridge loses its stability.

By introducing a finite bridge, small values of k corresponding to large values of λ can be screened out. The largest value of λ possible in a finite bridge is $2L_0$. Larger values of λ, the long wavelengths that destabilize the liquid jet, are cut off and need not be taken into account in determining the stability of a liquid bridge.

The infinite jet is not like this; it is stable against short wavelength disturbances, unstable against long wavelength disturbances. Its diameter, $2R_0$, then determines the critical wavelength, $2\pi R_0$, dividing these two wavelength ranges. The finite jet introduces a second geometrical variable, its length, and this can be used as a control variable to cut off unstable, long wavelength disturbances. Now, a jet may break up well beyond its critical point, but a bridge will break up at critical.

1.3 Some Unanswered Questions and the Direction Our Work Will Take

These calculations leave some interesting questions unanswered. One is this: Why is it fair to represent a displacement in the form $r = R + \epsilon\cos kz$? Although this might not be extraordinary in the case of an infinite jet, in the case of a finite jet, it surely cannot be correct if the end conditions are other than contact at right angles. While the answer must await the work in the fourth essay, it is important, as it bears on whether or not it is fair to identify a displacement by its wavelength as we have done in all of the above. In the meantime, two discussion notes can be read. In the first, the calculations in this essay are repeated, taking the displacement of the surface to be no more specific than $r = R + \epsilon f(z)$. In the second, the requirement that a displacement be carried out at constant volume, so important in the above calculation, is put to a severe test by introducing a second surface.

Now, the reader may have noticed that the viscosity of the fluid making up the jet is nowhere in evidence. But then Rayleigh's work principle is not an energy method. It does not propose to draw conclusions from a balance of the various forms of energy appearing in a problem. Instead, it directs attention to potential energy and, in particular, to whether or not potential energy is released by the displacement of an equilibrium state. The reader seeking a simple illustration of this principle might think about a pendulum bob being displaced from either of its two equilibrium states. Displaced from its upper rest position, gravitational potential energy is released; displaced from its lower rest position, gravitational potential energy must be supplied. The jet is like this, though it stores potential energy via its surface area, not via gravity. Some displacements of a jet decrease its surface area and thereby release potential energy. Some do not. Whether a displacement decreases or increases a jet's surface area depends entirely on the wavelength of the displacement. The critical wave number, then, classifies the possible displacements of a jet into those that cause the jet to release potential energy and those that do not. As this classification is entirely geometric, the critical wave number cannot depend on the viscosity of the fluid making up the jet; hence, inviscid jet calculations reveal the facts about the critical wave number correctly.[5]

Although this essay serves to introduce the reader to a problem where the domain on which the problem is defined is not fixed, it then presents a solution to the problem where not knowing the domain seems to be of little importance. Yet, our main aim is to solve these kinds of problems by turning them into problems on a known reference domain. Writing the equations on the reference domain requires a little explanation. This is the subject of the next essay. Redoing the problem of this essay by solving the perturbation equations is then the subject of the third essay.

---

[5] Although it might not be a simple matter to identify the class of problems to which Rayleigh's work principle can be applied, the archetypical problem is a pencil standing on its point. All displacements release gravitational potential energy and the pencil falls, no matter the viscosity of its surroundings.

1.4   DISCUSSION NOTES

1.4.1   The Displacement $r = R + \epsilon f(z)$

To determine the stability of a liquid bridge by Rayleigh's work principle, it would seem that no simple reason could be advanced for taking the displacement of its surface to be of the special form

$$r = R + \epsilon \cos kz$$

To see how much might be gained if the displacement is made arbitrary, we repeat the calculations of the essay taking the displacement to be of the form

$$r = R + \epsilon f(z)$$

The requirement that must be satisfied by the liquid bridge where its surface intersects the solid walls at $z = 0$ and at $z = L_0$ is left unspecified.

As before, the surface area of a displaced bridge, denoted A, is determined by the formula

$$A = \int_0^{L_0} 2\pi r \frac{ds}{dz} \, dz$$

where, to second order in ϵ, ds/dz can be replaced by $1 + \frac{1}{2}[f']^2\epsilon^2$. Then, carrying out the calculation to second order in ϵ, the surface area is

$$A = 2\pi R L_0 \ + \ 2\pi\epsilon \int_0^{L_0} f \, dz \ + \ \pi R\epsilon^2 \int_0^{L_0} [f']^2 \, dz \qquad (1.6)$$

The value of R in this formula must be determined to meet the requirement that the volume of the liquid making up the bridge remain fixed as its surface is displaced. This volume, denoted V, can be obtained via

$$V = \int_0^{L_0} \pi r^2 \, dz$$

whence it is

$$V = \pi R^2 L_0 + 2\pi R\epsilon \int_0^{L_0} f \, dz + \pi\epsilon^2 \int_0^{L_0} f^2 \, dz$$

As V must remain at the value $\pi R_0^2 L_0$, this formula determines R in terms of R_0, ϵ and f as a root of the quadratic equation

$$R^2 + \left[\frac{2\epsilon}{L_0} \int_0^{L_0} f \, dz \right] R + \left[\frac{\epsilon^2}{L_0} \int_0^{L_0} f^2 \, dz - R_0^2 \right] = 0$$

Now, there are two roots to this equation and if ϵ is small, one is positive, the other is negative. To order ϵ^2, the positive root is

$$R = -\frac{\epsilon}{L_0} \int_0^{L_0} f \, dz + R_0 \left[1 + \frac{\epsilon^2}{2R_0^2 L_0^2} \left[\int_0^{L_0} f \, dz \right]^2 - \frac{\epsilon^2}{2R_0^2 L_0} \int_0^{L_0} f^2 \, dz \right]$$

and this can be substituted into equation (1.6), making that equation satisfy the fixed-volume requirement. The result is

$$A = 2\pi R_0 L_0 + \pi \epsilon^2 \left[\frac{1}{R_0 L_0} \left[\int_0^{L_0} f \, dz \right]^2 + R_0 \int_0^{L_0} [f']^2 \, dz - \frac{1}{R_0} \int_0^{L_0} f^2 \, dz \right]$$

whence the increase in surface area on displacement is $\pi \epsilon^2$ multiplied by

$$\frac{1}{R_0 L_0} \left[\int_0^{L_0} f \, dz \right]^2 + R_0 \int_0^{L_0} [f']^2 \, dz - \frac{1}{R_0} \int_0^{L_0} f^2 \, dz$$

The first thing to notice is that the increase in the surface area is zero to order ϵ; it differs from zero only at order ϵ^2 and this is due entirely to the constant-volume requirement.

The second thing to notice is that if $f(z) = \cos kz$, this result is just the result found in the essay, except that in the essay $\int_0^{L_0} f(z) \, dz$ is zero when k is restricted so that only an integer number of half-wavelengths fit on the interval $(0, L_0)$. In any case, the term $\frac{1}{R_0 L_0} \left[\int_0^{L_0} f \, dz \right]^2$ is stabilizing.

If f is required to satisfy homogeneous Dirichlet, homogeneous Neumann (and this is the case in the essay) or homogeneous Robin end conditions at $z = 0$ and $z = L_0$, then[6]

$$\int_0^{L_0} [f']^2 \, dz \geq \lambda_1^2 \int_0^{L_0} f^2 \, dz$$

where λ_1^2 is the smallest positive eigenvalue of the differential operator $-d^2/dz^2$ corresponding to the end conditions satisfied by f.[7]

This inequality tells us that the increase in the surface area must always exceed

$$\pi \epsilon^2 \left[\frac{1}{R_0 L_0} \left[\int_0^{L_0} f \, dz \right]^2 + \left[R_0 \lambda_1^2 - \frac{1}{R_0} \right] \left[\int_0^{L_0} f^2 \, dz \right] \right]$$

and this tells us how to build a stable liquid bridge. Indeed, satisfying the condition

$$R_0 \lambda_1^2 - \frac{1}{R_0} > 0$$

---

[6]See equation 36.11, p. 164, Chapter 7, H. F. Weinberger. *A First Course in Partial Differential Equations.* Wiley, New York, 1965.

[7]There is a technical difficulty in the case of fixed end conditions. It has to do with the fact that the radius of the jet must change from R_0 to R if a displacement is to maintain the volume of the jet fixed. It is not important.

is sufficient to make a liquid bridge stable to all displacements. As λ_1^2 is a multiple of $1/L_0^2$, this is a geometric condition determining L_0 in terms of R_0 or vice versa.[8]

To discover an odd fact about Rayleigh's work principle, let $f(z)$ take the special form

$$f(z) = A \cos k_1 z + B \cos k_2 z$$

where

$$k_1 = \frac{\pi}{L_0} m \quad \text{and} \quad k_2 = \frac{\pi}{L_0} n$$

and where the integers m and n can take any of the values 1, 2, $\cdots$. Then $f(z)$ is the sum of two terms where each term satisfies the requirement that an integer number of half wavelengths fit on the interval $(0, \ L_0)$.

To determine the increase in the surface area of a liquid bridge due to a displacement of this kind, turn again to

$$\pi \epsilon^2 \left[\frac{1}{R_0 L_0} \left[\int_0^{L_0} f \ dz \right]^2 + R_0 \int_0^{L_0} [f']^2 \ dz - \frac{1}{R_0} \int_0^{L_0} f^2 \ dz \right]$$

and observe that now there obtains

$$\int_0^{L_0} f \ dz = 0$$

$$\int_0^{L_0} [f']^2 \ dz = A^2 k_1^2 \frac{L_0}{2} + B^2 k_2^2 \frac{L_0}{2}$$

and

$$\int_0^{L_0} f^2 \ dz = A^2 \frac{L_0}{2} + B^2 \frac{L_0}{2}$$

Hence the surface area increase is

$$\frac{1}{2} \pi \frac{\epsilon^2 L_0}{R_0} \left[A^2 [R_0^2 k_1^2 - 1] + B^2 [R_0^2 k_2^2 - 1] \right]$$

and it is the sum of the area increase due to each term taken by itself.

This seems to tell us two things. First, the stability of the bridge is not determined by each term acting alone, but by the two terms acting together. Second, a disturbance, which would be unstable acting by itself, can be stabilized by the presence of another displacement.

It turns out that neither of these conclusions is true. However, a perturbation calculation must be carried out to explain this and that is presented in the third essay.

---

[8]The eigenvalue λ_1^2 depends on the end conditions and in case they are Neumann–Neumann, it is π^2/L_0^2. In other cases (viz., Neumann–Dirichlet and Dirichlet–Dirichlet), it is nearly $9\pi^2/4L_0^2$ and $4\pi^2/L_0^2$.

1.4.2   The Annular Jet

This problem is introduced to see how far Rayleigh's work principle goes in predicting the stability of a liquid jet and, at the same time, to shed some light on the requirement that displacements be carried out at constant volume.

Let an infinitely long liquid jet of radius R_0 have its central core of radius R_0^* removed, leaving behind what might be called an annular jet. If the liquid fills all space outside the hollow core (i.e., if $R_0 = \infty$), Chandrasekhar calls what results a hollow jet and this simply inverts what we have been calling a liquid thread. We refer to the configuration where $0 < R_0^* < R_0 < \infty$ as an annular liquid jet. Our hope is that the requirement of fixed volume on displacement is now not enough to classify displacements as stable or unstable according to their wavelengths.

The infinite annular jet lies between the two surfaces $r = R_0^*$ and $r = R_0 > R_0^*$. To determine whether or not a disturbance increases its surface area, two displacements must be introduced and these can be taken to be

$$r = R^* + \epsilon^* \cos kz$$

and

$$r = R + \epsilon \cos kz$$

where, again, the values of R^* and R must be determined to meet the condition that the volume of the annular jet remain fixed when its reference configuration is disturbed.

The surface area per unit length of a disturbed annular jet can be obtained to second order in both ϵ^* and ϵ as

$$2\pi R^* + 2\pi R + \frac{1}{2}\pi R^* \epsilon^{*2} k^2 + \frac{1}{2}\pi R \epsilon^2 k^2$$

and to decide by how much this is an increase over the surface area per unit length in the reference configuration (i.e., over $2\pi R_0^* + 2\pi R_0$), the values of R^* and R need to be determined in terms of R_0^*, R_0, ϵ^* and ϵ. Yet, there is only one requirement that remains to be satisfied by which this can be done. It is the requirement that the volume remain fixed when an annular jet is disturbed, and although this was enough to determine R in terms of R_0 and ϵ in the case of an ordinary jet, it is now not enough. Indeed, the volume per unit length of a displaced annular jet is

$$\pi R^2 - \pi R^{*2} + \frac{1}{2}\pi \epsilon^2 - \frac{1}{2}\pi \epsilon^{*2}$$

and the requirement that this must be

$$\pi R_0^2 - \pi R_0^{*2}$$

does not determine both R^* and R. Even if R^* and R could be eliminated, both ϵ^* and ϵ would remain in our formula for the surface area increase and so all of its ambiguity would not be removed.

So our attempt to decide whether or not an annular liquid jet is stable to a disturbance imposed on it, by deciding whether or not the disturbance increases the surface area of the jet, seems to end without producing a conclusion.

Although it is possible to resolve this question by carrying out a perturbation calculation, and this is done in a discussion note following the third essay, we can anticipate the answer by trying to make the above calculation conclusive. To do this, we can require the displacement of each surface, by itself, to be carried out at constant volume, which is a special way of making the volume enclosed remain constant. This produces two conditions, namely

$$\pi R^{*2} + \frac{1}{2}\pi\epsilon^{*2} = \pi R_0^{*2}$$

and

$$\pi R^2 + \frac{1}{2}\pi\epsilon^2 = \pi R_0^2$$

which lead, to second order, to

$$R^* = R_0^* \left[1 - \frac{\epsilon^{*2}}{4R_0^{*2}} \right]$$

and

$$R = R_0 \left[1 - \frac{\epsilon^2}{4R_0^2} \right]$$

Using these values of R^* and R, the increase in the surface area per unit length turns out to be

$$\frac{1}{2}\pi\frac{\epsilon^{*2}}{R_0^*}[k^2 R_0^{*2} - 1] + \frac{1}{2}\pi\frac{\epsilon^2}{R_0}[k^2 R_0^2 - 1]$$

This is still not definitive due to the presence of both ϵ^* and ϵ, but it tells us that if

$$k^2 R_0^2 < 1$$

the surface area decreases and the annular jet is unstable, whereas if

$$k^2 R_0^{*2} > 1$$

the surface area increases and the jet is stable. The annular jet is then stable to displacements such that $k^2 > 1/R_0^{*2}$, but unstable to displacements such that $k^2 < 1/R_0^2$. The seemingly indeterminate range between $1/R_0^2$ and $1/R_0^{*2}$ must be taken to be unstable, as it would be if the inside surface were displaced while the outside surface were not. This being the case, the inside surface is in control of the stability of an annular jet, a not unexpected conclusion.

In the indeterminate range where

$$\frac{1}{R_0^2} < k^2 < \frac{1}{R_0^{*2}}$$

it might seem that a large stable displacement of the outside surface could stabilize a small unstable displacement of the inside surface, but it remains to be seen whether ϵ^* and ϵ are independent or whether they must satisfy some, as yet unknown, requirement.

What we have learned in the essay and in the discussion note is this: Imposing a displacement on the surface of a jet does not, by itself, lead to a condition for instability. A side condition is required and the side condition used in the essay is that a displacement may not alter the volume of the jet. If only one surface is of interest, this is sufficient to make the increase in the surface area of the jet definite. However, if two surfaces are present and both can be displaced, holding the jet volume fixed is not, by itself, enough. It leaves the increase in the surface area ambiguous, as it does not eliminate both R^* and R and it leaves ϵ^* and ϵ independent. Some progress can be made by making only constant-volume displacements at each surface, but no physical reason can be advanced to explain why such a limitation should obtain.

2

An Explanation of the Basic Formulas Required to Carry Out the Work of the Remaining Essays

2.1 Introduction and Plan

Let u denote the solution to a problem defined on a domain D, where equations sufficient in number are specified on D, and on its boundary, to determine u. Among problems like this are those where the domain D itself presents a major challenge. These problems are of two kinds. Those where D is not specified and must be determined as part of the solution, and those where D is specified but is irregular.[1] The number of equations to be satisfied can be different in the two cases.

In either case, estimates of u, or of u and D, can be obtained by solving a new problem defined on a specified and regular domain. This second domain, denoted by D_0, is called the reference domain. It may be determined by the original problem or it may be set so as to simplify our work.

Our job is to discover how to change the problem on D to make up for the change in its domain to D_0. Specifically, it is to determine u in terms of the solutions to problems defined on D_0. The first part of our job is to decide what problems on D_0 might help us.

To explain how to do this and to illustrate exactly how all of the details of the calculations work out, let the problem be two dimensional and work in terms of Cartesian coordinates. Then, in terms of an origin O and axes

---

[1]By the word *irregular* we do not mean to imply not smooth, but instead, not convenient.

Ox and Oy at right angles, the points of D_0 will be denoted by coordinates x_0, y_0 and those of D by coordinates x, y.

The main idea is to imagine a family of domains D_ϵ growing out of the reference domain D_0 and to imagine that u must be determined on each of these, one being the domain of interest. The points (x, y) of the domain D_ϵ are then determined in terms of the points (x_0, y_0) of the reference domain D_0 by the mapping

$$x = f(x_0, y_0, \epsilon) \tag{2.1}$$

and

$$y = g(x_0, y_0, \epsilon) \tag{2.2}$$

By this, each point of D_ϵ acquires an ancestor in D_0 and determining the mapping of D_0 into D_ϵ may be part of solving the problem.

Now, equations (2.1) and (2.2) define a mapping which is a little more general than what is required in order to explain the main idea of this essay. It is enough to let D_ϵ be unspecified or irregular in only one part of its boundary and to let D_0 be specified and regular where D_ϵ is unspecified or irregular. Then, let the domains D_ϵ grow out of the domain D_0 as part of the boundary of D_0, denoted $y_0 = Y_0(x_0)$, passes through a sequence of configurations $y = Y(x, \epsilon)$ forming the corresponding parts of the boundaries of the domains D_ϵ. It is enough to simplify the mapping of D_0 into D_ϵ to

$$x = x_0$$

and

$$y = g(x_0, y_0, \epsilon)$$

where this mapping must carry the points $(x_0, y_0 = Y_0(x_0))$ of the boundary of D_0 into the corresponding points $(x, y = Y(x, \epsilon))$ of the boundary of D_ϵ, namely

$$x = x_0$$

and

$$y = Y(x, \epsilon) = Y(x_0, \epsilon) = g(x_0, Y_0(x_0), \epsilon)$$

This simple mapping is the basis of most of our work. It is illustrated in Figure 2.1.

The plan of the essay is this:

• First, we work out the expansion along the mapping of a variable u on the domain D_ϵ in terms of variables u_0, u_1, $\cdots$ on the reference domain.

• Second, we determine the equations to be satisfied by u_0, u_1, $\cdots$ on the reference domain and on its boundary in terms of the equations known to be satisfied by u on the present domain and on its boundary. This is

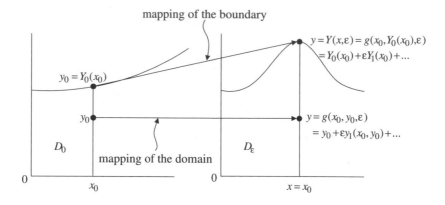

The variable ε may denote the amplitude by which the boundary
of D_ε is displaced from the boundary of D_0. The definition of ε is
in our hands. The applications presented in Appendix A illustrate
a variety of choices of ε.

Figure 2.1. *The Mapping of D_0 into D_ϵ*

done by requiring the equations satisfied by u on D_ϵ and on its boundary
to hold for all values of ϵ.

• Third, we explain[2] how to determine u at a point $(x,\ y)$ in the present
domain in terms of u_0, u_1, $\cdots$, once they are determined at the ancestor
point $(x_0,\ y_0)$ in the reference domain. An explanation is required due
to the fact that the mapping cannot be determined, other than at the
boundary. But the mapping must be determined at the boundary if D_ϵ
is to be constructed from D_0, and this fact can be used to determine u
anywhere in D_ϵ.

---

[2]The reader may ask: Why does this need to be explained? The reason is this: No
mapping of the interior points of the reference domain into the interior points of the
present domain can be determined. No extra domain equations will be available by
which this can be done. At the boundary, things are a little different. In fact, it is by
specifying its boundary that a domain is specified. Whether the boundary of the present
domain is specified in advance or whether it is part of the solution, it is represented
by a mapping of the boundary of a reference domain. In both cases, this mapping can
be obtained. In the second case, an extra equation, holding at the boundary, will be
available to do this. The second example in Appendix A illustrates this. In this example,
the solution to the one-dimensional heat conduction equation must satisfy one condition
at its fixed left-hand boundary, but it must satisfy two conditions at its moving right-
hand boundary. The first example in Appendix A illustrates the estimation of u at an
interior point.

To put this plan to work, the function g is expanded in powers of ϵ as

$$g(x_0,\ y_0,\ \epsilon) = g(x_0,\ y_0,\ \epsilon = 0) + \epsilon \frac{\partial g}{\partial \epsilon}(x_0,\ y_0,\ \epsilon = 0)$$

$$+ \frac{1}{2}\epsilon^2 \frac{\partial^2 g}{\partial \epsilon^2}(x_0,\ y_0,\ \epsilon = 0) + \cdots$$

where $g(x_0,\ y_0,\ \epsilon = 0) = y_0$ and where the derivatives of g, evaluated in the reference domain, are taken holding x_0 and y_0 fixed. Then, in terms of the notation

$$y_1(x_0,\ y_0) = \frac{\partial g}{\partial \epsilon}(x_0,\ y_0,\ \epsilon = 0) \tag{2.3}$$

$$y_2(x_0,\ y_0) = \frac{\partial^2 g}{\partial \epsilon^2}(x_0,\ y_0,\ \epsilon = 0) \tag{2.4}$$

etc.

the mapping can be written

$$x = x_0 \tag{2.5}$$

and

$$y = g(x_0,\ y_0,\ \epsilon) = y_0 + \epsilon y_1(x_0,\ y_0) + \frac{1}{2}\epsilon^2 y_2(x_0,\ y_0) + \cdots \tag{2.6}$$

and this is the form in which it is used.

Now, the boundary of the reference domain must be carried into the boundary of the present domain by our mapping. Hence, the function $Y(x,\ \epsilon)$, which describes the boundary of the current domain, inherits its expansion in powers of ϵ by putting $y_0 = Y_0(x_0)$ into the expansion of the domain mapping. Then, at the boundary, the mapping

$$x = x_0$$

and

$$y = Y(x,\ \epsilon) = Y_0(x_0) + \epsilon y_1(x_0,\ Y_0(x_0)) + \frac{1}{2}\epsilon^2 y_2(x_0,\ Y_0(x_0)) + \cdots$$

can be written

$$x = x_0 \tag{2.7}$$

and

$$y = Y(x,\ \epsilon) = Y_0(x_0) + \epsilon Y_1(x_0) + \frac{1}{2}\epsilon^2 Y_2(x_0) + \cdots \tag{2.8}$$

by introducing the notation

$$Y_1(x_0) = y_1(x_0,\ Y_0(x_0)) \tag{2.9}$$

$$Y_2(x_0) = y_2(x_0, \ Y_0(x_0)) \tag{2.10}$$

etc.

It is Y_1, Y_2, $\cdots$ that must be determined or assigned to specify the domain D_ϵ in terms of the domain D_0.

2.2 The Expansion of a Domain Variable and Its Derivatives Along the Mapping

There are two kinds of variables of interest to us: those defined on the domain, and hence on its boundary as well, and those defined only on the boundary of the domain. These latter variables ordinarily depend on the shape of the boundary and acquire their expansion via the expansion of the function determining this shape. The curvature of the bounding surface, the normal and the normal speed are examples of variables defined only on the boundary, but many variables of interest to us are defined on the domain as well as on its boundary and it is to these variables that we turn first.

Let u, as it depends on x, y and ϵ, denote the solution to a family of problems on a family of domains, denoted D_ϵ. The domains D_ϵ grow out of a reference domain D_0. A point $(x, \ y)$ of the domain D_ϵ is connected to its ancestor $(x_0, \ y_0)$ of D_0 by equations (2.5) and (2.6). Let $u(x_0, \ y_0, \ \epsilon = 0)$ be denoted by $u_0(x_0, \ y_0)$. It is understood that the problem satisfied by u can be solved on D_0 to produce u_0 and D_0 too if it is not known. Our plan is to determine u at the point $(x, \ y)$ in D_ϵ in terms of u_0, and whatever else is required, at the point $(x_0, \ y_0)$ in D_0.

To find out what is required, we expand $u(x, \ y, \ \epsilon)$ in powers of ϵ along the mapping and write this expansion

$$u(x,y,\epsilon) = u(x = x_0, \ y = y_0, \ \epsilon = 0) + \epsilon \frac{du}{d\epsilon}(x = x_0, \ y = y_0, \ \epsilon = 0)$$

$$+ \frac{1}{2}\epsilon^2 \frac{d^2u}{d\epsilon^2}(x = x_0, \ y = y_0, \ \epsilon = 0) + \cdots \tag{2.11}$$

where $d/d\epsilon$ denotes the derivative of a function depending on x, y and ϵ taken along the mapping [i.e., taken holding x_0 and y_0 fixed where x and y depend on x_0, y_0 and ϵ via the mapping, viz., via equations (2.5) and (2.6)].

Then, with u_0 defined by

$$u_0(x_0, \ y_0) = u(x = x_0, \ y = y_0, \ \epsilon = 0)$$

work out $\dfrac{du}{d\epsilon}$ and $\dfrac{d^2u}{d\epsilon^2}$.

To obtain a formula for $\dfrac{du}{d\epsilon}(x = x_0,\ y = y_0,\ \epsilon = 0)$, differentiate u along the mapping taking x and y to depend on ϵ, holding x_0 and y_0 fixed. By the chain rule, there results

$$\frac{du}{d\epsilon}(x,\ y,\ \epsilon) = \frac{\partial u}{\partial \epsilon}(x,\ y,\ \epsilon) + \frac{\partial u}{\partial y}(x,\ y,\ \epsilon)\frac{\partial g}{\partial \epsilon}(x_0,\ y_0,\ \epsilon) \qquad (2.12)$$

where $y = g(x_0,\ y_0,\ \epsilon)$ and where the term $\dfrac{\partial u}{\partial x}(x,\ y,\ \epsilon)\dfrac{\partial f}{\partial \epsilon}(x_0,\ y_0,\ \epsilon)$, that might have been expected had equation (2.1) been used, does not appear due to the fact that equation (2.5), namely

$$x = f(x_0,\ y_0,\ \epsilon) = x_0$$

is being used instead, where x_0 is being held fixed. Then, set ϵ to zero in equation (2.12) and use equation (2.3) to obtain

$$\frac{du}{d\epsilon}(x = x_0,\ y = y_0,\ \epsilon = 0)$$

$$= \frac{\partial u}{\partial \epsilon}(x_0,\ y_0,\ \epsilon = 0) + y_1(x_0,\ y_0)\frac{\partial u}{\partial y}(x_0,\ y_0,\ \epsilon = 0) \qquad (2.13)$$

In equation (2.13), introduce u_1 via

$$u_1(x_0,\ y_0) = \frac{\partial u}{\partial \epsilon}(x = x_0,\ y = y_0,\ \epsilon = 0)$$

and observe that

$$\frac{\partial u}{\partial y}(x = x_0,\ y = y_0,\ \epsilon = 0) = \frac{\partial u_0}{\partial y_0}(x_0,\ y_0)$$

to get

$$\frac{du}{d\epsilon}(x = x_0,\ y = y_0,\ \epsilon = 0) = u_1(x_0,\ y_0) + y_1(x_0,\ y_0)\frac{\partial u_0}{\partial y_0}(x_0,\ y_0) \qquad (2.14)$$

where y_1 specifies the mapping to first order and where u_1, like u_0, remains to be determined.

To work out the second derivative, differentiate equation (2.12) to obtain

$$\frac{d}{d\epsilon}\left[\frac{du}{d\epsilon}\right] = \frac{d}{d\epsilon}\left[\frac{\partial u}{\partial \epsilon} + \frac{\partial u}{\partial y}\frac{\partial g}{\partial \epsilon}\right] = \frac{d}{d\epsilon}\left[\frac{\partial u}{\partial \epsilon}\right] + \frac{d}{d\epsilon}\left[\frac{\partial u}{\partial y}\right]\frac{\partial g}{\partial \epsilon} + \frac{\partial u}{\partial y}\frac{\partial^2 g}{\partial \epsilon^2}$$

and substitute

$$\frac{d}{d\epsilon}\left[\frac{\partial u}{\partial \epsilon}\right] = \frac{\partial}{\partial \epsilon}\left[\frac{\partial u}{\partial \epsilon}\right] + \frac{\partial}{\partial y}\left[\frac{\partial u}{\partial \epsilon}\right]\frac{\partial g}{\partial \epsilon}$$

and

$$\frac{d}{d\epsilon}\left[\frac{\partial u}{\partial y}\right] = \frac{\partial}{\partial \epsilon}\left[\frac{\partial u}{\partial y}\right] + \frac{\partial}{\partial y}\left[\frac{\partial u}{\partial y}\right]\frac{\partial g}{\partial \epsilon}$$

to obtain

$$\frac{d^2u}{d\epsilon^2} = \frac{\partial^2 u}{\partial \epsilon^2} + 2\frac{\partial^2 u}{\partial y \partial \epsilon}\frac{\partial g}{\partial \epsilon} + \frac{\partial^2 u}{\partial y^2}\left[\frac{\partial g}{\partial \epsilon}\right]^2 + \frac{\partial u}{\partial y}\frac{\partial^2 g}{\partial \epsilon^2} \tag{2.15}$$

Then, in equation (2.15), set $x = x_0$, $y = y_0$ and $\epsilon = 0$, introduce u_2 via

$$u_2(x_0,\ y_0) = \frac{\partial^2 u}{\partial \epsilon^2}(x = x_0,\ y = y_0,\ \epsilon = 0)$$

observe that

$$\frac{\partial^2 u}{\partial y^2}(x = x_0,\ y = y_0,\ \epsilon = 0) = \frac{\partial^2 u_0}{\partial y_0^2}(x_0,\ y_0)$$

and use equations (2.3) and (2.4) to obtain

$$\frac{d^2u}{d\epsilon^2}(x = x_0,\ y = y_0,\ \epsilon = 0) = u_2(x_0,\ y_0) + 2y_1(x_0,\ y_0)\frac{\partial u_1}{\partial y_0}(x_0,\ y_0)$$

$$+ y_1^2(x_0,\ y_0)\frac{\partial^2 u_0}{\partial y_0^2}(x_0,\ y_0) + y_2(x_0,\ y_0)\frac{\partial u_0}{\partial y_0}(x_0,\ y_0) \tag{2.16}$$

where y_2 specifies the mapping to second order and where u_2 remains to be determined.

Additional derivatives needed to write $u(x,\ y,\ \epsilon)$ to any order in ϵ can be determined by carrying this work further; indeed, their forms can even be guessed, but equations (2.14) and (2.16) are enough to write u to second order in ϵ as

$$u(x,\ y,\ \epsilon) = u_0 + \epsilon\left[u_1 + y_1\frac{\partial u_0}{\partial y_0}\right]$$

$$+ \frac{1}{2}\epsilon^2\left[u_2 + 2y_1\frac{\partial u_1}{\partial y_0} + y_1^2\frac{\partial^2 u_0}{\partial y_0^2} + y_2\frac{\partial u_0}{\partial y_0}\right] + \cdots \tag{2.17}$$

where every variable on the right-hand side is evaluated at $(x_0,\ y_0)$.

To determine u on the domain D_ϵ, then, it would seem as though u_0, u_1, u_2, $\cdots$ as well as y_1, y_2, $\cdots$ would need to be determined on the domain D_0. It turns out that while u_0, u_1, u_2, $\cdots$ can be determined, y_1, y_2, $\cdots$ cannot, and, although this presents a technical difficulty, it is helpful to be guided by the idea that what is required is a mapping-independent prediction of the values of u at the interior points of the domain D_ϵ.

One aim, then, is to derive the equations that must be satisfied by the functions u_0, u_1, u_2, $\cdots$ on the reference domain D_0 from the equations satisfied by u on the present domain D_ϵ. Now, derivatives of u appear in most of the equations of interest; hence, in addition to the expansion of u, expansions of its derivatives (e.g., expansions of $\partial u/\partial x$ and $\partial u/\partial y$), must be obtained. These can be used along with the expansion of u in determining the equations satisfied by u_0, u_1, u_2, $\cdots$ on the reference domain.

Taking $\partial u / \partial y$ first, we must differentiate the right-hand side of equation (2.17) with respect to y, holding x and ϵ fixed, where the functions appearing there depend on x_0, y_0 and ϵ. To do this, equations (2.5) and (2.6) can be used to obtain x_0 and y_0 in terms of x, y and ϵ, and then the chain rule of differentiation can be used. In this use of the chain rule, functions of x_0, y_0 and ϵ are to be differentiated with respect to y, holding x and ϵ fixed; earlier, to obtain equations (2.14) and (2.16), functions of x, y and ϵ were differentiated with respect to ϵ, holding x_0 and y_0 fixed.

The chain rule, then, requires functions of x_0 and y_0 to be differentiated according to

$$\frac{\partial}{\partial y}[\ ] = \frac{\partial}{\partial x_0}[\ ]\frac{\partial x_0}{\partial y} + \frac{\partial}{\partial y_0}[\ ]\frac{\partial y_0}{\partial y}$$

but $\partial x_0 / \partial y$ is zero, due to the fact that $x = f(x_0, y_0, \epsilon) = x_0$, and by this, the chain rule reduces to

$$\frac{\partial}{\partial y}[\ ] = \frac{\partial}{\partial y_0}[\ ]\frac{\partial y_0}{\partial y}$$

Hence, we must differentiate the right-hand side of equation (2.17) with respect to y_0, holding x_0 and ϵ fixed, and then multiply this by $\partial y_0 / \partial y$ which comes from equation (2.6), holding x and ϵ fixed. Doing this leads to

$$\frac{\partial u}{\partial y} = \left[\frac{\partial u_0}{\partial y_0}\frac{\partial y_0}{\partial y}\right] + \epsilon\left[\frac{\partial u_1}{\partial y_0}\frac{\partial y_0}{\partial y} + y_1\frac{\partial^2 u_0}{\partial y_0^2}\frac{\partial y_0}{\partial y} + \frac{\partial y_1}{\partial y}\frac{\partial u_0}{\partial y_0}\right]$$

$$+ \frac{1}{2}\epsilon^2\left[\frac{\partial u_2}{\partial y_0}\frac{\partial y_0}{\partial y} + 2y_1\frac{\partial^2 u_1}{\partial y_0^2}\frac{\partial y_0}{\partial y} + 2\frac{\partial y_1}{\partial y}\frac{\partial u_1}{\partial y_0} + y_1^2\frac{\partial^3 u_0}{\partial y_0^3}\frac{\partial y_0}{\partial y}\right.$$

$$\left. + 2y_1\frac{\partial y_1}{\partial y}\frac{\partial^2 u_0}{\partial y_0^2} + y_2\frac{\partial^2 u_0}{\partial y_0^2}\frac{\partial y_0}{\partial y} + \frac{\partial y_2}{\partial y}\frac{\partial u_0}{\partial y_0}\right] + \cdots \tag{2.18}$$

where the chain rule has been used to differentiate u_0, $\partial u_0/\partial y_0$, $\partial^2 u_0/\partial y_0^2$, u_1, $\partial u_1/\partial y_0$ and u_2, but it has not been used to differentiate y_1 and y_2. These derivatives have been indicated as $\partial y_1/\partial y$ and $\partial y_2/\partial y$ for reasons to appear presently.

Now, equations (2.5) and (2.6) can be used to determine $\partial y_0/\partial y$ as

$$\frac{\partial y_0}{\partial y} = 1 - \epsilon\frac{\partial y_1}{\partial y} - \frac{1}{2}\epsilon^2\frac{\partial y_2}{\partial y} - \cdots$$

and this reduces equation (2.18), the expansion of $\partial u/\partial y$, to its final form:

$$\frac{\partial u}{\partial y}(x, y, \epsilon) = \frac{\partial u_0}{\partial y_0} + \epsilon\left[\frac{\partial u_1}{\partial y_0} + y_1\frac{\partial^2 u_0}{\partial y_0^2}\right]$$

$$+ \frac{1}{2}\epsilon^2\left[\frac{\partial u_2}{\partial y_0} + 2y_1\frac{\partial^2 u_1}{\partial y_0^2} + y_1^2\frac{\partial^3 u_0}{\partial y_0^3} + y_2\frac{\partial^2 u_0}{\partial y_0^2}\right] + \cdots \tag{2.19}$$

The important thing to observe in this formula is that all the terms in equation (2.18) in which derivatives of y_1, y_2, etc. appear (e.g., $\dfrac{\partial y_1}{\partial y}\dfrac{\partial u_0}{\partial y_0}$), are lost in taking the last step, with the result that the expansion of $\partial u/\partial y$ looks just like the expansion of u, as might have been anticipated, namely

$$u = u_0 + \epsilon\left[u_1 + y_1\frac{\partial u_0}{\partial y_0}\right] + \frac{1}{2}\epsilon^2\left[u_2 + 2y_1\frac{\partial u_1}{\partial y_0} + y_1^2\frac{\partial^2 u_0}{\partial y_0^2} + y_2\frac{\partial u_0}{\partial y_0}\right] + \cdots$$

and

$$\frac{\partial u}{\partial y} = \frac{\partial u_0}{\partial y_0} + \epsilon\left[\frac{\partial u_1}{\partial y_0} + y_1\frac{\partial^2 u_0}{\partial y_0^2}\right]$$

$$+ \frac{1}{2}\epsilon^2\left[\frac{\partial u_2}{\partial y_0} + 2y_1\frac{\partial^2 u_1}{\partial y_0^2} + y_1^2\frac{\partial^3 u_0}{\partial y_0^3} + y_2\frac{\partial^2 u_0}{\partial y_0^2}\right] + \cdots$$

and this would lead us to expect the expansion of $\dfrac{\partial^2 u}{\partial y^2}$ to be

$$\frac{\partial^2 u}{\partial y^2} = \frac{\partial^2 u_0}{\partial y_0^2} + \epsilon\left[\frac{\partial^2 u_1}{\partial y_0^2} + y_1\frac{\partial^3 u_0}{\partial y_0^3}\right]$$

$$+ \frac{1}{2}\epsilon^2\left[\frac{\partial^2 u_2}{\partial y_0^2} + 2y_1\frac{\partial^3 u_1}{\partial y_0^3} + y_1^2\frac{\partial^4 u_0}{\partial y_0^4} + y_2\frac{\partial^3 u_0}{\partial y_0^3}\right] + \cdots \qquad (2.20)$$

which, in fact, is correct. In all these formulas, the variables on the left-hand side are evaluated at x, y and ϵ whereas all the variables on the right-hand side are evaluated at x_0 and y_0.

Turning to $\partial u/\partial x$, we must differentiate the right-hand side of equation (2.17) with respect to x, holding y and ϵ fixed. Now, x is equal to x_0, but y is being held fixed; hence, y_0 cannot be fixed. The result is that $\partial/\partial x$, where y is held fixed, operating on the right-hand side cannot be replaced by $\partial/\partial x_0$, where y_0 is being held fixed, even though $x = x_0$. The mapping must be used to determine y_0 as a function of x at fixed values of y and ϵ. Again, all of the variables on the right-hand side of the expansion of $u(x, y, \epsilon)$, outside of ϵ itself, which is being held fixed, depend only on x_0 and y_0. In view of this, our calculation of $\partial u/\partial x$ again requires the use of the chain rule, now in the form

$$\frac{\partial}{\partial x}[\ ] = \frac{\partial}{\partial x_0}[\ ]\frac{\partial x_0}{\partial x} + \frac{\partial}{\partial y_0}[\ ]\frac{\partial y_0}{\partial x}$$

Once more using equations (2.5) and (2.6) to express x_0 and y_0 in terms of x and y, there now obtains

$$\frac{\partial x_0}{\partial x} = 1$$

and

$$0 = \frac{\partial y_0}{\partial x} + \epsilon \frac{\partial y_1}{\partial x} + \frac{1}{2}\epsilon^2 \frac{\partial y_2}{\partial x} + \cdots$$

whence the chain rule takes the form

$$\frac{\partial}{\partial x}[\,] = \frac{\partial}{\partial x_0}[\,] + \frac{\partial}{\partial y_0}[\,]\frac{\partial y_0}{\partial x}$$

where

$$\frac{\partial y_0}{\partial x} = -\epsilon\frac{\partial y_1}{\partial x} - \frac{1}{2}\epsilon^2\frac{\partial y_2}{\partial x} - \cdots \qquad (2.21)$$

Working out $\partial/\partial x$ applied to the right-hand side of equation (2.17) then leads to

$$\frac{\partial u}{\partial x} = \left[\frac{\partial u_0}{\partial x_0} + \frac{\partial u_0}{\partial y_0}\frac{\partial y_0}{\partial x}\right] + \epsilon\left[\left[\frac{\partial u_1}{\partial x_0} + \frac{\partial u_1}{\partial y_0}\frac{\partial y_0}{\partial x}\right] + y_1\left[\frac{\partial^2 u_0}{\partial x_0 \partial y_0} + \frac{\partial^2 u_0}{\partial y_0^2}\frac{\partial y_0}{\partial x}\right]\right.$$

$$\left. + \frac{\partial y_1}{\partial x}\frac{\partial u_0}{\partial y_0}\right] + \frac{1}{2}\epsilon^2\left[\left[\frac{\partial u_2}{\partial x_0} + \frac{\partial u_2}{\partial y_0}\frac{\partial y_0}{\partial x}\right] + 2y_1\left[\frac{\partial^2 u_1}{\partial x_0 \partial y_0} + \frac{\partial^2 u_1}{\partial y_0^2}\frac{\partial y_0}{\partial x}\right]\right.$$

$$+ 2\frac{\partial y_1}{\partial x}\frac{\partial u_1}{\partial y_0} + y_1^2\left[\frac{\partial^3 u_0}{\partial x_0 \partial y_0^2} + \frac{\partial^3 u_0}{\partial y_0^3}\frac{\partial y_0}{\partial x}\right] + 2y_1\frac{\partial y_1}{\partial x}\frac{\partial^2 u_0}{\partial y_0^2}$$

$$\left. + y_2\left[\frac{\partial^2 u_0}{\partial x_0 \partial y_0} + \frac{\partial^2 u_0}{\partial y_0^2}\frac{\partial y_0}{\partial x}\right] + \frac{\partial y_2}{\partial x}\frac{\partial u_0}{\partial y_0}\right] + \cdots$$

where the chain rule has been used to differentiate u_0, $\partial u_0/\partial y_0$, $\partial^2 u_0/\partial y_0^2$, u_1, $\partial u_1/\partial y_0$ and u_2 but it has not been used to differentiate y_1 and y_2. Again, these derivatives are simply denoted $\partial y_1/\partial x$ and $\partial y_2/\partial x$.

The final formula for $\partial u/\partial x$ obtains on eliminating $\partial y_0/\partial x$ via equation (2.21) and it is

$$\frac{\partial u}{\partial x} = \frac{\partial u_0}{\partial x_0} + \epsilon\left[\frac{\partial u_1}{\partial x_0} + y_1\frac{\partial^2 u_0}{\partial x_0 \partial y_0}\right]$$

$$+ \frac{1}{2}\epsilon^2\left[\frac{\partial u_2}{\partial x_0} + 2y_1\frac{\partial^2 u_1}{\partial x_0 \partial y_0} + y_1^2\frac{\partial^3 u_0}{\partial x_0 \partial y_0^2} + y_2\frac{\partial^2 u_0}{\partial x_0 \partial y_0}\right] + \cdots \quad (2.22)$$

Again, the derivatives of y_1, y_2, etc. are lost in the last step and the formula for $\partial u/\partial x$, like the formula for $\partial u/\partial y$, takes the form of the formula for u itself. This, of course, makes it possible to guess other formulas (e.g., the formula for $\partial^2 u/\partial x^2$). It is worthwhile observing that each time $\partial y_0/\partial x$ is eliminated, by substituting $-\epsilon\dfrac{\partial y_1}{\partial x} - \dfrac{1}{2}\epsilon^2\dfrac{\partial y_2}{\partial x} - \cdots$ in its place, terms to all higher orders in ϵ are introduced. It is by this that the derivatives of y_1, y_2, etc. cancel out; for example, just as $\dfrac{\partial u_0}{\partial y_0}\dfrac{\partial y_0}{\partial x}$ cancels $\epsilon\dfrac{\partial y_1}{\partial x}\dfrac{\partial u_0}{\partial y_0}$ and

$\dfrac{1}{2}\epsilon^2\dfrac{\partial y_2}{\partial x}\dfrac{\partial u_0}{\partial y_0}$ in the foregoing calculation, so too it will cancel $\dfrac{1}{6}\epsilon^3\dfrac{\partial y_3}{\partial x}\dfrac{\partial u_0}{\partial y_0}$ at the third order.

Equations (2.17), (2.19), (2.20) and (2.22) provide for the expansion of a domain variable, yet two kinds of variables can appear in the boundary equations: domain variables evaluated at the boundary and variables defined only at the boundary.

The expansion of a domain variable evaluated at the boundary simply requires the mapping to be evaluated at the boundary in the formulas already obtained. The expansion of a variable defined only at the boundary stems from the expansion of the function determining the shape of the boundary. The function Y defines the shape of the boundary in the present configuration. Its expansion along the mapping appears as equation (2.8).

Let n denote a function defined only at the boundary of the present domain. Now, n is ordinarily a function of Y and we denote its values by $n(Y)$. The expansion of n, then, must come out of the expansion of Y, and to expand n in powers of ϵ along the mapping, write

$$n(Y) = n(\epsilon = 0) + \epsilon\frac{dn}{d\epsilon}(\epsilon = 0) + \frac{1}{2}\epsilon^2\frac{d^2n}{d\epsilon^2}(\epsilon = 0) + \cdots$$

and calculate $\dfrac{dn}{d\epsilon}, \dfrac{d^2n}{d\epsilon^2}$, etc. As n does not depend explicitly on ϵ, the first derivative is[3]

$$\frac{dn}{d\epsilon} = n_Y\left\{\frac{dY}{d\epsilon}\right\}$$

and the second is

$$\frac{d^2n}{d\epsilon^2} = \frac{d}{d\epsilon}\frac{dn}{d\epsilon} = n_Y\left\{\frac{d^2Y}{d\epsilon^2}\right\} + n_{YY}\left\{\frac{dY}{d\epsilon}\right\}\left\{\frac{dY}{d\epsilon}\right\}$$

where n_Y is a linear operator and n_{YY} is a bilinear operator.[4] Both depend

---

[3] As n is a function of Y, and Y is, itself, a function, some way of indicating the derivative of n with respect to Y is required. The notation $n_Y\{\}$ does this, where n_Y operates on whatever appears inside $\{\}$. In fact, n_Y can be obtained as

$$n_Y\{X\} = \lim_{\epsilon\to 0}\frac{n(Y + \epsilon X) - n(Y)}{\epsilon}$$

where n_Y may again be a function of Y. It is a linear function of X, hence, the use of $\{\}$. In terms of n_Y, there obtains

$$n(Y + \epsilon X) = n(Y) + \epsilon n_Y\{X\} + R$$

where $R/\epsilon \to 0$ as $\epsilon \to 0$.

on Y. Using these formulas and

$$\frac{dY}{d\epsilon} = Y_1 + \epsilon Y_2 + \cdots$$

$$\frac{d^2Y}{d\epsilon^2} = Y_2 + \cdots$$

etc.

and evaluating the derivatives of n at $\epsilon = 0$ produces the expansion of $n(Y)$. It is, on writing $n(Y_0)$ in place of $n(\epsilon = 0)$,

$$n(Y) = n(Y_0) + \epsilon n_Y\{Y_1\} + \frac{1}{2}\epsilon^2\Big[n_Y\{Y_2\} + n_{YY}\{Y_1\}\{Y_1\}\Big] + \cdots \quad (2.23)$$

where n_Y and n_{YY} must be evaluated at Y_0.

The expansion of domain variables and boundary variables can be used to carry out the second part of our plan for this essay and it is to this that we now turn.

2.3 The Equations Satisfied by u_0, u_1, u_2, etc. on the Reference Domain and at Its Boundary

The most direct way to obtain the equations satisfied by u_0, u_1, u_2, etc. on the reference domain, and at its boundary, is to substitute the expansions of u and its derivatives into the equations that must be satisfied on the present domain and at its boundary. Then, as the right-hand sides of these expansions depend only on x_0, y_0 and ϵ, on setting the coefficients of each power of ϵ to zero, a set of equations in u_0, u_1, u_2, etc. appear.

Taking the domain equations first, we work out two examples, partly to illustrate the method, but mainly to show that the mapping does not survive[5] in the equations satisfied by u_0, u_1, u_2, etc. This is a favorable result, not only because it is the expected result, but also because the mapping of the interior of the reference domain to the interior of the present domain

---

[4]In a definite problem, n must be specified as a definite function of Y. Its expansion can then be obtained most easily by differentiating this specified function; for example, if $\vec{n}$ denotes the normal to the curve $y = Y(x, \epsilon)$, then

$$\vec{n} = \frac{-\frac{\partial Y}{\partial x}\vec{i} + \vec{j}}{\sqrt{\left[\frac{\partial Y}{\partial \tau}\right]^2 + 1}}$$

and the derivatives of $\vec{n}$ can be obtained in terms of the derivatives of Y. But if n is not specified as a definite function of Y, the notation n_Y, n_{YY}, etc. must be introduced. It is not used in the essays.

[5]A proof of this, not by example, is presented in the first discussion note.

cannot be determined. The derivatives of y_1, y_2, etc. were already lost in expanding the derivatives of u, so they cannot appear in the equations for u_0, u_1, u_2, etc.; now y_1, y_2, etc. themselves are lost.

First, let u and v satisfy

$$\frac{\partial u}{\partial x} + \frac{\partial v}{\partial y} = 0$$

on the current domain. Then, substituting the expansions of $\frac{\partial u}{\partial x}(x,\ y,\ \epsilon)$ and $\frac{\partial v}{\partial y}(x,\ y,\ \epsilon)$ into this equation and setting the coefficients of each power of ϵ to zero, there obtains, on the reference domain,

$$\frac{\partial u_0}{\partial x_0} + \frac{\partial v_0}{\partial y_0} = 0 \tag{2.24}$$

$$\frac{\partial u_1}{\partial x_0} + \frac{\partial v_1}{\partial y_0} + y_1 \left[\frac{\partial^2 u_0}{\partial x_0 \partial y_0} + \frac{\partial^2 v_0}{\partial y_0^2} \right] = 0$$

$$\frac{\partial u_2}{\partial x_0} + \frac{\partial v_2}{\partial y_0} + 2y_1 \left[\frac{\partial^2 u_1}{\partial x_0 \partial y_0} + \frac{\partial^2 v_1}{\partial y_0^2} \right] + y_1^2 \left[\frac{\partial^3 u_0}{\partial x_0 \partial y_0^2} + \frac{\partial^3 v_0}{\partial y_0^3} \right]$$

$$+ y_2 \left[\frac{\partial^2 u_0}{\partial x_0 \partial y_0} + \frac{\partial^2 v_0}{\partial y_0^2} \right] = 0$$

etc.

The first equation reduces the second and the third to

$$\frac{\partial u_1}{\partial x_0} + \frac{\partial v_1}{\partial y_0} = 0 \tag{2.25}$$

and

$$\frac{\partial u_2}{\partial x_0} + \frac{\partial v_2}{\partial y_0} + 2y_1 \left[\frac{\partial^2 u_1}{\partial x_0 \partial y_0} + \frac{\partial^2 v_1}{\partial y_0^2} \right] = 0$$

and the new second equation, equation (2.25), reduces the new third equation to

$$\frac{\partial u_2}{\partial x_0} + \frac{\partial v_2}{\partial y_0} = 0 \tag{2.26}$$

The mapping does not appear in equations (2.24), (2.25) and (2.26) nor in any higher-order equations.

To take one more example, let u satisfy[6]

$$u\frac{\partial u}{\partial y} + \epsilon u + \epsilon^2 + 1 = 0$$

on the current domain. Then substituting the expansions of $u(x,\ y,\ \epsilon)$ and $\frac{\partial u}{\partial y}(x,\ y,\ \epsilon)$ into this, there obtains

$$\left[u_0 + \epsilon\left[u_1 + y_1\frac{\partial u_0}{\partial y_0} \right] + \frac{1}{2}\epsilon^2\left[u_2 + 2y_1\frac{\partial u_1}{\partial y_0} + y_1^2\frac{\partial^2 u_0}{\partial y_0^2} + y_2\frac{\partial u_0}{\partial y_0} \right] + \cdots \right]$$

$$\times \left[\frac{\partial u_0}{\partial y_0} + \epsilon\left[\frac{\partial u_1}{\partial y_0} + y_1\frac{\partial^2 u_0}{\partial y_0^2} \right] + \frac{1}{2}\epsilon^2\left[\frac{\partial u_2}{\partial y_0} + 2y_1\frac{\partial^2 u_1}{\partial y_0^2} + y_1^2\frac{\partial^3 u_0}{\partial y_0^3} \right.\right.$$

$$\left.\left. + y_2\frac{\partial^2 u_0}{\partial y_0^2} \right] + \cdots \right] + \epsilon\left[u_0 + \epsilon\left[u_1 + y_1\frac{\partial u_0}{\partial y_0} \right] + \cdots \right] + \epsilon^2 + 1 = 0$$

which leads, on requiring the coefficient of each power of ϵ to vanish, to

$$u_0\frac{\partial u_0}{\partial y_0} + 1 = 0 \tag{2.27}$$

$$u_0\left[\frac{\partial u_1}{\partial y_0} + y_1\frac{\partial^2 u_0}{\partial y_0^2} \right] + \left[u_1 + y_1\frac{\partial u_0}{\partial y_0} \right]\frac{\partial u_0}{\partial y_0} + u_0 = 0$$

$$\frac{1}{2}u_0\left[\frac{\partial u_2}{\partial y_0} + 2y_1\frac{\partial^2 u_1}{\partial y_0^2} + y_1^2\frac{\partial^3 u_0}{\partial y_0^3} + y_2\frac{\partial^2 u_0}{\partial y_0^2} \right] + \frac{1}{2}\left[u_2 + 2y_1\frac{\partial u_1}{\partial y_0} + y_1^2\frac{\partial^2 u_0}{\partial y_0^2} \right.$$

$$\left. + y_2\frac{\partial u_0}{\partial y_0} \right]\frac{\partial u_0}{\partial y_0} + \left[u_1 + y_1\frac{\partial u_0}{\partial y_0} \right]\left[\frac{\partial u_1}{\partial y_0} + y_1\frac{\partial^2 u_0}{\partial y_0^2} \right] + \left[u_1 + y_1\frac{\partial u_0}{\partial y_0} \right] + 1 = 0$$

etc.

Now, by using $\frac{\partial}{\partial y_0}\left[u_0\frac{\partial u_0}{\partial y_0} + 1 \right] = 0$, the first equation, equation (2.27), eliminates $y_1\left[u_0\frac{\partial^2 u_0}{\partial y_0^2} + \frac{\partial u_0}{\partial y_0}\frac{\partial u_0}{\partial y_0} \right]$ in the second, reducing it to

$$u_0\frac{\partial u_1}{\partial y_0} + u_1\frac{\partial u_0}{\partial y_0} + u_0 = 0 \tag{2.28}$$

while the third equation can be written

---

[6]The reader can try his hand at $u\frac{\partial u}{\partial y} + \epsilon y u + \epsilon^2 + 1 = 0$.

$$\frac{1}{2}u_0\frac{\partial u_2}{\partial y_0} + \frac{1}{2}u_2\frac{\partial u_0}{\partial y_0} + u_1\frac{\partial u_1}{\partial y_0} + u_1 + 1 + y_1\left[u_0\frac{\partial^2 u_1}{\partial y_0^2} + \frac{\partial u_1}{\partial y_0}\frac{\partial u_0}{\partial y_0}\right.$$

$$\left. + \frac{\partial u_0}{\partial y_0}\frac{\partial u_1}{\partial y_0} + u_1\frac{\partial^2 u_0}{\partial y_0^2} + \frac{\partial u_0}{\partial y_0}\right] + \frac{1}{2}y_1^2\left[u_0\frac{\partial^3 u_0}{\partial y_0^3} + \frac{\partial^2 u_0}{\partial y_0^2}\frac{\partial u_0}{\partial y_0}\right.$$

$$\left. +2\frac{\partial u_0}{\partial y_0}\frac{\partial^2 u_0}{\partial y_0^2}\right] + \frac{1}{2}y_2\left[u_0\frac{\partial^2 u_0}{\partial y_0^2} + \frac{\partial u_0}{\partial y_0}\frac{\partial u_0}{\partial y_0}\right] = 0$$

In this, the terms multiplying y_1 add to zero, as do the terms multiplying y_1^2 and those multiplying y_2. The sum of terms multiplying y_1 vanishes on the use of

$$\frac{\partial}{\partial y_0}\left[u_0\frac{\partial u_1}{\partial y_0} + \frac{\partial u_0}{\partial y_0}u_1 + u_0\right] = 0$$

while the sums of terms multiplying y_1^2 and y_2 vanish on the use of

$$\frac{\partial^2}{\partial y_0^2}\left[u_0\frac{\partial u_0}{\partial y_0} + 1\right] = 0 = \frac{\partial}{\partial y_0}\left[u_0\frac{\partial u_0}{\partial y_0} + 1\right]$$

whence the third equation reduces to

$$\frac{1}{2}u_0\frac{\partial u_2}{\partial y_0} + \frac{1}{2}\frac{\partial u_0}{\partial y_0}u_2 + u_1\frac{\partial u_1}{\partial y_0} + u_1 + 1 = 0 \qquad (2.29)$$

Again, a sequence of equations on the reference domain is obtained, equations (2.27), (2.28) and (2.29), and, again, they are free of the mapping. The mapping is lost in writing the equations on the reference domain. The earlier equations in the sequence eliminate the mapping in the later equations. The result, then, is an equation for u_0 at order zero. It is just the original equation for u written on the reference domain. Then comes an equation for u_1 in terms of u_0, succeeded by an equation for u_2 in terms of u_0 and u_1, etc.

To completely determine u_0, u_1, u_2, etc., equations must be worked out at the boundary of the reference domain. In these, the mapping of the boundary must remain if determining the present domain is to be part of the solution. Yet before deriving the equations at the boundary of the reference domain, two small matters remain to be explained before our work on the domain equations is complete. The first of these is a rule by which the correct equations on the reference domain obtain without the need to remove the mapping at each step by using equations obtained in earlier steps. This rule is just what anyone would use, unless it occurs to them to ask the following:

How can a sequence of functions on the reference domain add up to a function on the present domain in view of the fact that the mapping of the one domain into the other cannot be determined?

The rule does not carry over to the boundary equations. It is the boundary equations that retain the mapping and hence the trademark of problems on an unknown domain.

The second small matter comes out of the first. It is a formal justification of our domain rule. It appears as a discussion note to this essay.

2.3.1 The Rule

The rule is this: To get the correct equations satisfied by u_0, u_1, u_2, etc. on the reference domain, replace u wherever it is present in the equations on the current domain by

$$u_0 + \epsilon u_1 + \frac{1}{2}\epsilon^2 u_2 + \cdots$$

and then replace $\dfrac{\partial}{\partial y}$ by $\dfrac{\partial}{\partial y_0}$, $\dfrac{\partial}{\partial x}$ by $\dfrac{\partial}{\partial x_0}$, etc.

To see that this works, let u satisfy

$$u\frac{\partial u}{\partial y} + \epsilon u + \epsilon^2 + 1 = 0$$

on the current domain. Then, making the replacement there obtains

$$\left[u_0 + \epsilon u_1 + \frac{1}{2}\epsilon^2 u_2 + \cdots\right]\left[\frac{\partial u_0}{\partial y_0} + \epsilon\frac{\partial u_1}{\partial y_0} + \frac{1}{2}\epsilon^2\frac{\partial u_2}{\partial y_0} + \cdots\right]$$

$$+\epsilon\left[u_0 + \epsilon u_1 + \frac{1}{2}\epsilon^2 u_2 + \cdots\right] + \epsilon^2 + 1 = 0$$

and setting the coefficient of each power of ϵ to zero leads to

$$u_0\frac{\partial u_0}{\partial y_0} + 1 = 0$$

$$u_0\frac{\partial u_1}{\partial y_0} + u_1\frac{\partial u_0}{\partial y_0} + u_0 = 0$$

$$\frac{1}{2}u_0\frac{\partial u_2}{\partial y_0} + \frac{1}{2}u_2\frac{\partial u_0}{\partial y_0} + u_1\frac{\partial u_1}{\partial y_0} + u_1 + 1 = 0$$

etc.

This is the sequence of equations obtained earlier as equations (2.27), (2.28) and (2.29), but, by use of the rule explained here, obtaining the sequence requires a lot less work. Each equation is defined on the reference domain simply because we say it is. Again, these equations determine, in sequence, u_0, u_1, u_2, etc. and it would seem that our earlier work is of little use until it occurs to us that the sum

$$u_0 + \epsilon u_1 + \frac{1}{2}\epsilon^2 u_2 + \cdots$$

does not add up to u unless a second rule is introduced to explain how the sum is to be interpreted.

The two series

$$u(x,\ y,\ \epsilon) = u_0(x_0,\ y_0)$$

$$+\epsilon\left[u_1(x_0,\ y_0) + y_1(x_0,\ y_0)\frac{\partial u_0}{\partial y_0}(x_0,\ y_0)\right] + \cdots \tag{2.30}$$

and

$$y = y_0 + \epsilon y_1(x_0,\ y_0) + \cdots$$

where $x = x_0$, are free of this defect. However, the functions y_1, y_2, etc., which define the mapping of the reference domain into the present domain, cannot be determined.

If $u(x,\ y,\ \epsilon)$ is of interest, and many times it is not, as the essays illustrate, some unanswered questions remain. Yet, these questions can be answered, although their answers must await further work. Equation (2.30) must be used because we are certain what it adds up to. It only remains to learn how to use it while not being able to determine the mapping.

2.3.2   Equations at the Boundary

To complete the sequence of problems that must be solved in order to obtain u_0, u_1, u_2, etc., we must add equations at the boundary to the equations on the reference domain. Again, to illustrate the method, we work out an example. Let u be a domain variable and let n be a boundary variable where the values of n along the boundary depend on the shape of the boundary. The dependence of n on Y must be specified, but Y may remain to be determined.

At the boundary of the present domain, let u and n satisfy

$$n(Y)\frac{\partial u}{\partial y}(x,\ Y,\ \epsilon) = 0 \tag{2.31}$$

where $y = Y(x,\ \epsilon)$ defines the shape of the boundary, where the expansion of Y is

$$Y(x,\ \epsilon) = Y_0(x_0) + \epsilon Y_1(x_0) + \frac{1}{2}\epsilon^2 Y_2(x_0) + \cdots$$

and where n inherits its ϵ dependence through Y. Then, substitute into equation (2.31) the expansions of $n(Y)$ and $\frac{\partial u}{\partial y}(x,\ Y,\ \epsilon)$.

The expansion of $\partial u/\partial y$ at the boundary can be written by evaluating its domain expansion [viz., equation (2.19)], at $y = Y(x, \epsilon)$ and hence at $y_0 = Y_0(x_0)$. It is

$$\frac{\partial u}{\partial y}(x, Y, \epsilon) = \frac{\partial u_0}{\partial y_0} + \epsilon \left[\frac{\partial u_1}{\partial y_0} + Y_1 \frac{\partial^2 u_0}{\partial y_0^2} \right]$$

$$+ \frac{1}{2}\epsilon^2 \left[\frac{\partial u_2}{\partial y_0} + 2Y_1 \frac{\partial^2 u_1}{\partial y_0^2} + Y_1^2 \frac{\partial^3 u_0}{\partial y_0^3} + Y_2 \frac{\partial^2 u_0}{\partial y_0^2} \right] + \cdots$$

where all the variables on the right-hand side must be evaluated at $y_0 = Y_0(x_0)$, and hence they all depend only on x_0, and where the right-hand side is already defined on the boundary of the reference domain. The expansion of $n(Y)$ has been obtained as equation (2.23), namely

$$n(Y) = n(Y_0) + \epsilon n_Y\{Y_1\} + \frac{1}{2}\epsilon^2 \left[n_Y\{Y_2\} + n_{YY}\{Y_1\}\{Y_1\} \right] + \cdots$$

where n_Y and n_{YY} depend also on Y_0. Substituting these expansions into equation (2.31) then leads to a sequence of equations on the boundary of the reference domain, one at each order in ϵ. The sequence begins

$$n(Y_0)\frac{\partial u_0}{\partial y_0}(x_0, Y_0) = 0$$

and

$$n(Y_0)\left[\frac{\partial u_1}{\partial y_0}(x_0, Y_0) + Y_1 \frac{\partial^2 u_0}{\partial y_0^2}(x_0, Y_0) \right] + n_Y\{Y_1\}\frac{\partial u_0}{\partial y_0}(x_0, Y_0) = 0$$

where it is left to the reader to write the next equation. Each equation holds on the boundary of the reference domain [viz., on $y_0 = Y_0(x_0)$] and the derivatives n_Y, n_{YY}, etc. must be evaluated at $Y = Y_0$. This often leads to simple formulas if Y_0 is simple (e.g., if Y_0 is a constant).

As Y_0 must be known, the first equation completes the problem for u_0. Of course, the problem for u_0 is just the nonlinear problem on the reference domain. Then, the second equation completes the problem for u_1, or at least it completes the problem for u_1 if Y_1 is known. Often Y_1 is not known and a second boundary equation must then be specified if Y_1 is to be determined. This leads to another equation in u_1 and Y_1 at the boundary of the reference domain and the two variables must be determined together, likewise u_2 and Y_2, etc. As the functions Y_1, Y_2, etc. become known, so too the shape of the present domain to first, second, etc. orders in ϵ.

It is important to observe that the derivation of the boundary equations is just like the derivation of the domain equations, save that the mapping cannot be eliminated from the boundary equations. It is important that the mapping be eliminated in the domain equations, for it cannot be determined on the domain; it is equally important that it cannot be eliminated from the boundary equations, otherwise the shape of the present domain could not be determined.

In the third endnote the no-flow equation at the boundary of a present domain, namely

$$\vec{n} \cdot \vec{v} = u$$

is turned into a sequence of equations at the boundary of a reference domain. These equations are used many times in the essays.

2.4   How to Determine $u(x,\ y,\ \epsilon)$

It remains only to explain how to determine the value of u at a point $(x,\ y)$ of the present domain.

Let the functions Y_0, Y_1, Y_2, etc. be known, where Y_0 specifies the boundary of the reference domain and where Y_0, Y_1, Y_2, etc. determine the boundary of the present domain via

$$Y(x,\ \epsilon) = Y_0(x_0) + \epsilon Y_1(x_0) + \frac{1}{2}\epsilon^2 Y_2(x_0) + \cdots$$

The points $(x_0,\ y_0)$ lie on the boundary of the reference domain if $y_0 = Y_0(x_0)$, whereas they lie inside the boundary if $y_0 < Y_0(x_0)$. Likewise, the points $(x,\ y)$ lie on the boundary of the present domain if $x = x_0$, $y = Y(x,\ \epsilon)$, whereas they lie inside the boundary if $x = x_0$, $y < Y(x,\ \epsilon)$. Off the boundary, the mapping cannot be determined and hence neither can the correspondence between interior points $(x_0,\ y_0)$ and interior points $(x,\ y)$.

Let the functions u_0, u_1, u_2, etc. be known on the reference domain. Then, to determine the value of u at a point $(x,\ y)$ of the present domain in terms of u_0, u_1, u_2, etc., we must use the expansions

$$u(x,\ y,\ \epsilon) = u_0(x_0,\ y_0)$$

$$+\epsilon \left[u_1(x_0,\ y_0) + y_1(x_0,\ y_0)\frac{\partial u_0}{\partial y_0}(x_0,\ y_0) \right] + \cdots$$

where

$$x = x_0$$

and

$$y = y_0 + \epsilon y_1(x_0,\ y_0) + \frac{1}{2}\epsilon^2 y_2(x_0,\ y_0) + \cdots$$

But the functions y_1, y_2, etc. cannot be determined, save at the boundary where

$$y_1\big(x_0,\ Y_0(x_0)\big) = Y_1(x_0), \text{ etc.}$$

To begin to evaluate u at points of the present domain, notice that u can be determined at the points of its boundary by using equations (2.7), (2.8) and (2.17) at these boundary points, namely

$$x = x_0$$

$$Y = Y_0(x_0) + \epsilon Y_1(x_0) + \frac{1}{2}\epsilon^2 Y_2(x_0) + \cdots$$

and

$$u(x, Y, \epsilon) = u_0(x_0, Y_0(x_0))$$

$$+\epsilon \left[u_1(x_0, Y_0(x_0)) + Y_1(x_0)\frac{\partial u_0}{\partial y_0}(x_0, Y_0(x_0)) \right] + \cdots$$

Likewise, the values of all the derivatives of u can be determined at points of the boundary by using the formulas for their expansions along the mapping, restricted to the boundary. By this u, and all of its derivatives can be determined along the boundary of the present domain.

The interior points of the present domain present a problem in the use of our expansions that does not come up in their use at the boundary points. To none of the interior points of the present domain can an ancestor point of the reference domain be assigned. This is only a technical difficulty. To see how to overcome it (e.g., at second order), write equations (2.17) and (2.6), the expansions of u and y, in the form

$$u(x, y, \epsilon) = u_0 + \epsilon u_1 + [\epsilon y_1]\frac{\partial u_0}{\partial y_0} + \frac{1}{2}\epsilon^2 u_2$$

$$+\epsilon[\epsilon y_1]\frac{\partial u_1}{\partial y_0} + \frac{1}{2}[\epsilon y_1]^2\frac{\partial^2 u_0}{\partial y_0^2} + \frac{1}{2}\epsilon^2 y_2\frac{\partial u_0}{\partial y_0}$$

and

$$[\epsilon y_1] = y - y_0 - \frac{1}{2}\epsilon^2 y_2 - \cdots$$

and then substitute the second into the first to obtain

$$u(x, y, \epsilon) = u_0(x_0, y_0) + \frac{\partial u_0}{\partial y_0}(x_0, y_0)[y - y_0] + \frac{1}{2}\frac{\partial^2 u_0}{\partial y_0^2}(x_0, y_0)[y - y_0]^2$$

$$+\epsilon \left[u_1(x_0, y_0) + \frac{\partial u_1}{\partial y_0}(x_0, y_0)[y - y_0] \right] + \frac{1}{2}\epsilon^2 \left[u_2(x_0, y_0) \right]$$

Now,

$$u_0(x_0, y_0) + \frac{\partial u_0}{\partial y_0}(x_0, y_0)[y - y_0] + \frac{1}{2}\frac{\partial^2 u_0}{\partial y_0^2}(x_0, y_0)[y - y_0]^2$$

is an estimate to second order of $u_0(x_0, y)$. Likewise,

$$\epsilon \left[u_1(x_0, y_0) + \frac{\partial u_1}{\partial y_0}(x_0, y_0)[y - y_0] \right]$$

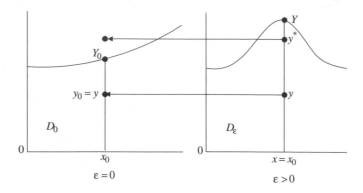

The point $(x = x_0, y)$ of the present domain can be carried straight back to a point $(x_0, y_0 = y)$ of the reference domain. The point $(x = x_0, y^*)$ cannot.

Figure 2.2. *Matching Points of the Present Domain to Points of the Reference Domain*

and

$$\frac{1}{2}\epsilon^2 u_2(x_0,\ y_0)$$

are second-order estimates of $\epsilon u_1(x_0,\ y)$ and $\frac{1}{2}\epsilon^2 u_2(x_0,\ y)$, whence, to second order, $u(x,\ y,\ \epsilon)$ can be estimated by

$$u(x,\ y,\ \epsilon) = u_0(x,\ y) + \epsilon u_1(x,\ y) + \frac{1}{2}\epsilon^2 u_2(x,\ y)$$

This solves the problem of predicting $u(x,\ y,\ \epsilon)$ inside the present domain, at least as long as the point $(x_0 = x,\ y_0 = y)$ lies inside the reference domain [i.e., at least as long as the point $(x,\ y)$ of the present domain is also a point of the reference domain].

This is illustrated in Figure 2.2, which depicts a point $(x,\ y)$ in the present domain, which is also a point of the reference domain, as well as a point $(x,\ y^*)$ in the present domain, which is not a point of the reference domain.

This resumming of the expansion of u along the mapping establishes its value at most, if not all, of the points of the present domain, as the calculation to second order can be carried to higher orders.

It remains only to decide what to do in the case of points like $(x,\ y^*)$ in Figure 2.2, which when carried straight back, do not hit points of the reference domain. These points do not present a difficult problem inasmuch as they must lie near the boundary and u, $\partial u / \partial y$, $\partial^2 u / \partial y^2$, etc. can all be determined at nearby boundary points $(x,\ Y)$ via

$$u(x,\ Y,\ \epsilon) = u_0(x_0,\ Y_0) + \epsilon \left[u_1(x_0,\ Y_0) + Y_1(x_0)\frac{\partial u_0}{\partial y_0}(x_0,\ Y_0) \right] + \cdots$$

$$\frac{\partial u}{\partial y}(x,\ Y,\ \epsilon) = \frac{\partial u_0}{\partial y_0}(x_0,\ Y_0)$$

$$+\epsilon \left[\frac{\partial u_1}{\partial y_0}(x_0,\ Y_0) + Y_1(x_0)\frac{\partial^2 u_0}{\partial y_0^2}(x_0,\ Y_0) \right] + \cdots$$

$$\frac{\partial^2 u}{\partial y^2}(x,\ Y,\ \epsilon) = \frac{\partial^2 u_0}{\partial y_0^2}(x_0,\ Y_0)$$

$$+\epsilon \left[\frac{\partial^2 u_1}{\partial y_0^2}(x_0,\ Y_0) + Y_1(x_0)\frac{\partial^3 u_0}{\partial y_0^3}(x_0,\ Y_0) \right] + \cdots$$

etc.

where $x = x_0$ and $Y = Y_0(x_0) + \epsilon Y_1(x_0) + \cdots$

Then, $u(x,\ y,\ \epsilon)$ can be obtained at a point near the boundary via

$$u(x,\ y^*,\ \epsilon) = u(x,\ Y,\ \epsilon) + \frac{\partial u}{\partial y}(x,\ Y,\ \epsilon)[y^* - Y]$$

$$+\frac{1}{2}\frac{\partial^2 u}{\partial y^2}(x,\ Y,\ \epsilon)[y^* - Y]^2 + \cdots$$

and, by this, u can be obtained throughout the present domain.

The ideas presented in this essay are illustrated in Appendix A, where several problems are worked out. In the first problem, estimation of u on the current domain is emphasized.

The remaining essays require the formulas for the expansions of u and its derivatives. Some of this is repeated in the next essay, less in subsequent essays. Three endnotes are presented, one on the expansion of vector-valued functions and their derivatives, one on the expansion of integrals and one on the no-flow boundary condition which appears in most of the essays. These can be read as the need arises.

2.5   ENDNOTES

2.5.1   *Vector Formulas*

Sometimes, vector formulas are useful. To derive the expansions of vector-valued functions in a way that will be especially useful in the third, fourth and fifth essays, let a problem to be solved be most easily formulated in a cylindrical coordinate system. This is the case in the third essay.

In this endnote, the mapping of the reference domain into the present domain is not taken to be special in any way. Let $\vec{r}$ denote the position vector to the point $(r,\ \theta,\ z)$ in the present domain and let $\vec{r}_0$ denote the position vector to the point $(r_0,\ \theta_0,\ z_0)$ in the reference domain. Denote a mapping of the reference domain into the present domain by

$$\vec{r} = \vec{r}(\vec{r}_0,\ \epsilon)$$

or by[7]

$$r = r(r_0,\ \theta_0,\ z_0,\ \epsilon)$$

$$\theta = \theta(r_0,\ \theta_0,\ z_0,\ \epsilon)$$

and

$$z = z(r_0,\ \theta_0,\ z_0,\ \epsilon)$$

Let the mapping be expanded in powers of ϵ as

$$\vec{r} = \vec{r}_0 + \epsilon\vec{r}_1 + \frac{1}{2}\epsilon^2\vec{r}_2 + \cdots$$

or as

$$r = r_0 + \epsilon r_1 + \frac{1}{2}\epsilon^2 r_2 + \cdots$$

$$\theta = \theta_0 + \epsilon\theta_1 + \frac{1}{2}\epsilon^2\theta_2 + \cdots$$

and

$$z = z_0 + \epsilon z_1 + \frac{1}{2}\epsilon^2 z_2 + \cdots$$

where $\vec{r}_1,\ r_1,\ \theta_1,\ z_1,\ \vec{r}_2,\ r_2,\ \theta_2,\ z_2,\ \cdots$ depend on $\vec{r}_0$ or on $r_0,\ \theta_0$ and z_0 and where

$$\vec{r}_1 = \frac{\partial\vec{r}}{\partial\epsilon}(\epsilon = 0),\ \vec{r}_2 = \frac{\partial^2\vec{r}}{\partial\epsilon^2}(\epsilon = 0),\ \cdots$$

---

[7]The earlier mapping, namely

$$x = f(x_0,\ y_0,\ \epsilon) = x_0\ \text{ and }\ y = g(x_0,\ y_0,\ \epsilon)$$

was special; this mapping is not.

$$r_1 = \frac{\partial r}{\partial \epsilon}(\epsilon = 0), \ r_2 = \frac{\partial^2 r}{\partial \epsilon^2}(\epsilon = 0), \ \cdots$$

$$\theta_1 = \frac{\partial \theta}{\partial \epsilon}(\epsilon = 0), \ \theta_2 = \frac{\partial^2 \theta}{\partial \epsilon^2}(\epsilon = 0), \ \cdots$$

and

$$z_1 = \frac{\partial z}{\partial \epsilon}(\epsilon = 0), \ z_2 = \frac{\partial^2 z}{\partial \epsilon^2}(\epsilon = 0), \ \cdots$$

By differentiating

$$\vec{r} = r\vec{i}_r(\theta) + z\vec{i}_z$$

once, twice, etc., with respect to ϵ and then setting ϵ to zero, $\vec{r}_1$ can be obtained in terms of r_1, θ_1 and z_1, $\vec{r}_2$ can be obtained in terms of r_1, θ_1, z_1, r_2, θ_2, and z_2, etc. For example, the formula for $\vec{r}_1$ is

$$\vec{r}_1 = r_1\vec{i}_{r_0}(\theta_0) + r_0\theta_1\vec{i}_{\theta_0}(\theta_0) + z_1\vec{i}_{z_0}$$

Now, the expansion along the mapping of a vector-valued function defined at a point $\vec{r}$ in the present configuration can be written

$$\vec{v}(\vec{r}, \ \epsilon) = \vec{v}(\vec{r}_0, \ \epsilon = 0) + \epsilon\frac{d\vec{v}}{d\epsilon}(\vec{r}_0, \ \epsilon = 0)$$

$$+\frac{1}{2}\epsilon^2\frac{d^2\vec{v}}{d\epsilon^2}(\vec{r}_0, \ \epsilon = 0) \ + \ \cdots \tag{2.32}$$

where $d/d\epsilon$ denotes the derivative of a function of $\vec{r}$ and ϵ holding $\vec{r}_0$ fixed and where $\vec{r}$ depends on $\vec{r}_0$ and on ϵ via the mapping. The derivatives are evaluated at $\epsilon = 0$ and hence at $\vec{r} = \vec{r}_0$.

To turn equation (2.32) into a useful formula:
(i) denote $\vec{v}(\vec{r}_0, \ \epsilon = 0)$ by $\vec{v}_0(\vec{r}_0)$
(ii) use the chain rule in the form

$$\frac{d}{d\epsilon}[\ ] = \frac{\partial}{\partial \epsilon}[\ ] + \frac{d\vec{r}}{d\epsilon} \cdot \nabla[\ ]$$

to work out $\dfrac{d}{d\epsilon}\vec{v}(\vec{r}, \ \epsilon)$, $\dfrac{d^2}{d\epsilon^2}\vec{v}(\vec{r}, \ \epsilon)$, etc.

(iii) set ϵ to zero and denote $\dfrac{\partial \vec{v}}{\partial \epsilon}(\vec{r}_0, \ \epsilon = 0)$ by $\vec{v}_1(\vec{r}_0)$, $\dfrac{\partial^2 \vec{v}}{\partial \epsilon^2}(\vec{r}_0, \ \epsilon = 0)$ by $\vec{v}_2(\vec{r}_0)$, etc.

Then, to second order, there obtains

$$\vec{v}(\vec{r}, \ \epsilon) = \vec{v}_0(\vec{r}_0) + \epsilon\left[\vec{v}_1(\vec{r}_0) + \vec{r}_1 \cdot \nabla_0 \vec{v}_0(\vec{r}_0)\right] + \frac{1}{2}\epsilon^2\left[\vec{v}_2(\vec{r}_0)\right.$$

$$\left. +2\vec{r}_1 \cdot \nabla_0 \vec{v}_1(\vec{r}_0) + \vec{r}_1\vec{r}_1 : \nabla_0\nabla_0\vec{v}_0(\vec{r}_0) + \vec{r}_2 \cdot \nabla_0\vec{v}_0(\vec{r}_0)\right] + \cdots \ (2.33)$$

where

$$\nabla_0 = \vec{i}_{r_0}\frac{\partial}{\partial r_0} + \vec{i}_{\theta_0}\frac{1}{r_0}\frac{\partial}{\partial \theta_0} + \vec{i}_{z_0}\frac{\partial}{\partial z_0}$$

and where

$$\vec{r}_1 \cdot \nabla_0 = r_1\frac{\partial}{\partial r_0} + \theta_1\frac{\partial}{\partial \theta_0} + z_1\frac{\partial}{\partial z_0}$$

This is our basic formula. It predicts $\vec{v}$ at a point $\vec{r}$ in the current domain in terms of $\vec{v}_0$, $\vec{v}_1$, etc. at a point $\vec{r}_0$ in the reference domain. It is by means of formulas like this that a problem on the current domain can be turned into a sequence of problems on the reference domain. Henceforth, our formulas will be written only to first order.

Now, derivatives of vectors on the present domain are also vectors on the present domain. Our basic formula for expanding $\vec{v}$ along a mapping can just as well be used to expand its derivatives. Equation (2.33) would then lead us to write expansions of the derivatives of $\vec{v}$ as

$$\frac{\partial}{\partial r}\vec{v}(\vec{r}, \ \epsilon) = \frac{\partial}{\partial r_0}\vec{v}_0(\vec{r}_0) + \epsilon\left[\frac{\partial}{\partial r_0}\vec{v}_1(\vec{r}_0) + \vec{r}_1 \cdot \nabla_0\frac{\partial}{\partial r_0}\vec{v}_0(\vec{r}_0)\right] + \cdots$$

$$\frac{\partial}{\partial \theta}\vec{v}(\vec{r}, \ \epsilon) = \frac{\partial}{\partial \theta_0}\vec{v}_0(\vec{r}_0) + \epsilon\left[\frac{\partial}{\partial \theta_0}\vec{v}_1(\vec{r}_0) + \vec{r}_1 \cdot \nabla_0\frac{\partial}{\partial \theta_0}\vec{v}_0(\vec{r}_0)\right] + \cdots$$

and

$$\frac{\partial}{\partial z}\vec{v}(\vec{r}, \ \epsilon) = \frac{\partial}{\partial z_0}\vec{v}_0(\vec{r}_0) + \epsilon\left[\frac{\partial}{\partial z_0}\vec{v}_1(\vec{r}_0) + \vec{r}_1 \cdot \nabla_0\frac{\partial}{\partial z_0}\vec{v}_0(\vec{r}_0)\right] + \cdots$$

These formulas can be obtained by the same kind of calculations that led to $\partial u/\partial y$ and $\partial u/\partial x$ in the essay.

To expand $\nabla\vec{v}$ along the mapping, notice first that the base vectors can be expanded using our basic formula for expanding vectors. Taking into account that the base vectors do not depend explicitly on ϵ, but depend on ϵ only implicitly through the dependence of $\vec{r}$ on ϵ, their expansions are

$$\vec{i}_r(\vec{r}) = \vec{i}_{r_0}(\vec{r}_0) + \epsilon\left[\vec{r}_1 \cdot \nabla_0\vec{i}_{r_0}(\vec{r}_0)\right] + \cdots$$

$$\vec{i}_\theta(\vec{r}) = \vec{i}_{\theta_0}(\vec{r}_0) + \epsilon\left[\vec{r}_1 \cdot \nabla_0\vec{i}_{\theta_0}(\vec{r}_0)\right] + \cdots$$

and

$$\vec{i}_z(\vec{r}) = \vec{i}_{z_0}(\vec{r}_0) + \epsilon\left[\vec{r}_1 \cdot \nabla_0\vec{i}_{z_0}(\vec{r}_0)\right] + \cdots$$

As the base vectors in cylindrical coordinates depend at most on θ, these formulas reduce to

$$\vec{i}_r = \vec{i}_{r_0} + \epsilon\theta_1\vec{i}_{\theta_0} + \cdots$$

$$\vec{i}_\theta = \vec{i}_{\theta_0} - \epsilon\theta_1\vec{i}_{r_0} + \cdots$$

and

$$\vec{i}_z = \vec{i}_{z_0}$$

Then, to expand $\nabla\vec{v}$ along the mapping, the expansions of the base vectors and the derivatives of $\vec{v}$ can be used to evaluate the terms $\vec{i}_r\,\dfrac{\partial}{\partial r}\vec{v}(\vec{r},\ \epsilon)$, $\vec{i}_\theta\,\dfrac{1}{r}\dfrac{\partial}{\partial\theta}\vec{v}(\vec{r},\ \epsilon)$, etc. Doing this and using

$$[\vec{r}_1 \cdot \nabla_0]\vec{i}_{r_0} = \theta_1\vec{i}_{\theta_0}, \quad [\vec{r}_1 \cdot \nabla_0]\vec{i}_{\theta_0} = -\theta_1\vec{i}_{r_0} \quad \text{and} \quad [\vec{r}_1 \cdot \nabla_0]\frac{1}{r_0} = -\frac{r_1}{r_0^2}$$

there obtains

$$\vec{i}_r(\vec{r})\frac{\partial}{\partial r}\vec{v}(\vec{r},\ \epsilon) = \vec{i}_{r_0}\frac{\partial\vec{v}_0}{\partial r_0} + \epsilon\left[\vec{i}_{r_0}\frac{\partial\vec{v}_1}{\partial r_0} + \vec{r}_1 \cdot \nabla_0\left[\vec{i}_{r_0}\frac{\partial\vec{v}_0}{\partial r_0}\right]\right] + \cdots$$

$$\vec{i}_\theta(\vec{r})\frac{1}{r}\frac{\partial}{\partial\theta}\vec{v}(\vec{r},\ \epsilon) = \vec{i}_{\theta_0}\frac{1}{r_0}\frac{\partial\vec{v}_0}{\partial\theta_0} + \epsilon\left[\vec{i}_{\theta_0}\frac{1}{r_0}\frac{\partial\vec{v}_1}{\partial\theta_0} + \vec{r}_1 \cdot \nabla_0\left[\vec{i}_{\theta_0}\frac{1}{r_0}\frac{\partial\vec{v}_0}{\partial\theta_0}\right]\right] + \cdots$$

and

$$\vec{i}_z\frac{\partial}{\partial z}\vec{v}(\vec{r},\ \epsilon) = \vec{i}_{z_0}\frac{\partial\vec{v}_0}{\partial z_0} + \epsilon\left[\vec{i}_{z_0}\frac{\partial\vec{v}_1}{\partial z_0} + \vec{r}_1 \cdot \nabla_0\left[\vec{i}_{z_0}\frac{\partial\vec{v}_0}{\partial z_0}\right]\right] + \cdots$$

whereupon $\nabla\vec{v}$ can be expanded as

$$\nabla\vec{v}(\vec{r},\ \epsilon) = \nabla_0\,\vec{v}_0(\vec{r}_0) + \epsilon\left[\nabla_0\,\vec{v}_1(\vec{r}_0) + \vec{r}_1 \cdot \nabla_0\,\nabla_0\,\vec{v}_0(\vec{r}_0)\right] + \cdots \quad (2.34)$$

These expansions of $\vec{v}$ and $\nabla\vec{v}$, equations (2.33) and (2.34), will be used in the third essay in just the same way that the expansions of u, $\partial u/\partial x$ and $\partial u/\partial y$ were used earlier in this essay, namely to turn the equations on the present domain into a sequence of equations on the reference domain.

It is worth noticing the following:

• The expansion of $\nabla\vec{v}$ could have been obtained formally from the expansion of $\vec{v}$ by putting $\nabla\vec{v}$ in place of $\vec{v}$, $\nabla_0\vec{v}_0$ in place of $\vec{v}_0$, etc.

• The functions on the left-hand side of equations (2.33) and (2.34) depend on $\vec{r}$ and ϵ; the functions on the right-hand side depend only on $\vec{r}_0$. The mapping shows up on the right-hand side via $\vec{r}_1$, which depends on $\vec{r}_0$.

• The expansions are not restricted to cylindrical coordinate systems. In fact, they are form invariant under a change to another coordinate system and they could have been derived much more easily in a Cartesian coordinate system.

- The variable $\vec{v}$ need not be a vector.
- The expansions of $\nabla \cdot \vec{v}$ and $\nabla^2 \vec{v}$ can be obtained as

$$\nabla \cdot \vec{v} = \nabla_0 \cdot \vec{v}_0 + \epsilon\left[\nabla_0 \cdot \vec{v}_1 + \vec{r}_1 \cdot \nabla_0[\nabla_0 \cdot \vec{v}_0]\right] + \cdots$$

and

$$\nabla^2 \vec{v} = \nabla_0^2 \vec{v}_0 + \epsilon\left[\nabla_0^2 \vec{v}_1 + \vec{r}_1 \cdot \nabla_0 \nabla_0^2 \vec{v}_0\right] + \cdots$$

2.5.2   The Expansion of Integrals Along the Mapping

Let V denote the volume of the present domain. Then, in many cases, there are physical requirements on some state variable, denoted by u, which can be stated in terms of its integral over V. In this endnote, the form of this requirement is determined to first order in ϵ.

Let I denote the integral of u over V, namely

$$I = \int_V u \; dV$$

Now, u depends on $\vec{r}$ and ϵ and V depends on ϵ; hence, I depends on ϵ. The mapping $\vec{r} = \vec{r}_0 + \epsilon \vec{r}_1 + \cdots$ carries the points $\vec{r}_0$ of the reference domain into the points $\vec{r}$ of the present domain, and by this, it ties V to V_0.

Our job is to expand I along the mapping in powers of ϵ. To do this, write

$$I = I_0 + \epsilon I_1 + \cdots$$

where

$$I_0 = I(\epsilon = 0)$$

$$I_1 = \frac{dI}{d\epsilon}(\epsilon = 0)$$

etc.

and where

$$I_0 = \int_{V_0} u_0 \; dV_0$$

To determine I_1, use Leibnitz' rule to write

$$\frac{dI}{d\epsilon} = \int_V \frac{\partial u}{\partial \epsilon} \; dV + \int_{\partial V} dA \; \vec{n} \cdot \frac{d\vec{r}}{d\epsilon} u$$

Then, set $\epsilon = 0$ and use $\dfrac{d\vec{r}}{d\epsilon}(\epsilon = 0) = \vec{r}_1$, where, at the boundary of the reference configuration, $\vec{r}_1 = \vec{R}_1$, to obtain

$$I_1 = \frac{dI}{d\epsilon}(\epsilon = 0) = \int_{V_0} u_1 \, dV_0 + \int_{\partial V_0} dA_0 \vec{n}_0 \cdot \vec{R}_1 u_0$$

and, by this, the expansion of I to two terms, namely

$$I = \int_{V_0} u_0 \, dV_0 + \epsilon \left[\int_{V_0} u_1 \, dV_0 + \int_{\partial V_0} dA_0 \vec{n}_0 \cdot \vec{R}_1 u_0 \right] \qquad (2.35)$$

This formula will be used in many essays. Its use can be illustrated in meeting the requirement that the volume of a liquid bridge remain fixed as its surface is displaced.

Turning to this, set $u = 1$, and hence $u_0 = 1$, $u_1 = 0$, etc., in equation (2.35) to obtain

$$V = V_0 + \epsilon \int_{\partial V_0} dA_0 \vec{n}_0 \cdot \vec{R}_1 + \cdots$$

Then, the fixed-volume requirement to first order is

$$\int_{\partial V_0} dA_0 \vec{n}_0 \cdot \vec{R}_1 = 0$$

In the case of a liquid bridge, where the surface being displaced is a cylinder of radius R_0, this is simply

$$\int_0^{L_0} \int_0^{2\pi} R_1 \, d\theta_0 \, dz_0 = 0$$

where $\vec{R}_1 = R_1 \vec{i}_{r_0} + Z_1 \vec{i}_{z_0}$. This formula is used in the fourth essay.

Now, the mapping of the domain does not appear in equation (2.35); only the mapping of the boundary appears via $\vec{R}_1$. To see how the mapping is lost, the calculation can be carried out another way. To do this, let u depend on x, y and ϵ and let the mapping be written

$$x = x_0$$

and

$$y = y_0 + \epsilon y_1(x_0, y_0) + \cdots$$

Then, to evaluate $\displaystyle\iint_V u \, dx \, dy$, substitute the expansion of u along the mapping and at the same time transform the integral over the present domain to an integral over the reference domain to get

$$\iint_V u(x, y, \epsilon) \, dx \, dy = \iint_{V_0} \left[u_0 + \epsilon \left[u_1 + y_1 \frac{\partial u_0}{\partial y_0} \right] + \cdots \right] \frac{\partial(x, y)}{\partial(x_0, y_0)} \, dx_0 \, dy_0$$

where

$$\frac{\partial(x, \, y)}{\partial(x_0, \, y_0)} = 1 + \epsilon \frac{\partial y_1}{\partial y_0} + \cdots$$

To first order in ϵ, this is

$$\iint\limits_{V} u \, dx \, dy = \iint\limits_{V_0} u_0 \, dx_0 \, dy_0 + \epsilon \iint\limits_{V_0} \left[u_1 + \frac{\partial [y_1 u_0]}{\partial y_0} \right] dx_0 \, dy_0$$

and, carrying out the last integration on the right-hand side, there obtains

$$\iint\limits_{V} u \, dx \, dy = \iint\limits_{V_0} u_0 \, dx_0 \, dy_0$$

$$+ \epsilon \left[\iint\limits_{V_0} u_1 \, dx_0 \, dy_0 + \epsilon \int_{y_0 = Y(x_0)} Y_1 u_0 \, dx_0 \right] \qquad (2.36)$$

The mapping of the domain does not appear in this formula; it is lost in the last step, leaving only the mapping of the boundary. Now, let $u = 1$; then equation (2.36) reduces to

$$\iint\limits_{V} dx \, dy = \iint\limits_{V_0} dx_0 \, dy_0 + \epsilon \int_{y_0 = Y(x_0)} Y_1 \, dx_0$$

This is a formula for the volume (area) of a perturbed domain.

2.5.3 Expanding the Normal Velocity of a Fluid at a Moving Surface

As an illustration of the use of the formulas presented in this essay, the perturbed form of the no-flow boundary condition that turns up many times in subsequent essays will be obtained.

Let a physical process take place in a domain defined by a boundary denoted by

$$y = Y(x, \, t, \, \epsilon)$$

and suppose that fluid does not cross this boundary. Then, one equation of interest there is

$$\vec{n} \cdot \vec{v} = u$$

where $\vec{n}$ is the unit normal to the surface and u is its normal speed. This equation can be written[8]

$$-v_x \frac{\partial Y}{\partial x} + v_y = \frac{\partial Y}{\partial t} \qquad (2.37)$$

and it holds at the points of the boundary of the present domain.

Now, let there be a reference domain defined by a boundary denoted by $y_0 = Y_0(x_0, t_0)$. This reference domain can be mapped into the present domain and the form we take is

$$x = x_0$$

$$y = g(x_0, y_0, t_0, \epsilon) = y_0 + \epsilon y_1(x_0, y_0, t_0) + \frac{1}{2}\epsilon^2 y_2(x_0, y_0, t_0) + \cdots$$

and

$$t = t_0$$

This mapping must carry the boundary of the reference domain into the boundary of the present domain, and due to this, $Y(x, t, \epsilon)$ acquires the expansion

$$Y(x, t, \epsilon) = g(x_0, Y_0(x_0, t_0), t_0, \epsilon) = Y_0 + \epsilon Y_1 + \frac{1}{2}\epsilon^2 Y_2 + \cdots$$

where

$$Y_1(x_0, t_0) = y_1(x_0, Y_0, t_0)$$

$$Y_2(x_0, t_0) = y_2(x_0, Y_0, t_0)$$

etc.

Now, the expansions of the domain variables v_x and v_y, evaluated at the boundary, can be written

$$v_x(x, Y, t, \epsilon) = v_{x_0} + \epsilon\left[v_{x_1} + Y_1\frac{\partial v_{x_0}}{\partial y_0}\right] + \cdots$$

and

$$v_y(x, Y, t, \epsilon) = v_{y_0} + \epsilon\left[v_{y_1} + Y_1\frac{\partial v_{y_0}}{\partial y_0}\right] + \cdots$$

and substituting these along with the expansion[9] of Y into equation (2.37) leads, to zeroth and first orders in ϵ and at the boundary of the reference

---

[8]Formulas for the normal to a surface and for its normal speed are presented in Appendices B and C.

[9]As Y depends on x and t, where $x = x_0$ and $t = t_0$, the expansions of $\partial Y/\partial x$ and $\partial Y/\partial t$ present no problem.

domain, to

$$-v_{x_0}\frac{\partial Y_0}{\partial x_0} + v_{y_0} = \frac{\partial Y_0}{\partial t_0}$$

and

$$-v_{x_0}\frac{\partial Y_1}{\partial x_0} - \left[v_{x_1} + Y_1\frac{\partial v_{x_0}}{\partial y_0}\right]\frac{\partial Y_0}{\partial x_0} + v_{y_1} + Y_1\frac{\partial v_{y_0}}{\partial y_0} = \frac{\partial Y_1}{\partial t_0}$$

The first of these, of course, is just the original nonlinear equation evaluated at the boundary of the reference domain. The second is its perturbation. It is rarely seen in a form this general.

Some special cases are of interest. The case where Y_0 is a constant, independent of x_0 and t_0, comes up most often. Then, the first-order equation reduces to

$$-v_{x_0}\frac{\partial Y_1}{\partial x_0} + v_{y_1} + Y_1\frac{\partial v_{y_0}}{\partial y_0} = \frac{\partial Y_1}{\partial t_0}$$

Yet, even this is seldom seen. If, in addition, v_{x_0} and v_{y_0} vanish (i.e., if the fluid is at rest on the reference domain) it reduces further to

$$v_{y_1} = \frac{\partial Y_1}{\partial t_0}$$

Now, this equation is often used by Chandrasekhar, and it could be written down intuitively. Indeed, in this form, it must be so self-evident that not even a hint of its origin can be found in the writings of the great workers in fluid mechanics, not in the writing of Rayleigh nor Taylor nor Chandrasekhar.

2.6   DISCUSSION NOTES

2.6.1   A Justification of the Rule

In the essay, a rule was introduced whereby the perturbation equations on the reference domain could be obtained from the nonlinear equations on the current domain in a very simple way. It was illustrated by example. In this discussion note, the rule will be justified and the reader will not then think that the examples presented there were in any way special.

To begin, let u satisfy

$$A(u) = 0$$

on the current domain D_ϵ. The solution u depends on y, as well as on ϵ, and A denotes an operator on u that may, itself, depend on ϵ. The solution depends on ϵ, essentially because the domain depends on ϵ and incidently[10] because A may depend on ϵ. To illustrate this, the equation

$$u\frac{\partial u}{\partial y} + \epsilon u + \epsilon^2 + 1 = 0$$

can be written

$$A(u) = 0$$

where

$$A = \{\ \}\frac{\partial}{\partial y}\{\ \} + \epsilon\{\ \} + \epsilon^2 + 1$$

Now, the equation

$$A(u) = 0 \qquad \text{on } D_\epsilon$$

must be satisfied for any value of ϵ. Then, not only must it be true that

$$A(u) = 0 \qquad \text{on } D_\epsilon$$

but also that

$$\frac{d}{d\epsilon}A(u) = 0 \qquad \text{on } D_\epsilon$$

$$\frac{d^2}{d\epsilon^2}A(u) = 0 \qquad \text{on } D_\epsilon$$

etc.

where $d/d\epsilon$ is, as before, the total derivative along the mapping that ties all the domains, D_ϵ, together.

---

[10]Of course, A may depend on x and y as well.

Evaluating this sequence of equations at $\epsilon = 0$ produces a sequence of equations on the reference domain and this will turn out to be just the sequence produced by using our shorthand rule. But again, no way to reconstruct $u(x, y, \epsilon)$ comes out of doing this.

To find out what equations turn up on the reference domain, we must carry out the calculation of $\dfrac{d}{d\epsilon}A$, $\dfrac{d^2}{d\epsilon^2}A$, etc. To begin, we calculate $\dfrac{d}{d\epsilon}A(u)$, taking into account that u depends on ϵ. To do this, write

$$\frac{d}{d\epsilon}A(u) = A_\epsilon(u) + A_u\left\{\frac{du}{d\epsilon}\right\} \tag{2.38}$$

where A_ϵ, like A itself, is an operator on u and where A_u is a linear operator on $du/d\epsilon$, possibly depending also on u and ϵ. Then, substituting

$$\frac{du}{d\epsilon} = \frac{\partial u}{\partial \epsilon} + \frac{\partial y}{\partial \epsilon}\frac{\partial u}{\partial y}$$

into equation (2.38) leads to

$$\frac{d}{d\epsilon}A(u) = A_\epsilon(u) + A_u\left\{\frac{\partial u}{\partial \epsilon}\right\} + A_u\left\{\frac{\partial y}{\partial \epsilon}\frac{\partial u}{\partial y}\right\} \tag{2.39}$$

Now, $\partial y/\partial \epsilon$ depends at most on x_0, y_0 and ϵ and, although A_u may depend on ϵ, it does not operate on ϵ. Hence, $\partial y/\partial \epsilon$ is a multiplicative factor and equation (2.39) can be written

$$\frac{d}{d\epsilon}A(u) = A_\epsilon(u) + A_u\left\{\frac{\partial u}{\partial \epsilon}\right\} + \frac{\partial y}{\partial \epsilon}A_u\left\{\frac{\partial u}{\partial y}\right\} \tag{2.40}$$

To obtain the first two equations on the reference domain, we must evaluate

$$A(u) = 0 \qquad \text{on } D_\epsilon$$

and

$$\frac{d}{d\epsilon}A(u) = 0 \qquad \text{on } D_\epsilon$$

at $\epsilon = 0$. The first presents no problem, and equation (2.40) can be used to evaluate the second. Then, denoting u, $\partial u/\partial \epsilon$ and y at $\epsilon = 0$ by u_0, u_1 and y_0 and observing that $\dfrac{\partial y}{\partial \epsilon}(\epsilon = 0)$ is just y_1 while $\dfrac{\partial u}{\partial y}$ at $\epsilon = 0$ is $\dfrac{\partial u_0}{\partial y_0}$, there results

$$A(u_0) = 0 \qquad \text{on } D_0 \tag{2.41}$$

and

$$A_\epsilon(u_0) + A_u\{u_1\} + y_1 A_u\left\{\frac{\partial u_0}{\partial y_0}\right\} = 0 \qquad \text{on } D_0 \tag{2.42}$$

where A, A_ϵ and A_u are evaluated at $\epsilon = 0$ and A_ϵ and A_u are evaluated at $u = u_0$.

To eliminate y_1 in equation (2.42), differentiate equation (2.41) with respect to y_0 to obtain

$$0 = \frac{\partial}{\partial y_0} A(u_0) = A_u \left\{ \frac{\partial u_0}{\partial y_0} \right\}$$

where, again, A_u is evaluated at $\epsilon = 0$ and $u = u_0$ and where this result reduces equation (2.42) to

$$A_\epsilon(u_0) + A_u\{u_1\} = 0 \quad \text{on } D_0$$

The first two equations in the sequence are then

$$A(u_0) = 0 \quad \text{on } D_0 \tag{2.43}$$

and

$$A_u\{u_1\} = -A_\epsilon(u_0) \quad \text{on } D_0 \tag{2.44}$$

As an example, let $A(u)$ denote

$$u \frac{\partial u}{\partial y} + \epsilon u + \epsilon^2 + 1$$

then at $\epsilon = 0$ we obtain

$$A(u_0) = u_0 \frac{\partial u_0}{\partial y_0} + 1$$

$$A_\epsilon(u_0) = u_0$$

and

$$A_u\{u_1\} = u_0 \frac{\partial u_1}{\partial y_0} + u_1 \frac{\partial u_0}{\partial y_0}$$

and the first two equations turn out to be

$$u_0 \frac{\partial u_0}{\partial y_0} + 1 = 0$$

and

$$u_0 \frac{\partial u_1}{\partial y_0} + u_1 \frac{\partial u_0}{\partial y_0} = -u_0$$

just as they did in the essay.

To go on, we must calculate $\dfrac{d^2}{d\epsilon^2} A(u)$, and to do this, write

$$\frac{d^2}{d\epsilon^2} A(u) = \frac{d}{d\epsilon} \left[\frac{d}{d\epsilon} A(u) \right] = \frac{d}{d\epsilon} A_\epsilon(u) + \frac{d}{d\epsilon} A_u \left\{ \frac{du}{d\epsilon} \right\}$$

and then carry out the indicated differentiations to get

$$\frac{d^2}{d\epsilon^2} A(u) = A_{\epsilon\epsilon}(u) + A_{u\epsilon}\left\{\frac{du}{d\epsilon}\right\} + A_{\epsilon u}\left\{\frac{du}{d\epsilon}\right\}$$

$$+ A_{uu}\left\{\frac{du}{d\epsilon}\right\}\left\{\frac{du}{d\epsilon}\right\} + A_u\left\{\frac{d^2 u}{d\epsilon^2}\right\} \tag{2.45}$$

where $A_{\epsilon\epsilon}$ is an operator on u, where $A_{u\epsilon}$ and $A_{\epsilon u}$ are linear operators on $du/d\epsilon$, henceforth taken to be the same and possibly depending also on ϵ and u, and where A_{uu} is a symmetric bilinear operator on $du/d\epsilon$, again possibly depending on ϵ and u. Then, substituting

$$\frac{du}{d\epsilon} = \frac{\partial u}{\partial \epsilon} + \frac{\partial y}{\partial \epsilon}\frac{\partial u}{\partial y}$$

and

$$\frac{d^2 u}{d\epsilon^2} = \frac{\partial^2 u}{\partial \epsilon^2} + 2\frac{\partial y}{\partial \epsilon}\frac{\partial^2 u}{\partial y \partial \epsilon} + \left[\frac{\partial y}{\partial \epsilon}\right]^2 \frac{\partial^2 u}{\partial y^2} + \frac{\partial^2 y}{\partial \epsilon^2}\frac{\partial u}{\partial y}$$

into equation (2.45) and noticing that $\frac{\partial y}{\partial \epsilon}$, $\left[\frac{\partial y}{\partial \epsilon}\right]^2$ and $\frac{\partial^2 y}{\partial \epsilon^2}$ depend only on x_0, y_0 and ϵ, whence they are treated by the linear operators A_u, $A_{u\epsilon}$ and A_{uu} as multiplicative factors, there obtains

$$\frac{d^2}{d\epsilon^2} A(u) = A_{\epsilon\epsilon}(u) + 2A_{u\epsilon}\left\{\frac{\partial u}{\partial \epsilon}\right\} + 2\frac{\partial y}{\partial \epsilon} A_{u\epsilon}\left\{\frac{\partial u}{\partial y}\right\} + A_u\left\{\frac{\partial^2 u}{\partial \epsilon^2}\right\}$$

$$+ 2\frac{\partial y}{\partial \epsilon} A_u\left\{\frac{\partial^2 u}{\partial y \partial \epsilon}\right\} + \left[\frac{\partial y}{\partial \epsilon}\right]^2 A_u\left\{\frac{\partial^2 u}{\partial y^2}\right\} + \frac{\partial^2 y}{\partial \epsilon^2} A_u\left\{\frac{\partial u}{\partial y}\right\}$$

$$+ A_{uu}\left\{\frac{\partial u}{\partial \epsilon}\right\}\left\{\frac{\partial u}{\partial \epsilon}\right\} + 2\frac{\partial y}{\partial \epsilon} A_{uu}\left\{\frac{\partial u}{\partial \epsilon}\right\}\left\{\frac{\partial u}{\partial y}\right\}$$

$$+ \left[\frac{\partial y}{\partial \epsilon}\right]^2 A_{uu}\left\{\frac{\partial u}{\partial y}\right\}\left\{\frac{\partial u}{\partial y}\right\} \tag{2.46}$$

The third equation on the reference domain then turns up on evaluating

$$\frac{d^2}{d\epsilon^2} A(u) = 0 \quad \text{on } D_\epsilon$$

at $\epsilon = 0$. Using equation (2.46) to do this, denoting $\frac{\partial^2 u}{\partial \epsilon^2}$ at $\epsilon = 0$ by u_2, using $\frac{\partial^2 y}{\partial \epsilon^2}(\epsilon = 0) = y_2$ and observing that $\frac{\partial^2 u}{\partial y \partial \epsilon}$ and $\frac{\partial^2 u}{\partial y^2}$ at $\epsilon = 0$ are $\frac{\partial u_1}{\partial y_0}$ and $\frac{\partial^2 u_0}{\partial y_0^2}$, the result, after a little arranging, is

$$A_{\epsilon\epsilon}(u_0) + 2A_{u\epsilon}\{u_1\} + A_u\{u_2\} + A_{uu}\{u_1\}\{u_1\}$$

$$+2y_1\left[A_{u\epsilon}\left\{\frac{\partial u_0}{\partial y_0}\right\} + A_u\left\{\frac{\partial u_1}{\partial y_0}\right\} + A_{uu}\{u_1\}\left\{\frac{\partial u_0}{\partial y_0}\right\}\right]$$

$$+y_1^2\left[A_u\left\{\frac{\partial^2 u_0}{\partial y_0^2}\right\} + A_{uu}\left\{\frac{\partial u_0}{\partial y_0}\right\}\left\{\frac{\partial u_0}{\partial y_0}\right\}\right] + y_2 A_u\left\{\frac{\partial u_0}{\partial y_0}\right\} = 0 \quad (2.47)$$

where all the linear operators must be evaluated at $\epsilon = 0$, $x = x_0$, $y = y_0$ and $u = u_0$.

It remains only to eliminate the mapping. To do this, use equations (2.43) and (2.44). First, use

$$A(u_0) = 0$$

to get

$$\frac{\partial}{\partial y_0} A(u_0) = 0 = \frac{\partial^2}{\partial y_0^2} A(u_0)$$

whence there obtains

$$A_u\left\{\frac{\partial u_0}{\partial y_0}\right\} = 0 = A_{uu}\left\{\frac{\partial u_0}{\partial y_0}\right\}\left\{\frac{\partial u_0}{\partial y_0}\right\} + A_u\left\{\frac{\partial^2 u_0}{\partial y_0^2}\right\}$$

which eliminates the y_2 and y_1^2 terms in equation (2.47). Then, use

$$A_\epsilon(u_0) + A_u\{u_1\} = 0$$

to get

$$\frac{\partial}{\partial y_0}\left[A_\epsilon(u_0) + A_u\{u_1\}\right]$$

$$= A_{u\epsilon}\left\{\frac{\partial u_0}{\partial y_0}\right\} + A_{uu}\{u_1\}\left\{\frac{\partial u_0}{\partial y_0}\right\} + A_u\left\{\frac{\partial u_1}{\partial y_0}\right\} = 0$$

which eliminates the y_1 term in equation (2.47).

The third equation on the reference domain is then

$$A_{\epsilon\epsilon}(u_0) + 2A_{u\epsilon}\{u_1\} + A_u\{u_2\} + A_{uu}\{u_1\}\{u_1\} = 0 \quad (2.48)$$

and our sequence of equations on the reference domain, carried to second order in ϵ, turns out to be equations (2.43), (2.44) and (2.48) that is,

$$A(u_0) = 0$$

$$A_u\{u_1\} = -A_\epsilon(u_0)$$

$$A_u\{u_2\} = -A_{\epsilon\epsilon}(u_0) - 2A_{u\epsilon}\{u_1\} - A_{uu}\{u_1\}\{u_1\}$$

All of the operators must be evaluated at $\epsilon = 0$, $x = x_0$ and $y = y_0$. The linear operators must also be evaluated at $u = u_0$.

The special case where A is independent of ϵ is important, and in this case, the sequence of equations on D_0 is

$$A(u_0) = 0$$

$$A_u\{u_1\} = 0$$

$$A_u\{u_2\} = -A_{uu}\{u_1\}\{u_1\}$$

etc.

All of the operators in the sequence are nonlinear operators on u_0 and A_u is the Fréchet derivative of A evaluated at u_0. As expected, it appears at first order. The Fréchet derivative of the linear operator $A_u\{\ \}$, with respect to whatever variable is present inside $\{\ \}$, is also $A_u\{\ \}$. Hence, A_u reasserts itself at all orders above the first, and at each order it operates on the unknown function present there for the first time.

Now, turning to the example that is being carried along to illustrate what is going on, let

$$A(u) = u\frac{\partial u}{\partial y} + \epsilon u + \epsilon^2 + 1$$

whence at $\epsilon = 0$ we obtain

$$A_{\epsilon\epsilon}(u_0) = 2$$

$$A_{u\epsilon}\{u_1\} = A_{\epsilon u}\{u_1\} = u_1$$

and

$$A_{uu}\{u_1\}\{u_1\} = 2u_1\frac{\partial u_1}{\partial y_0}$$

and the third equation turns out, also as before, to be

$$u_0\frac{\partial u_2}{\partial y_0} + u_2\frac{\partial u_0}{\partial y_0} = -2 - 2u_1 - 2u_1\frac{\partial u_1}{\partial y_0}$$

These calculations justify the simple rule presented in the essay. By using this formal derivation to write the equations on the reference domain, the expansions of u and its derivatives need not be written out explicitly. That being so, the derivatives of the functions y_1, y_2, $\cdots$ never turn up in the derivation, and we can hardly imagine that their elimination in the expansion of u, and in the expansion of its derivatives, is taking place.

Turning now to the equations that must be satisfied at the boundary, let u and n, a domain variable and a boundary variable, satisfy

$$B(u,\, n) = 0$$

at the boundary of the present domain. Let B depend explicitly on ϵ, let u be an ordinary function of x, y and ϵ, where now $y = Y(x,\, \epsilon)$, and let n

depend on the function Y. The expansion of Y along the mapping is

$$Y(x, \epsilon) = g(x_0, Y_0(x_0), \epsilon) = Y_0 + \epsilon Y_1 + \frac{1}{2}\epsilon^2 Y_2 + \cdots$$

where

$$Y_1 = \frac{\partial g}{\partial \epsilon}(x_0, y_0 = Y_0(x_0), \epsilon = 0)$$

$$Y_2 = \frac{\partial^2 g}{\partial \epsilon^2}(x_0, y_0 = Y_0(x_0), \epsilon = 0)$$

etc.

To determine the equations that must hold at the boundary of the reference domain, observe that if

$$B(u, n) = 0$$

for all values of ϵ, then so too

$$\frac{d}{d\epsilon}B(u, n) = 0$$

$$\frac{d^2}{d\epsilon^2}B(u, n) = 0$$

etc.

for all values of ϵ, where the derivative $d/d\epsilon$ is again taken along the mapping. The equations on the boundary of the reference domain then obtain on setting $\epsilon = 0$. The first is

$$B(u_0, n(Y_0)) = 0 \tag{2.49}$$

To get the second, calculate $\dfrac{d}{d\epsilon}B(u, n)$. This is

$$\frac{d}{d\epsilon}B(u, n) = B_\epsilon(u, n) + B_u\left\{\frac{du}{d\epsilon}\right\} + B_n\left\{\frac{dn}{d\epsilon}\right\}$$

where B_u and B_n are linear operators, which, like B_ϵ and B itself, depend on ϵ, u and n.

Then, using

$$Y = g(x_0, Y_0(x_0), \epsilon)$$

and

$$\frac{du}{d\epsilon} = \frac{\partial u}{\partial \epsilon} + \frac{\partial g}{\partial \epsilon}\frac{\partial u}{\partial y}$$

and the fact that $\dfrac{\partial g}{\partial \epsilon}$ depends at most on x_0, y_0 and ϵ, $B_u\left\{\dfrac{du}{d\epsilon}\right\}$ can be written

$$B_u\left\{\frac{du}{d\epsilon}\right\} = B_u\left\{\frac{\partial u}{\partial \epsilon}\right\} + \frac{\partial g}{\partial \epsilon}B_u\left\{\frac{\partial u}{\partial y}\right\}$$

This and

$$B_n\left\{\frac{dn}{d\epsilon}\right\} = B_n\left\{n_Y\left\{\frac{dY}{d\epsilon}\right\}\right\}$$

reduce $\dfrac{d}{d\epsilon}B(u,\ n)$ to

$$\frac{d}{d\epsilon}B(u,\ n) = B_\epsilon(u,\ n) + B_u\left\{\frac{\partial u}{\partial \epsilon}\right\} + \frac{\partial g}{\partial \epsilon}B_u\left\{\frac{\partial u}{\partial y}\right\} + B_n\left\{n_Y\left\{\frac{dY}{d\epsilon}\right\}\right\}$$

whence evaluating $\dfrac{d}{d\epsilon}B(u,\ n) = 0$ at $\epsilon = 0$ produces the second equation on the boundary of the reference domain. It is

$$B_\epsilon\big(u_0,\ n(Y_0)\big) + B_u\{u_1\} + Y_1 B_u\left\{\frac{\partial u_0}{\partial y_0}\right\} + B_n\left\{n_Y\{Y_1\}\right\} = 0 \quad (2.50)$$

where B_u and B_n must be evaluated at $u = u_0$ and $n = n(Y_0)$, where n_Y must be evaluated at $Y = Y_0$ and where all operators must be evaluated at $\epsilon = 0$.

If $B(u,\ n) = n\dfrac{\partial u}{\partial y}$, the example worked out in the essay, then

$$B_\epsilon = 0$$

$$B_u\{\ \} = n(Y_0)\frac{\partial}{\partial y_0}\{\ \}$$

and

$$B_n\{\ \} = \{\ \}\frac{\partial u_0}{\partial y_0}$$

and equations (2.49) and (2.50), the first two equations on the boundary of the reference domain, turn out to be

$$n(Y_0)\frac{\partial u_0}{\partial y_0}(x_0,\ Y_0) = 0$$

and

$$n(Y_0)\frac{\partial u_1}{\partial y_0}(x_0,\ Y_0) + Y_1 n(Y_0)\frac{\partial^2 u_0}{\partial y_0^2}(x_0,\ Y_0) + n_Y\{Y_1\}\frac{\partial u_0}{\partial y_0}(x_0,\ Y_0) = 0$$

and this is what we obtained earlier by direct substitution of the expansions of $n(Y)$ and $\partial u/\partial y$.

The equation to second order in ϵ on the boundary of the reference domain obtains on calculating $\dfrac{d^2}{d\epsilon^2}B(u,\ n)$ then evaluating it at $\epsilon = 0$ and setting the result to zero. To get $\dfrac{d^2}{d\epsilon^2}B(u,\ n)$, write

$$\frac{d^2}{d\epsilon^2}B(u,\ n) = \frac{d}{d\epsilon}\frac{d}{d\epsilon}B(u,\ n) = \frac{d}{d\epsilon}\left[B_\epsilon(u,\ n) + B_u\left\{\frac{du}{d\epsilon}\right\} + B_n\left\{\frac{dn}{d\epsilon}\right\}\right]$$

and then use

$$\frac{d}{d\epsilon}B_\epsilon(u,\ n) = B_{\epsilon\epsilon}(u,\ n) + B_{u\epsilon}\left\{\frac{du}{d\epsilon}\right\} + B_{n\epsilon}\left\{\frac{dn}{d\epsilon}\right\}$$

$$\frac{d}{d\epsilon}B_u\left\{\frac{du}{d\epsilon}\right\}$$

$$= B_{\epsilon u}\left\{\frac{du}{d\epsilon}\right\} + B_{uu}\left\{\frac{du}{d\epsilon}\right\}\left\{\frac{du}{d\epsilon}\right\} + B_{nu}\left\{\frac{du}{d\epsilon}\right\}\left\{\frac{dn}{d\epsilon}\right\} + B_u\left\{\frac{d^2u}{d\epsilon^2}\right\}$$

and

$$\frac{d}{d\epsilon}B_n\left\{\frac{dn}{d\epsilon}\right\}$$

$$= B_{\epsilon n}\left\{\frac{dn}{d\epsilon}\right\} + B_{un}\left\{\frac{dn}{d\epsilon}\right\}\left\{\frac{du}{d\epsilon}\right\} + B_{nn}\left\{\frac{dn}{d\epsilon}\right\}\left\{\frac{dn}{d\epsilon}\right\} + B_n\left\{\frac{d^2n}{d\epsilon^2}\right\}$$

along with symmetry, to write $\dfrac{d^2}{d\epsilon^2}B(u,\ n)$ as

$$\frac{d^2}{d\epsilon^2}B(u,\ n)$$

$$= B_{\epsilon\epsilon}(u,\ n) + 2B_{\epsilon u}\left\{\frac{du}{d\epsilon}\right\} + 2B_{\epsilon n}\left\{\frac{dn}{d\epsilon}\right\} + B_{uu}\left\{\frac{du}{d\epsilon}\right\}\left\{\frac{du}{d\epsilon}\right\}$$

$$+2B_{un}\left\{\frac{du}{d\epsilon}\right\}\left\{\frac{dn}{d\epsilon}\right\} + B_u\left\{\frac{d^2u}{d\epsilon^2}\right\} + B_{nn}\left\{\frac{dn}{d\epsilon}\right\}\left\{\frac{dn}{d\epsilon}\right\} + B_n\left\{\frac{d^2n}{d\epsilon^2}\right\}$$

Now, the factors $\dfrac{\partial g}{\partial\epsilon}$, etc. that are present in $\dfrac{du}{d\epsilon}$, $\dfrac{d^2u}{d\epsilon^2}$, etc., for example,

$$\frac{du}{d\epsilon} = \frac{\partial u}{\partial\epsilon} + \frac{\partial g}{\partial\epsilon}\frac{\partial u}{\partial y}$$

depend at most on x_0 and ϵ and so certain of the terms in the formula for $\dfrac{d^2}{d\epsilon^2}B(u,\ n)$ can be written in a better form before ϵ is set to zero, for example

$$B_{un}\left\{\frac{du}{d\epsilon}\right\}\left\{\frac{dn}{d\epsilon}\right\} = B_{un}\left\{\frac{\partial u}{\partial\epsilon}\right\}\left\{\frac{dn}{d\epsilon}\right\} + \frac{\partial g}{\partial\epsilon}B_{un}\left\{\frac{\partial u}{\partial y}\right\}\left\{\frac{dn}{d\epsilon}\right\}$$

Writing $B_{\epsilon u}\left\{\dfrac{du}{d\epsilon}\right\}$, $B_{uu}\left\{\dfrac{du}{d\epsilon}\right\}\left\{\dfrac{du}{d\epsilon}\right\}$, $B_{un}\left\{\dfrac{du}{d\epsilon}\right\}\left\{\dfrac{dn}{d\epsilon}\right\}$ and $B_u\left\{\dfrac{d^2u}{d\epsilon^2}\right\}$ in this way, before setting ϵ to zero, using

$$\frac{dn}{d\epsilon} = n_Y\left\{\frac{dY}{d\epsilon}\right\}$$

etc.

where $Y = Y_0 + \epsilon Y_1 + \dfrac{1}{2}\epsilon^2 Y_2 + \cdots$, and then setting $\epsilon = 0$ produces the second-order equation on the boundary of the reference domain. It is

$$B_{\epsilon\epsilon}\big(u_0,\ n(Y_0)\big) + 2B_{\epsilon u}\{u_1\} + 2Y_1 B_{\epsilon u}\left\{\frac{\partial u_0}{\partial y_0}\right\} + 2B_{\epsilon n}\Big\{n_Y\{Y_1\}\Big\}$$

$$+ B_{uu}\{u_1\}\{u_1\} + 2Y_1 B_{uu}\{u_1\}\left\{\frac{\partial u_0}{\partial y_0}\right\} + Y_1^2 B_{uu}\left\{\frac{\partial u_0}{\partial y_0}\right\}\left\{\frac{\partial u_0}{\partial y_0}\right\}$$

$$+ 2B_{un}\{u_1\}\Big\{n_Y\{Y_1\}\Big\} + 2Y_1 B_{un}\left\{\frac{\partial u_0}{\partial y_0}\right\}\Big\{n_Y\{Y_1\}\Big\} + B_u\{u_2\}$$

$$+ 2Y_1 B_u\left\{\frac{\partial u_1}{\partial y_0}\right\} + Y_1^2 B_u\left\{\frac{\partial^2 u_0}{\partial y_0^2}\right\} + Y_2 B_u\left\{\frac{\partial u_0}{\partial y_0}\right\}$$

$$+ B_{nn}\Big\{n_Y\{Y_1\}\Big\}\Big\{n_Y\{Y_1\}\Big\} + B_n\Big\{n_Y\{Y_2\}\Big\} + B_n\Big\{n_{YY}\{Y_1\}\{Y_1\}\Big\} = 0$$

The readers can use this to check their second-order equation, obtained by direct substitution of the expansions of n and u, in the case where

$$B(u,\ n) = n\frac{\partial u}{\partial y}$$

2.6.2 A Justification of the Rule Under a Little More General Mapping

The conclusions of the previous discussion note hold just as well under a little more general mapping. This is no surprise, but the details are worked out in this discussion note.

Let the mapping be

$$x = f(x_0,\ y_0,\ \epsilon) = x_0 + \epsilon x_1(x_0,\ y_0) + \frac{1}{2}\epsilon^2 x_2(x_0,\ y_0) + \cdots$$

and

$$y = g(x_0,\ y_0,\ \epsilon) = y_0 + \epsilon y_1(x_0,\ y_0) + \frac{1}{2}\epsilon^2 y_2(x_0,\ y_0) + \cdots$$

in place of the simpler mapping used in the earlier work.

Then, to carry out the expansion of u and its derivatives, again write

$$u(x, y, \epsilon) = u(\epsilon = 0) + \epsilon \frac{du}{d\epsilon}(\epsilon = 0) + \cdots$$

but now use

$$\frac{du}{d\epsilon} = \frac{\partial u}{\partial \epsilon} + \frac{\partial x}{\partial \epsilon}\frac{\partial u}{\partial x} + \frac{\partial y}{\partial \epsilon}\frac{\partial u}{\partial y}$$

at $\epsilon = 0$ to produce the expansion of u as

$$u(x, y, \epsilon) = u_0(x_0, y_0) + \epsilon\left[u_1(x_0, y_0) + x_1(x_0, y_0)\frac{\partial u_0}{\partial x_0}(x_0, y_0)\right.$$

$$\left. +y_1(x_0, y_0)\frac{\partial u_0}{\partial y_0}(x_0, y_0)\right] + \cdots \tag{2.51}$$

Then, use the chain rule in the form

$$\frac{\partial}{\partial x}[\,] = \frac{\partial}{\partial x_0}[\,]\frac{\partial x_0}{\partial x} + \frac{\partial}{\partial y_0}[\,]\frac{\partial y_0}{\partial x}$$

to get

$$\frac{\partial u}{\partial x} = \left[\frac{\partial u_0}{\partial x_0}\frac{\partial x_0}{\partial x} + \frac{\partial u_0}{\partial y_0}\frac{\partial y_0}{\partial x}\right]$$

$$+\epsilon\left[\left[\frac{\partial u_1}{\partial x_0}\frac{\partial x_0}{\partial x} + \frac{\partial u_1}{\partial y_0}\frac{\partial y_0}{\partial x}\right] + x_1\left[\frac{\partial^2 u_0}{\partial x_0^2}\frac{\partial x_0}{\partial x} + \frac{\partial^2 u_0}{\partial y_0 \partial x_0}\frac{\partial y_0}{\partial x}\right]\right.$$

$$\left. +\frac{\partial x_1}{\partial x}\frac{\partial u_0}{\partial x_0} + y_1\left[\frac{\partial^2 u_0}{\partial x_0 \partial y_0}\frac{\partial x_0}{\partial x} + \frac{\partial^2 u_0}{\partial y_0^2}\frac{\partial y_0}{\partial x}\right] + \frac{\partial y_1}{\partial x}\frac{\partial u_0}{\partial y_0}\right] + \cdots$$

which on use of

$$\frac{\partial x_0}{\partial x} = 1 - \epsilon\frac{\partial x_1}{\partial x} - \frac{1}{2}\epsilon^2\frac{\partial x_2}{\partial x} - \cdots$$

and

$$\frac{\partial y_0}{\partial x} = 0 - \epsilon\frac{\partial y_1}{\partial x} - \frac{1}{2}\epsilon^2\frac{\partial y_2}{\partial x} - \cdots$$

produces the expansion of $\dfrac{\partial u}{\partial x}$ as

$$\frac{\partial u}{\partial x} = \frac{\partial u_0}{\partial x_0} + \epsilon\left[\frac{\partial u_1}{\partial x_0} + x_1\frac{\partial^2 u_0}{\partial x_0^2} + y_1\frac{\partial^2 u_0}{\partial x_0 \partial y_0}\right] + \cdots \tag{2.52}$$

Likewise, the expansion of $\dfrac{\partial u}{\partial y}$ is

$$\frac{\partial u}{\partial y} = \frac{\partial u_0}{\partial y_0} + \epsilon\left[\frac{\partial u_1}{\partial y_0} + x_1\frac{\partial^2 u_0}{\partial x_0 \partial y_0} + y_1\frac{\partial^2 u_0}{\partial y_0^2}\right] + \cdots \tag{2.53}$$

The derivatives of x_1, x_2, $\cdots$ and y_1, y_2, $\cdots$ do not appear in these expansions and hence they do not appear in the equations on the reference domain derived therefrom. In addition, on using these expansions to derive the equations on the reference domain, the mapping itself (viz., x_1, x_2, $\cdots$, y_1, y_2, $\cdots$) does not appear in the resulting equations.

This conclusion can be established by requiring u to satisfy

$$A(u) = 0 \qquad \text{on } D_\epsilon$$

Then, also

$$\frac{d}{d\epsilon} A(u) = 0 \qquad \text{on } D_\epsilon$$

$$\frac{d^2}{d\epsilon^2} A(u) = 0 \qquad \text{on } D_\epsilon$$

$$\text{etc.}$$

must hold and the equations on the reference domain obtain by setting $\epsilon = 0$. Working out $\frac{d}{d\epsilon} A(u)$ as

$$\frac{d}{d\epsilon} A(u) = A_\epsilon(u) + A_u \left\{ \frac{du}{d\epsilon} \right\}$$

where

$$\frac{du}{d\epsilon} = \frac{\partial u}{\partial \epsilon} + \frac{\partial x}{\partial \epsilon} \frac{\partial u}{\partial x} + \frac{\partial y}{\partial \epsilon} \frac{\partial u}{\partial y}$$

and using

$$\frac{\partial x}{\partial \epsilon} = x_1 + \epsilon x_2 + \cdots$$

and

$$\frac{\partial y}{\partial \epsilon} = y_1 + \epsilon y_2 + \cdots$$

the first two equations on the reference domain, D_0, turn out to be

$$A(u_0) = 0$$

and

$$A_\epsilon(u_0) + A_u\{u_1\} + x_1 A_u \left\{ \frac{\partial u_0}{\partial x_0} \right\} + y_1 A_u \left\{ \frac{\partial u_0}{\partial y_0} \right\} = 0$$

where the first can be used to reduce the second to

$$A_u\{u_1\} = -A_\epsilon(u_0)$$

and the mapping is seen to disappear.

3

The Stability of a Liquid Jet by a Perturbation Calculation

3.1 The Nonlinear Equations

Although Rayleigh's work principle makes the conclusions reached in the first essay very likely, we would still like to establish these results directly by introducing a small displacement to the surface of an equilibrium jet and then determining whether or not it grows.

The fate of such a small displacement is described by the perturbation equations and it is the main business of this essay to write these equations and then to solve them.

The method by which the perturbation equations can be written is explained in the second essay.

To begin, introduce cylindrical coordinates r, θ and z and denote the surface of the jet by

$$r = R(\theta, \ z, \ t)$$

where the function R remains to be determined. Let the motions of the jet take place in free space (i.e., in an ambient fluid whose density and viscosity are too small to be taken into account). Then, the velocity and the pressure of a constant-density, inviscid jet, in a gravity-free environment, must satisfy the equations

$$\rho \frac{\partial \vec{v}}{\partial t} + \rho \vec{v} \cdot \nabla \vec{v} = -\nabla p \tag{3.1}$$

and

$$\nabla \cdot \vec{v} = 0 \tag{3.2}$$

in the region occupied by the jet, namely on the domain $r < R(\theta, z, t)$.

To these equations on the domain, must be added the equations

$$p = -\gamma \, 2H \tag{3.3}$$

and

$$\vec{n} \cdot \vec{v} = u \tag{3.4}$$

which must be satisfied at the boundary, $r = R(\theta, z, t)$. These equations introduce $\vec{n}$, H and u, where $\vec{n}$ is the unit outward normal to the jet surface, H is the mean curvature of this surface and u is its normal speed.[1] They express the requirement that the pressure at the surface of the jet differ from the ambient pressure, taken to be zero, by an amount proportional to the surface tension of the liquid, denoted γ, and the requirement that the liquid making up the jet not cross its own surface.

The normal, the mean curvature and the normal speed are determined by the formulas[2]

$$\vec{n} = \frac{\vec{i}_r - \dfrac{R_\theta}{R}\vec{i}_\theta - R_z\vec{i}_z}{\sqrt{1 + \dfrac{R_\theta^2}{R^2} + R_z^2}} \tag{3.5}$$

$$2H = \frac{\left[(1 + R_z^2)[-R^2 - 2R_\theta^2 + RR_{\theta\theta}] - 2R_\theta R_z[RR_{z\theta} - R_\theta R_z]\right]}{\left[R^2[1 + R_z^2] + R_\theta^2\right]^{\frac{3}{2}}}$$

$$+ \frac{\left[(1 + R_z^2)[R^2 + R_\theta^2]RR_{zz}\right]}{\left[R^2(1 + R_z^2) + R_\theta^2\right]^{\frac{3}{2}}} \tag{3.6}$$

and

$$u = \frac{R_t}{\sqrt{1 + \dfrac{R_\theta^2}{R^2} + R_z^2}} \tag{3.7}$$

These are obtained in Appendices B and C.

---

[1] Although u may be positive or negative, it is called the normal surface speed rather than the normal surface velocity.

[2] The symbol R_θ denotes $\dfrac{\partial R}{\partial \theta}$, etc.

There is a simple solution to the nonlinear equations which corresponds to a jet lying at rest as an infinite column of liquid in the shape of a right circular cylinder. It is

$$R = R_0$$

$$\vec{v} = \vec{0}$$

and

$$p = \frac{\gamma}{R_0}$$

where R_0 is a constant specifying the radius of the jet at rest.

This is the base solution and it is our aim to determine its stability. To do this, we plan to impose a small displacement on the surface of the jet and try to discover whether this displacement grows and perhaps moves the jet to a new configuration, or whether it does not. The shape of the surface is specified at only one instant of time. After that, it is part of the solution to the problem and it must be determined. It is denoted by the function R, which then defines the domain on which the equations must be solved. We seek to determine the conditions under which R remains near R_0 as time increases. These will be the conditions for the equilibrium jet to be stable.

We introduce a parameter, denoted ϵ, which allows us to look at a displacement as one member of a continuous family of displacements. Then, all of our variables, including R, depend on ϵ and we take this into account. In the case of the jet surface, we now write

$$r = R(\theta,\, z,\, t,\, \epsilon)$$

The parameter ϵ need not be defined explicitly, but it may denote the amplitude of the initial displacement of the surface of the jet.

The base problem corresponds to $\epsilon = 0$ and its configuration, namely

$$r \leq R_0$$

will be called the base or reference configuration. The variables r, θ, z and t in the base configuration will be denoted r_0, θ_0, z_0 and t_0 and the base velocity and pressure will be denoted $\vec{v}_0$ and p_0.

Any other problem corresponds to a value of $\epsilon > 0$ and its domain, namely

$$r \leq R(\theta,\, z,\, t,\, \epsilon)$$

will be called the current domain. The variables r, θ, z, t, $\vec{v}$ and p pertain to this and so too the nonlinear equations.

The plan is to turn the nonlinear problem on the current domain into a sequence of problems on the reference domain. This sequence corresponds to an expansion in powers of ϵ and it starts out with the base problem at zeroth order, being succeeded in turn by the perturbation problem at first order. Our interest is in the perturbation problem.

To produce these equations, introduce a mapping of the reference configuration into the current configuration. In this problem, it is enough to let

$$r = g(r_0, \theta_0, z_0, t_0, \epsilon) \tag{3.8}$$

$$\theta = \theta_0$$

$$z = z_0$$

and

$$t = t_0$$

where $g(\epsilon = 0) = r_0$. Then, expanding the mapping in powers of ϵ, the formulas in the second essay come into play.

Only the function g need be expanded and we write equation (3.8) as

$$r = r_0 + \epsilon r_1 + \frac{1}{2}\epsilon^2 r_2 + \cdots \tag{3.9}$$

where r_1, r_2, $\cdots$ depend on r_0, θ_0, z_0 and t_0, being abbreviations for derivatives of g, namely

$$r_1 = \frac{\partial g}{\partial \epsilon}(\epsilon = 0), \ r_2 = \frac{\partial^2 g}{\partial \epsilon^2}(\epsilon = 0), \ \text{etc.}$$

Now, the surface of the jet in the reference domain must be mapped into the surface of the jet in the current domain. To meet this requirement, the expansion of R must be

$$R = R_0 + \epsilon R_1 + \frac{1}{2}\epsilon^2 R_2 + \cdots \tag{3.10}$$

where

$$R_1 = r_1(r_0 = R_0), \ R_2 = r_2(r_0 = R_0), \ \text{etc.}$$

and where R_1, R_2, $\cdots$ depend on θ_0, z_0 and t_0.

3.2   The Perturbation Equations

To obtain the perturbation equations, the vector expansions presented in the first endnote to the second essay will be used, but only the special mapping introduced as equation (3.8) is needed, whence, in the notation of that endnote, we have $\theta_1 = 0 = z_1$ and $\vec{r}_1 \cdot \nabla_0 = r_1 \frac{\partial}{\partial r_0}$.

3.2.1   The Equations on the Reference Domain

Substituting the expansions

$$\vec{v} = \vec{v}_0 + \epsilon\left[\vec{v}_1 + r_1\frac{\partial\vec{v}_0}{\partial r_0}\right] + \cdots$$

and

$$p = p_0 + \epsilon\left[p_1 + r_1\frac{\partial p_0}{\partial r_0}\right] + \cdots$$

into equations (3.1) and (3.2), there obtains

$$\vec{0} = \rho\frac{\partial\vec{v}}{\partial t} + \rho\vec{v}\cdot\nabla\vec{v} + \nabla p$$

$$= \rho\left[\frac{\partial\vec{v}_0}{\partial t_0} + \epsilon\left[\frac{\partial\vec{v}_1}{\partial t_0} + r_1\frac{\partial}{\partial r_0}\frac{\partial\vec{v}_0}{\partial t_0}\right] + \cdots\right]$$

$$+ \rho\left[\vec{v}_0 + \epsilon\left[\vec{v}_1 + r_1\frac{\partial\vec{v}_0}{\partial r_0}\right] + \cdots\right]\cdot\left[\nabla_0\vec{v}_0 + \epsilon\left[\nabla_0\vec{v}_1 + r_1\frac{\partial}{\partial r_0}\nabla_0\vec{v}_0\right] + \cdots\right]$$

$$+ \nabla_0 p_0 + \epsilon\left[\nabla_0 p_1 + r_1\frac{\partial}{\partial r_0}\nabla_0 p_0\right] + \cdots$$

and

$$0 = \nabla\cdot\vec{v} = \nabla_0\cdot\vec{v}_0 + \epsilon\left[\nabla_0\cdot\vec{v}_1 + r_1\frac{\partial}{\partial r_0}\nabla_0\cdot\vec{v}_0\right] + \cdots$$

whence, to zeroth order, $\vec{v}_0$ and p_0 must satisfy

$$\vec{0} = \rho\frac{\partial\vec{v}_0}{\partial t_0} + \rho\vec{v}_0\cdot\nabla_0\vec{v}_0 + \nabla_0 p_0$$

and

$$0 = \nabla_0\cdot\vec{v}_0$$

while, to first order, $\vec{v}_1$ and p_1 must satisfy

$$\vec{0} = \rho\frac{\partial\vec{v}_1}{\partial t_0} + \rho\vec{v}_0\cdot\nabla_0\vec{v}_1 + \rho\vec{v}_1\cdot\nabla_0\vec{v}_0 + \nabla_0 p_1$$

$$+ r_1\frac{\partial}{\partial r_0}\left[\rho\frac{\partial\vec{v}_0}{\partial t_0} + \rho\vec{v}_0\cdot\nabla_0\vec{v}_0 + \nabla_0 p_0\right]$$

and

$$0 = \nabla_0\cdot\vec{v}_1 + r_1\frac{\partial}{\partial r_0}\nabla_0\cdot\vec{v}_0$$

The first-order equations then reduce to

$$\vec{0} = \rho\frac{\partial\vec{v}_1}{\partial t_0} + \rho\vec{v}_0\cdot\nabla_0\vec{v}_1 + \rho\vec{v}_1\cdot\nabla_0\vec{v}_0 + \nabla_0 p_1$$

and

$$0 = \nabla_0 \cdot \vec{v}_1$$

by using the zeroth-order equations. As the zeroth-order equations are satisfied by the base solution, namely by

$$\vec{v}_0 = \vec{0} \quad \text{and} \quad p_0 = \text{constant}$$

the perturbation equations become

$$\rho \frac{\partial \vec{v}_1}{\partial t_0} + \nabla_0 p_1 = \vec{0} \tag{3.11}$$

and

$$\nabla_0 \cdot \vec{v}_1 = 0 \tag{3.12}$$

It is worth observing that the mapping is not present in any of these equations.

3.2.2   The Equations on the Boundary of the Reference Domain

To turn the equations that hold on the boundary of the current domain into equations on the boundary of the reference domain, two kinds of variables must be expanded: domain variables evaluated at the boundary and surface variables defined only on the boundary. The domain variables do not present a problem, but the surface variables require a little work. These variables, the unit normal, the normal speed and the mean curvature, depend on the shape of the surface and acquire their expansions via the expansion of this shape.

Let the surface of the current jet be expanded along the mapping, according to equation (3.10), namely

$$R(\theta, z, t, \epsilon) = R_0 + \epsilon R_1(\theta_0, z_0, t_0) + \cdots \tag{3.13}$$

where R_0 is constant and does not depend on θ_0, z_0 or t_0. Turn first to the expansion of $\vec{n}$ and write this as

$$\vec{n} = \vec{n}_0 + \epsilon \vec{n}_1 + \cdots$$

where $\vec{n}_0$ and $\vec{n}_1$ need to be determined. To do this, substitute equation (3.13) into equation (3.5), after writing equation (3.5) in the form

$$\vec{n} = \frac{\vec{N}(R)}{D(R)}$$

to indicate that $\vec{n}$ depends explicitly on R alone due to the fact that $\theta = \theta_0$ along our mapping and, hence, $\vec{i}_r(\theta)$ and $\vec{i}_\theta(\theta)$ can be replaced by $\vec{i}_{r_0}$ and $\vec{i}_{\theta_0}$.

The expansion of $\vec{n}$, then, requires the expansions of $\vec{N}$ and D, and to first order, in the case where R_0 is constant, these are

$$\vec{N} = \vec{i}_{r_0} - \epsilon \left[\frac{1}{R_0} \frac{\partial R_1}{\partial \theta_0} \vec{i}_{\theta_0} - \frac{\partial R_1}{\partial z_0} \vec{i}_{z_0} \right]$$

and

$$D = 1$$

whence there obtains

$$\vec{n}_0 = \vec{i}_{r_0} \tag{3.14}$$

and

$$\vec{n}_1 = - \left[\vec{i}_{\theta_0} \frac{1}{R_0} \frac{\partial}{\partial \theta_0} + \vec{i}_{z_0} \frac{\partial}{\partial z_0} \right] R_1 \tag{3.15}$$

The expansion of u can be obtained by a similar calculation. Write

$$u = u_0 + \epsilon u_1 + \cdots$$

and observe that equation (3.7) can be written

$$u = \frac{1}{D(R)} \frac{\partial R}{\partial t}$$

Then, by substituting equation (3.13) into equation (3.7), again taking R_0 to be constant, there obtains

$$u = \epsilon \frac{\partial R_1}{\partial t_0}$$

whence

$$u_0 = 0 \tag{3.16}$$

and

$$u_1 = \frac{\partial R_1}{\partial t_0} \tag{3.17}$$

The expansions of $\vec{n}$ and u are enough to turn equation (3.4), namely

$$\vec{n} \cdot \vec{v} - u = 0$$

at the boundary of the current domain into a sequence of equations at the boundary of the reference domain. Substituting our expansions of $\vec{n}$, $\vec{v}$ and u into equation (3.4) and using equations (3.14), (3.15), (3.16) and (3.17) leads to

$$\left[\vec{i}_{r_0} - \epsilon \left[\vec{i}_{\theta_0} \frac{1}{R_0} \frac{\partial R_1}{\partial \theta_0} + \vec{i}_{z_0} \frac{\partial R_1}{\partial z_0} \right] + \cdots \right] \cdot \left[\vec{v}_0 + \epsilon \left[\vec{v}_1 + R_1 \frac{\partial \vec{v}_0}{\partial r_0} \right] + \cdots \right]$$

$$- \epsilon \frac{\partial R_1}{\partial t_0} + \cdots = 0$$

where $\vec{v}_0$, $\vec{v}_1$ and $\partial\vec{v}_0/\partial r_0$ must be evaluated at $r_0 = R_0$. To zeroth and first orders in ϵ, there obtains

$$\vec{i}_{r_0} \cdot \vec{v}_0 = 0$$

and

$$\vec{i}_{r_0} \cdot \left[\vec{v}_1 + R_1 \frac{\partial\vec{v}_0}{\partial r_0} \right] - \left[\vec{i}_{\theta_0} \frac{1}{R_0} \frac{\partial R_1}{\partial\theta_0} + \vec{i}_{z_0} \frac{\partial R_1}{\partial z_0} \right] \cdot \vec{v}_0 - \frac{\partial R_1}{\partial t_0} = 0$$

and, as $\vec{v}_0 = \vec{0}$, the first-order equation reduces to

$$\vec{i}_{r_0} \cdot \vec{v}_1 = v_{r_1} = \frac{\partial R_1}{\partial t_0} \tag{3.18}$$

and it is in this form that the no-flow condition is often used.

To turn equation (3.3) at the surface of the current domain into an equation at the surface of the reference domain requires the expansion of the mean curvature along the mapping. Writing this as

$$2H = 2H_0 + \epsilon 2H_1 + \cdots$$

and substituting equation (3.13) into equation (3.6) leads to a simple result in case R_0 is constant.[3] It is

$$2H_0 = -\frac{1}{R_0}$$

and

$$2H_1 = \left[\frac{1}{R_0^2} + \frac{1}{R_0^2} \frac{\partial^2}{\partial\theta_0^2} + \frac{\partial^2}{\partial z_0^2} \right] R_1$$

Then, substituting the expansions of p and $2H$ into equation (3.3), namely into

$$p + \gamma\, 2H(R) = 0$$

results in

$$p_0 + \epsilon \left[p_1 + R_1 \frac{\partial p_0}{\partial r_0} \right] + \cdots$$

$$+\gamma \left[-\frac{1}{R_0} + \epsilon \left[\frac{R_1}{R_0^2} + \frac{1}{R_0^2} \frac{\partial^2 R_1}{\partial\theta_0^2} + \frac{\partial^2 R_1}{\partial z_0^2} \right] + \cdots \right] = 0$$

which leads, to zeroth and first orders in ϵ, at $r_0 = R_0$, to

$$p_0 = \frac{\gamma}{R_0}$$

and

$$p_1 + R_1 \frac{\partial p_0}{\partial r_0} = -\gamma \left[\frac{R_1}{R_0^2} + \frac{1}{R_0^2} \frac{\partial^2 R_1}{\partial\theta_0^2} + \frac{\partial^2 R_1}{\partial z_0^2} \right]$$

---

[3]This is left to the reader.

Now, p_0 is constant, whence the first-order equation is just

$$p_1 = -\gamma \left[\frac{R_1}{R_0^2} + \frac{1}{R_0^2} \frac{\partial^2 R_1}{\partial \theta_0^2} + \frac{\partial^2 R_1}{\partial z_0^2} \right] \tag{3.19}$$

The perturbation problem is then made up of the first-order equations, equations (3.11), (3.12), (3.18) and (3.19). It is

$$\rho \frac{\partial \vec{v}_1}{\partial t_0} = -\nabla_0 \, p_1 \tag{3.20}$$

$$\left. \right\} \quad 0 < r_0 \leq R_0$$

$$\nabla_0 \cdot \vec{v}_1 = 0 \tag{3.21}$$

and

$$\vec{i}_{r_0} \cdot \vec{v}_1 = \frac{\partial R_1}{\partial t_0} \tag{3.22}$$

$$\left. \right\} \quad r_0 = R_0$$

$$p_1 = -\gamma \left[\frac{R_1}{R_0^2} + \frac{1}{R_0^2} \frac{\partial^2 R_1}{\partial \theta_0^2} + \frac{\partial^2 R_1}{\partial z_0^2} \right] \tag{3.23}$$

3.3   The Stability of Liquid Jets

The perturbation problem must be solved[4] to find $\vec{v}_1$ and p_1 on the reference domain and R_1 at its boundary. No part of the mapping survives, save R_1, which is needed to describe the small motions of the surface of the jet.

This problem is a linear, homogeneous, initial-value problem. To solve it, first eliminate $\vec{v}_1$ from equation (3.22) by using equation (3.20). Then, eliminate $\vec{v}_1$ from equation (3.20) by using equation (3.21). By doing this, there obtains a problem in p_1 and R_1, namely

$$\nabla_0^2 \, p_1 = 0, \quad 0 < r_0 \leq R_0$$

and

$$\vec{i}_{r_0} \cdot \nabla_0 \, p_1 = -\rho \frac{\partial^2 R_1}{\partial t_0^2}$$

$$\left. \right\} \quad r_0 = R_0$$

$$p_1 = -\gamma \left[\frac{R_1}{R_0^2} + \frac{1}{R_0^2} \frac{\partial^2 R_1}{\partial \theta_0^2} + \frac{\partial^2 R_1}{\partial z_0^2} \right]$$

---

[4]C.C. Lin and L.A. Segel, in their book *Mathematics Applied to Deterministic Problems in the Natural Sciences*, (MacMillan, New York 1974), on pages 515-529, take the readers through the thought process associated with carrying out a perturbation calculation. They raise all the important questions, but do so in an elementary way, while working out a concrete problem: the stability of a stratified fluid lying at rest in a gravitational field.

Then, turn this initial-value problem into an eigenvalue problem by substituting[5]

$$p_1 = \hat{p}_1(r_0)e^{\sigma t_0}e^{im\theta_0}e^{ikz_0}$$

and

$$R_1 = \hat{R}_1 e^{\sigma t_0}e^{im\theta_0}e^{ikz_0}$$

to obtain

$$\frac{1}{r_0}\frac{d}{dr_0}\left[r_0\frac{d\hat{p}_1}{dr_0}\right] - \left[\frac{m^2}{r_0^2} + k^2\right]\hat{p}_1 = 0, \quad 0 < r_0 \le R_0 \qquad (3.24)$$

and

$$\frac{d\hat{p}_1}{dr_0} = -\sigma^2\rho\hat{R}_1 \qquad (3.25)$$

$$\left.\begin{array}{c}\\[2ex]\\\end{array}\right\} \quad r_0 = R_0$$

$$\hat{p}_1 = -\gamma\left[\frac{1}{R_0^2} - \frac{m^2}{R_0^2} - k^2\right]\hat{R}_1 \qquad (3.26)$$

The eigenvalues are the values of σ at which this problem has a solution other than $\hat{p}_1 = 0 = \hat{R}_1$. They determine whether or not the jet is stable to a displacement in the form of an eigenfunction specified by the values of m and k. If the real part of σ is not positive, the jet is stable to the displacement; otherwise, it is unstable. It is σ^2 that the problem determines and this depends on m^2 and k^2.

To solve this eigenvalue problem, look first for the values of m^2 and k^2 such that $\sigma^2 = 0$ is a solution. Now, the bounded solution to equation (3.24) is[6]

$$\hat{p}_1 = AI_m(kr_0)$$

where A can be determined by equation (3.25). If σ^2 is set to zero, then $\hat{p}_1$ must satisfy

$$\frac{d\hat{p}_1}{dr_0}(r_0 = R_0) = 0$$

whence A, and therefore $\hat{p}_1$, must vanish. This leaves only equation (3.26), namely

$$0 = -\frac{\gamma}{R_0^2}[1 - m^2 - k^2 R_0^2]\hat{R}_1$$

---

[5]This is not unlike the way in which the reader would try to solve the constant coefficient differential equation

$$a_0\frac{d^n y}{dx^n} + a_1\frac{d^{n-1}y}{dx^{n-1}} + \cdots + a_n y = 0$$

by substituting $y = e^{mx}$ and trying to find m. Here, we substitute $e^{\sigma t_0}$ and try to find σ. Likewise we substitute $e^{im\theta_0}e^{ikz_0}$, but m and k are the inputs, whereas σ is the output.

[6]Equation (3.24) is Bessel's equation and its solutions are denoted I_m and K_m.

to be satisfied; and this can be satisfied, for $\hat{R}_1$ other than zero, if and only if $[1 - m^2 - k^2 R_0^2]$ is zero.

Now, periodicity requires the values of m to be $0, \pm 1, \pm 2, \cdots$, and, hence, if k^2 is greater than zero,[7] equation (3.26) can be satisfied if and only if $m = 0$ and, then, if and only if $k^2 R_0^2 = 1$. This determines the critical value of k^2. It is

$$k^2_{critical} R_0^2 = 1$$

and this is Rayleigh's criterion for the stability of a liquid jet, as we discovered it in the first essay.

Now, let σ^2 be other than zero and observe that $\hat{R}_1$ cannot be zero, otherwise both $\hat{R}_1$ and $\hat{p}_1$ must be zero. Then, $\hat{R}_1$ can be eliminated, and our problem, equations (3.24), (3.25) and (3.26), reduces to

$$\frac{1}{r_0}\frac{d}{dr_0}\left[r_0\frac{d\hat{p}_1}{dr_0}\right] - \left[\frac{m^2}{r_0^2} + k^2\right]\hat{p}_1 = 0, \quad 0 < r_0 < R_0$$

where $\hat{p}_1$ must be bounded at $r_0 = 0$, and

$$\sigma^2\rho\hat{p}_1 = \frac{\gamma}{R_0^2}[1 - m^2 - k^2 R_0^2]\frac{d\hat{p}_1}{dr_0}, \quad r_0 = R_0 \qquad (3.27)$$

Again, the bounded solution to equation (3.24) is

$$\hat{p}_1 = AI_m(kr_0)$$

and substituting this into equation (3.27) produces a formula for σ^2 in terms of m^2 and k^2. It is

$$\sigma^2 = \frac{\gamma}{\rho R_0^3}[1 - m^2 - k^2 R_0^2]\frac{kR_0 I'_m(kR_0)}{I_m(kR_0)} \qquad (3.28)$$

where $I'_m(x) = \dfrac{d}{dx}I_m(x)$ and where $\dfrac{xI'_m(x)}{I_m(x)}$ is positive for all x other than zero. Rayleigh obtained this equation.[8] It predicts the growth rate, $\pm\sqrt{\sigma^2}$, of a small displacement in the shape of the eigenfunction corresponding to assigned values of m^2 and k^2 and it tells us that σ^2 must be real.

Each value of σ^2 produces two values of σ (viz., $\pm\sqrt{\sigma^2}$). If $\vec{v}_1$, p_1 and R_1 denote the eigenfunction paired with $\sqrt{\sigma^2}$, then $\vec{v}_1$, $-p_1$ and $-R_1$ will be an independent eigenfunction paired with $-\sqrt{\sigma^2}$.

If m is any of $\pm 1, \pm 2, \cdots$ and if k^2 is positive, then σ^2 must be negative. Accordingly, σ must be purely imaginary. Displacements corresponding to these values of m and k^2 neither grow nor die, but cause the jet to execute small amplitude oscillations about its equilibrium configuration.

---

[7]If k^2 is zero, m equal to ± 1 is critical.

[8]He used it to explain the results of some rather remarkable experiments; cf., J.W.S. Rayleigh *Scientific Papers*, vol. i:377, 1899.

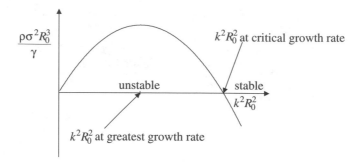

Figure 3.1. *The Graph of Equation (3.29)*

To determine the growth rate of unstable displacements, we must put m to zero in equation (3.28). Then, σ^2 as a function of k^2 is given by

$$\sigma^2 = \frac{\gamma}{\rho R_0^3}\,[1 - k^2 R_0^2]\,\frac{kR_0 I_0'(kR_0)}{I_0(kR_0)} \tag{3.29}$$

and we find, as Rayleigh did, that

$$k^2 R_0^2 > 1 \;\Rightarrow\; \sigma^2 < 0$$

$$k^2 R_0^2 = 1 \;\Rightarrow\; \sigma^2 = 0 \text{ whence } 1/R_0 \text{ is the critical wave number}$$

$$k^2 R_0^2 < 1 \;\Rightarrow\; \sigma^2 > 0$$

Rayleigh drew these conclusions while drawing attention to their physical origins. Equation (3.29) is illustrated graphically in Figure 3.1. This figure indicates that there is a fastest growing displacement corresponding to a long wavelength such that $k^2 R_0^2 < 1$. Rayleigh determined this wavelength in order to predict the results of jet-breakup experiments. It seemed to him that a jet ought to break up at the fastest growing displacement.

Now, the viscosity of the fluid making up the jet presented some problems for Rayleigh as he carried out many experiments, mostly using very viscous liquids. Yet, whether or not a jet is stable to a small displacement depends only on the wavelength of the displacement and the diameter of the jet, not at all on the viscosity of the liquid out of which it is formed. This can be explained by drawing a picture.

To draw a picture illustrating why a critical condition ought to be expected and why it ought not depend on the viscosity of the liquid making up the jet, let a jet in free space have an axisymmetric disturbance imposed on its surface. This results in a ripple of crests and troughs along the jet. The jet cross sections remain circular, but their diameters will be greater under a crest, lesser under a trough. Now, these circles increase the pressure in the jet above its ambient value and they do so more strongly under a trough than under a crest. Then, looking at a crest and its adjacent trough, notice that this produces an unstable arrangement of pressure tending to

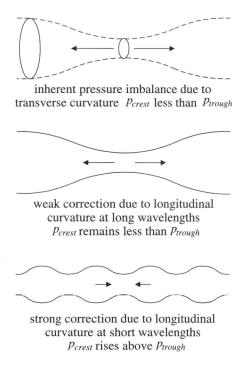

inherent pressure imbalance due to
transverse curvature P_{crest} less than P_{trough}

weak correction due to longitudinal
curvature at long wavelengths
P_{crest} remains less than P_{trough}

strong correction due to longitudinal
curvature at short wavelengths
P_{crest} rises above P_{trough}

Figure 3.2. *The Inherent Pressure Imbalance and Its Correction*

move fluid out of the trough into the crest, tending to reinforce the imposed displacement. This obtains no matter the wavelength of the disturbance; it is simply a consequence of the fact of a disturbance.

However, there is a second set of circles arising due to the diameter variation along the jet. These lie in a plane through its axis: half the circles lying under crests, the other half lying above troughs. The circles lying under a crest increase the pressure of the jet, those lying above a trough decrease the pressure, both vis-à-vis its ambient value. This tends to turn the unstable pressure imbalance around. It is stabilizing. It increases the pressure under a crest and decreases the pressure under a trough, and it does this by an amount that depends on the wavelength of the disturbance. Long wavelength displacements correct the inherent pressure imbalance very little, whereas short wavelength disturbances correct it a great deal. This is illustrated in Figure 3.2.

By this, the jet must be unstable to long wavelength displacements due to the adverse pressure imbalance imposed by the displacement; however the jet must be stable to short enough wavelength displacements which turn the pressure variation around. At some intermediate wavelength, these opposing causes of pressure variation come into balance and the pressure remains uniform. This is the critical wavelength. It corresponds to a solution to the perturbation equations wherein p_1 must be uniform. Should viscosity be included in the calculation, then, at the critical condition where p_1 is uniform, $\vec{v}_1$ will turn out to be zero and the viscosity of the fluid cannot make a difference.

This is the correct picture of what is going on whether or not the viscosity of the fluid is taken into account. Although the viscosity can determine how fast an unstable pressure variation can reinforce itself, it can not determine what is required to make the pressure uniform (i.e., the critical condition where the jet remains at rest). The balance in this problem is between the pressure under a crest and the pressure under a trough, and the balance is struck when these pressures are equal. This has only to do with the geometry of the surface. It has nothing to do with the viscosity of the fluid.

The reader should not get the idea that including viscosity in the calculation would be such a hard thing[9]; yet, instead of doing this, our next step will be to introduce a control variable to make it possible to run an experiment in which the jet can be seen to go unstable at the critical point. This is done in the next essay.

---

[9]See page 540 in Chandrasekhar's book (cf., the Preface).

3.4   DISCUSSION NOTES

3.4.1   The Effect of Streaming on the Stability of a Liquid Jet

The first and third essays draw attention to the physical mechanism underlying the instability of a liquid jet undergoing its motions in free space. The work in this discussion note adds a little more to the problem. Now, free space is replaced by an ambient fluid, say air, and then the jet is made to move uniformly at constant speed along its axis vis-à-vis its surroundings.

Neither the viscosity of the liquid nor that of the surrounding fluid is taken into account and only axisymmetric displacements are entertained. Gravity is turned off.

The notation remains as it was in the essay, but what was free space is now filled by a fluid and the new variables introduced thereby are distinguished by a superscript *.

There is no need to write the nonlinear equations, for they are as they were in the essay, save certain additions to take the ambient fluid into account and to allow for the fact that the jet is now moving at a uniform speed along its axis. There is a simple solution to these equations, denoted by the subscript zero. It is called the base solution and it is

$$\vec{v}_0^* = \vec{0}$$

$$p_0^* = p_{ref}$$

$$\vec{v}_0 = W_0 \vec{k}$$

and

$$p_0 = p_0^* + \frac{\gamma}{R_0}$$

where R_0 denotes the fixed radius of the jet in the base case and where W_0 denotes the speed at which the jet translates to the right. The reference domain on which the perturbation equations must be solved is then $0 < r_0 < R_0$.

The perturbation equations obtain in the usual way. They are

$$\frac{\partial \vec{v}_1^*}{\partial t} = -\frac{1}{\rho^*} \, \nabla p_1^* \qquad\qquad\qquad\qquad (3.30)$$

$$\nabla \cdot \vec{v}_1^* = 0 \qquad\qquad\qquad\qquad\qquad (3.31)$$

$$r_0 > R_0$$

$$\frac{\partial \vec{v}_1}{\partial t} + W_0 \frac{\partial \vec{v}_1}{\partial z_0} = -\frac{1}{\rho} \nabla p_1 \qquad (3.32)$$

$$r_0 < R_0$$

$$\nabla \cdot \vec{v}_1 = 0 \qquad (3.33)$$

and

$$v_{r_1} - \frac{\partial R_1}{\partial z_0} W_0 = \frac{\partial R_1}{\partial t_0} = v^*_{r_1} \qquad (3.34)$$

$$r_0 = R_0$$

$$p_1 = p^*_1 - \gamma\, 2H_1 \qquad (3.35)$$

To turn these equations into an eigenvalue problem, first substitute

$$v_{r_1} = \hat{v}_{r_1} e^{\sigma t_0} e^{ikz_0}$$

$$v_{\theta_1} = \hat{v}_{\theta_1} e^{\sigma t_0} e^{ikz_0}$$

$$v_{z_1} = \hat{v}_{z_1} e^{\sigma t_0} e^{ikz_0}$$

and

$$p_1 = \hat{p}_1 e^{\sigma t_0} e^{ikz_0}$$

into equations (3.32) and (3.33) to obtain

$$[\sigma + ikW_0]\hat{v}_{r_1} = -\frac{1}{\rho} \frac{d\hat{p}_1}{dr_0} \qquad (3.36)$$

$$[\sigma + ikW_0]\hat{v}_{\theta_1} = 0 \qquad (3.37)$$

$$[\sigma + ikW_0]\hat{v}_{z_1} = -ik\frac{1}{\rho}\, \hat{p}_1 \qquad (3.38)$$

and

$$\frac{d\hat{v}_{r_1}}{dr_0} + \frac{\hat{v}_{r_1}}{r_0} + ik\hat{v}_{z_1} = 0 \qquad (3.39)$$

where, again, only axisymmetric solutions are being sought.

Then, eliminate $\hat{v}_{z_1}$ from equations (3.38) and (3.39) to get

$$[\sigma + ikW_0]\left[\frac{d\hat{v}_{r_1}}{dr_0} + \frac{\hat{v}_{r_1}}{r_0}\right] + k^2 \frac{1}{\rho}\, \hat{p}_1 = 0$$

which can be substituted into equation (3.36) to obtain

$$\frac{d^2\hat{p}_1}{dr_0^2} + \frac{1}{r_0} \frac{d\hat{p}_1}{dr_0} - k^2 \hat{p}_1 = 0$$

By this, $\hat{p}_1$ and $\hat{v}_{r_1}$ can be obtained in terms of two constants, that is

$$\hat{p}_1 = AI_0(kr_0) + BK_0(kr_0) \qquad (3.40)$$

and

$$[\sigma + ikW_0]\hat{v}_{r_1} = -\frac{1}{\rho}\frac{d\hat{p}_1}{dr_0} = -\frac{1}{\rho}\Big[AkI_0'(kr_0) + BkK_0'(kr_0)\Big] \qquad (3.41)$$

Now, turn to equations (3.30) and (3.31) and repeat the calculation just carried out on equations (3.32) and (3.33). This leads to formulas for $\hat{p}_1^*$ and $\hat{v}_{r_1}^*$ in terms of two constants. They are

$$\hat{p}_1^* = A^* I_0(kr_0) + B^* K_0(kr_0) \qquad (3.42)$$

and

$$\sigma\hat{v}_{r_1}^* = -\frac{1}{\rho^*}\frac{d\hat{p}_1^*}{dr_0} = -\frac{1}{\rho^*}\Big[A^* kI_0'(kr_0) + B^* kK_0'(kr_0)\Big] \qquad (3.43)$$

The domain equations have now been satisfied but four constants remain to be determined. Two of these constants must be zero (viz., B must be zero in order that $\hat{v}_{r_1}$ and $\hat{p}_1$ remain finite at $r_0 = 0$ and A^* must be zero in order that $\hat{v}_{r_1}^*$ and $\hat{p}_1^*$ remain finite as r_0 grows large). Then, to determine the remaining two constants and at the same time determine the eigenvalues corresponding to a displacement of wave number k, turn to equations (3.34) and (3.35) and substitute

$$R_1 = \hat{R}_1 e^{\sigma t_0} e^{ikz_0}$$

to obtain, at $r_0 = R_0$,

$$[\sigma + ikW_0]\hat{v}_{r_1} = [\sigma + ikW_0]^2 \hat{R}_1 \qquad (3.44)$$

$$\sigma\hat{v}_{r_1}^* = \sigma^2 \hat{R}_1 \qquad (3.45)$$

and

$$\hat{p}_1 = \hat{p}_1^* - \frac{\gamma}{R_0^2}[1 - R_0^2 k^2]\hat{R}_1 \qquad (3.46)$$

where

$$2H_1 = \frac{R_1}{R_0^2} + R_{1z_0z_0}$$

has been used. Then, substitute equations (3.40), (3.41), (3.42) and (3.43) into equations (3.44), (3.45) and (3.46), with $B = 0 = A^*$, to obtain

$$-\frac{1}{\rho} AkI_0'(kR_0) = [\sigma + ikW_0]^2 \hat{R}_1$$

$$-\frac{1}{\rho^*} B^* kK_0'(kR_0) = \sigma^2 \hat{R}_1$$

and

$$AI_0(kR_0) = B^* K_0(kR_0) - \frac{\gamma}{R_0^2}[1 - R_0^2 k^2]\hat{R}_1$$

This is a set of three, linear, homogeneous, algebraic equations in the unknown constants A, B^* and $\hat{R}_1$. To get solutions other than A, B^* and $\hat{R}_1$ all zero, the eigenvalues must be the roots of the equation

$$\det \begin{pmatrix} \dfrac{1}{\rho} kI_0'(kR_0) & 0 & [\sigma + ikW_0]^2 \\[2ex] 0 & \dfrac{1}{\rho^*} kK_0'(kR_0) & \sigma^2 \\[2ex] I_0(kR_0) & -K_0(kR_0) & \dfrac{\gamma}{R_0^2}[1 - R_0^2 k^2] \end{pmatrix} = 0$$

which can be written

$$[\sigma + ikW_0]^2 - \sigma^2 \frac{\rho^*}{\rho} \frac{K_0(kR_0)}{K_0'(kR_0)} \frac{I_0'(kR_0)}{I_0(kR_0)}$$

$$- \frac{\gamma}{\rho R_0^3}[1 - k^2 R_0^2] \frac{kR_0 I_0'(kR_0)}{I_0(kR_0)} = 0 \tag{3.47}$$

In case W_0 is zero, equation (3.47) can be solved easily to get

$$\sigma^2 = \frac{\dfrac{\gamma}{\rho R_0^3}[1 - k^2 R_0^2] \dfrac{kR_0 I_0'(kR_0)}{I_0(kR_0)}}{1 - \dfrac{\rho^*}{\rho} \dfrac{I_0'(kR_0)}{I_0(kR_0)} \dfrac{K_0(kR_0)}{K_0'(kR_0)}}$$

which extends the results in the third essay to take into account the presence of an ambient fluid. Upon noticing that the factor

$$- \frac{I_0'(kR_0)}{I_0(kR_0)} \frac{K_0(kR_0)}{K_0'(kR_0)}$$

does not depart significantly from one for all values of k^2, $0 < k^2 < \infty$, it can be observed that the presence of the ambient fluid simply lowers the growth rate of the unstable displacements as well as the oscillation frequency of the stable displacements, but it does not turn stable displacements unstable, nor unstable displacements stable. All that needs to be done to get this formula is to go to the free-space formula, namely

$$\sigma^2 = \frac{\gamma}{\rho R_0^3}[1 - k^2 R_0^2] \frac{kR_0 I_0'(kR_0)}{I_0(kR_0)}$$

and replace ρ by

$$\rho + \rho^* \left[-\frac{I_0'(kR_0)}{I_0(kR_0)} \frac{K_0(kR_0)}{K_0'(kR_0)} \right]$$

which accounts for the additional inertia due to the presence of the surrounding fluid.

Something new is found in case the jet is in uniform axial motion, unless, of course, the motion takes place in free space, whereupon the results in the third essay are recovered. It turns out that the axial motion is destabilizing, turning stable displacements unstable. To see this, write equation (3.47) for σ in the form

$$\sigma^2 \left[1 - \frac{\rho^*}{\rho} \frac{I_0'(kR_0)}{I_0(kR_0)} \frac{K_0(kR_0)}{K_0'(kR_0)} \right] + 2ikW_0\sigma - k^2W_0^2$$

$$- \frac{\gamma}{\rho R_0^3} \left[1 - k^2R_0^2 \right] \frac{kR_0 I_0'(kR_0)}{I_0(kR_0)} = 0$$

then notice that this is a quadratic equation for σ and that one-quarter of its discriminant is

$$-k^2W_0^2 + \left[1 - \frac{\rho^*}{\rho} \frac{I_0'(kR_0)}{I_0(kR_0)} \frac{K_0(kR_0)}{K_0'(kR_0)} \right]$$

$$\times \left[k^2W_0^2 + \frac{\gamma}{\rho R_0^3} \left[1 - k^2R_0^2 \right] \frac{kR_0 I_0'(kR_0)}{I_0(kR_0)} \right]$$

If, to any value of k^2, the discriminant is positive, one of the two values of σ must have a positive real part whereupon the jet is unstable; if it is negative, both values of σ must be purely imaginary and the jet is stable. If the discriminant vanishes, the double root

$$\frac{-ikW_0}{\left[1 - \frac{\rho^*}{\rho} \frac{I_0'(kR_0)}{I_0(kR_0)} \frac{K_0(kR_0)}{K_0'(kR_0)} \right]}$$

turns up, which tells us that in the case of a neutral displacement the stationary wave of the third essay would now be traveling at the speed W_0, if it were not for the inertia of the surrounding fluid.

For small values of k^2, the discriminant is positive and the jet is unstable. For large values of k^2, it is negative and the jet is stable. The surface tension term controls the sign of the discriminant in both limits. It is stabilizing at large values of k^2, it is destabilizing at small values of k^2, and this is for the reasons explained in the essay. But the streaming does have an effect, and to see what it is, write one-quarter of the discriminant as

$$- \frac{\rho^*}{\rho} \frac{I_0'(kR_0)}{I_0(kR_0)} \frac{K_0(kR_0)}{K_0'(kR_0)} k^2 R_0^2 \frac{W_0^2}{R_0^2}$$

$$+ \left[1 - \frac{\rho^*}{\rho} \frac{I_0'(kR_0)}{I_0(kR_0)} \frac{K_0(kR_0)}{K_0'(kR_0)} \right] \frac{\gamma}{\rho R_0^3} \left[1 - k^2R_0^2 \right] \frac{kR_0 I_0'(kR_0)}{I_0(kR_0)}$$

and set this to zero to determine the value of $k_{critical}^2$. If ρ^* is zero, the result is $k_{critical}^2 R_0^2 = 1$. Then, if ρ^* is small, the result must be near this, and writing it as

$$k^2_{critical}R_0^2 = 1 + \delta^2$$

the correction, δ^2, turns out to be

$$\delta^2 = -\frac{\rho^*}{\rho}\frac{K_0(1)}{K_0'(1)}\frac{\rho R_0^3}{\gamma}\frac{W_0^2}{R_0^2}$$

By this, streaming (i.e., the translation of the jet through an ambient fluid) is seen to be destabilizing. The effect is to shift the critical wave number to the right, making otherwise stable displacements now unstable.

3.4.2   The Two-Fluid Jet

The term *two-fluid jet* is intended to remind the reader of the problem presented in the second discussion note to the first essay. It was introduced to test Rayleigh's work principle. In that essay, one fluid was confined by two surfaces, hence the name *annular jet*. Here, two fluids are required, as the hollow space there is now filled, and so the name *two-fluid jet*.

The perturbation equations are now to be solved for the case of a jet, made up of two fluids, executing its motions in free space where the ambient pressure and density need not be taken into account. The jet is infinitely long and the variables referring to the inside fluid are distinguished by a superscript *.

The nonlinear equations must take the inside fluid into account but require little explanation beyond the observation that the immiscibility requirement at the inside surface leads to two conditions

$$\vec{n}\cdot\vec{v} = u = \vec{n}\cdot\vec{v}^*$$

The base solution is simple. It is

$$\vec{v}_0 = \vec{0}, \quad p_0 = \frac{\gamma}{R_0}$$

and

$$\vec{v}_0^* = \vec{0}, \quad p_0^* - p_0 = \frac{\gamma^*}{R_0^*}$$

where R_0^* and R_0 denote the radii of the two cylinders which, in the equilibrium configuration, divide the inner and the outer fluids and the outer fluid and free space.

The perturbation equations, written now only for axisymmetric displacements, are then

$$\rho\frac{\partial \vec{v}_1}{\partial t_0} = -\nabla_0\, p_1$$

$$\nabla_0 \cdot \vec{v}_1 = 0$$

$$\left.\begin{array}{c} \\ \\ \\ \\ \end{array}\right\} \quad R_0^* < r < R_0$$

$$\vec{i}_{r_0} \cdot \vec{v}_1 = \frac{\partial R_1}{\partial t_0}$$

$$p_1 = -\frac{\gamma}{R_0^2}\left[R_1 + R_0^2 \frac{\partial^2 R_1}{\partial z_0^2}\right] \qquad \left.\right\} \quad r_0 = R_0$$

$$\vec{i}_{r_0} \cdot \vec{v}_1 = \frac{\partial R_1^*}{\partial t_0}$$

$$p_1^* - p_1 = -\frac{\gamma^*}{R_0^{*2}}\left[R_1^* + R_0^{*2}\frac{\partial^2 R_1^*}{\partial z_0^2}\right] \qquad \left.\right\} \quad r_0 = R_0^*$$

$$\vec{i}_{r_0} \cdot \vec{v}_1^* = \frac{\partial R_1^*}{\partial t_0}$$

and

$$\rho^* \frac{\partial \vec{v}_1^*}{\partial t_0} = -\nabla_0\, p_1^*$$

$$\left.\right\} \quad 0 < r_0 < R_0^*$$

$$\nabla_0 \cdot \vec{v}_1^* = 0$$

where p_1^* and $\vec{v}_1^*$ must remain bounded at $r_0 = 0$.

Eliminating $\vec{v}_1$ in favor of p_1 and $\vec{v}_1^*$ in favor of p_1^* and then substituting

$$p_1 = \hat{p}_1(r_0)e^{\sigma t_0}e^{ikz_0}$$

$$R_1 = \hat{R}_1 e^{\sigma t_0}e^{ikz_0}$$

$$p_1^* = \hat{p}_1^*(r_0)e^{\sigma t_0}e^{ikz_0}$$

and

$$R_1^* = \hat{R}_1^* e^{\sigma t_0}e^{ikz_0}$$

leads to the eigenvalue problem. It is

$$\frac{1}{r_0}\frac{d}{dr_0}\left[r_0 \frac{d\hat{p}_1}{dr_0}\right] - k^2\hat{p}_1 = 0, \quad R_0^* < r_0 < R_0 \qquad (3.48)$$

$$\frac{d\hat{p}_1}{dr_0} - -\rho\sigma^2\hat{R}_1 \qquad\qquad (3.49)$$

$$\left.\right\} \quad r_0 = R_0$$

$$\hat{p}_1 = -\frac{\gamma}{R_0^2}[1 - k^2R_0^2]\hat{R}_1 \qquad\qquad (3.50)$$

$$\frac{d\hat{p}_1}{dr_0} = -\rho\sigma^2\hat{R}_1^* \tag{3.51}$$

$$\hat{p}_1^* - \hat{p}_1 = -\frac{\gamma^*}{R_0^{*2}}[1 - k^2R_0^{*2}]\hat{R}_1^* \qquad\qquad r_0 = R_0^* \tag{3.52}$$

$$\frac{d\hat{p}_1^*}{dr_0} = -\rho^*\sigma^2\hat{R}_1^* \tag{3.53}$$

and

$$\frac{1}{r_0}\frac{d}{dr_0}\left[r_0\frac{d\hat{p}_1^*}{dr_0}\right] - k^2\hat{p}_1^* = 0, \quad 0 < r_0 < R_0^* \tag{3.54}$$

where, again, $\hat{p}_1^*$ must be bounded at $r_0 = 0$.

The solution to this problem can be obtained by first writing

$$\hat{p}_1 = AI_0(kr_0) + BK_0(kr_0)$$

and

$$\hat{p}_1^* = A^*I_0(kr_0)$$

which satisfy equations (3.48) and (3.54) and the fact that $\hat{p}_1^*$ must be bounded at $r_0 = 0$, and then by using the five boundary conditions to determine the three constants A, B, and A^* as well as the displacements $\hat{R}_1$ and $\hat{R}_1^*$.

Now, equation (3.53) can be used to eliminate A^* in favor of $\hat{R}_1^*$, namely

$$A^* = \frac{-\rho^*\sigma^2}{kI_0'(kR_0^*)}\hat{R}_1^*$$

To determine A, B, $\hat{R}_1$ and $\hat{R}_1^*$, equations (3.49), (3.50), (3.51) and (3.52) remain to be satisfied. Substituting for $\hat{p}_1$, $\hat{p}_1^*$ and A^*, they are

$$AkI_0'(kR_0) + BkK_0'(kR_0) = -\rho\sigma^2\hat{R}_1 \tag{3.55}$$

$$AI_0(kR_0) + BK_0(kR_0) = -\frac{\gamma}{R_0^2}[1 - k^2R_0^2]\hat{R}_1 \tag{3.56}$$

$$AkI_0'(kR_0^*) + BkK_0'(kR_0^*) = -\rho\sigma^2\hat{R}_1^* \tag{3.57}$$

and

$$AI_0(kR_0^*) + BK_0(kR_0^*) = \left[\frac{\gamma^*}{R_0^{*2}}[1 - k^2R_0^{*2}] - \rho^*\sigma^2\frac{I_0(kR_0^*)}{kI_0'(kR_0^*)}\right]\hat{R}_1^* \tag{3.58}$$

Equations (3.55) and (3.56) can be solved for A and B in terms of $\hat{R}_1$; likewise equations (3.57) and (3.58) can be solved for A and B in terms of $\hat{R}_1^*$. The results are

$$A = R_0\hat{R}_1\left[-\rho\sigma^2K_0(kR_0) + \frac{\gamma}{R_0^3}[1 - k^2R_0^2]kR_0K_0'(kR_0)\right] \tag{3.59}$$

and

$$B = R_0 \hat{R}_1 \left[\rho \sigma^2 I_0(kR_0) - \frac{\gamma}{R_0^3}[1 - k^2 R_0^2] kR_0 I_0'(kR_0) \right] \tag{3.60}$$

and

$$A = R_0^* \hat{R}_1^* \left[-\rho \sigma^2 K_0(kR_0^*) - \left[\frac{\gamma^*}{R_0^{*3}}[1 - k^2 R_0^{*2}] \right. \right.$$

$$\left. \left. - \rho^* \sigma^2 \frac{I_0(kR_0^*)}{kR_0^* I_0'(kR_0^*)} \right] kR_0^* K_0'(kR_0^*) \right] \tag{3.61}$$

and

$$B = R_0^* \hat{R}_1^* \left[\rho \sigma^2 I_0(kR_0^*) + \left[\frac{\gamma^*}{R_0^{*3}}[1 - k^2 R_0^{*2}] \right. \right.$$

$$\left. \left. - \rho^* \sigma^2 \frac{I_0(kR_0^*)}{kR_0^* I_0'(kR_0^*)} \right] kR_0^* I_0'(kR_0^*) \right] \tag{3.62}$$

If $\sigma^2 = 0 = [1 - k^2 R_0^2]$, these equations have the solution

$$A = 0 = B, \quad \hat{R}_1 \neq 0, \quad \hat{R}_1^* = 0$$

whereas if $\sigma^2 = 0 = [1 - k^2 R_0^{*2}]$, they have the solution

$$A = 0 = B, \quad \hat{R}_1 = 0, \quad \hat{R}_1^* \neq 0$$

This tells us that $k^2 = 1/R_0^2$ and $k^2 = 1/R_0^{*2}$ are both critical points.

Then, to determine σ^2, use equations (3.59) and (3.61) to eliminate A and equations (3.60) and (3.62) to eliminate B. By doing this, there obtains

$$R_0 \hat{R}_1 \left[-\rho \sigma^2 K_0(kR_0) + \frac{\gamma}{R_0^3}[1 - k^2 R_0^2] kR_0 K_0'(kR_0) \right]$$

$$= R_0^* \hat{R}_1^* \left[-\rho \sigma^2 K_0(kR_0^*) - \left[\frac{\gamma^*}{R_0^{*3}}[1 - k^2 R_0^{*2}] \right. \right.$$

$$\left. \left. - \rho^* \sigma^2 \frac{I_0(kR_0^*)}{kR_0^* I_0'(kR_0^*)} \right] kR_0^* K_0'(kR_0^*) \right]$$

and

$$R_0 \hat{R}_1 \left[\rho \sigma^2 I_0(kR_0) - \frac{\gamma}{R_0^3}[1 - k^2 R_0^2] kR_0 I_0'(kR_0) \right]$$

$$= R_0^* \hat{R}_1^* \left[\rho \sigma^2 I_0(kR_0^*) + \left[\frac{\gamma^*}{R_0^{*3}}[1 - k^2 R_0^{*2}] \right. \right.$$

$$\left. \left. - \rho^* \sigma^2 \frac{I_0(kR_0^*)}{kR_0^* I_0'(kR_0^*)} \right] kR_0^* I_0'(kR_0^*) \right]$$

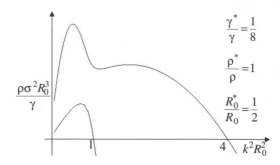

Figure 3.3. *The Prediction of Equation (3.63)*

which is a system of two, homogeneous, linear, algebraic equations in $\hat{R}_1$ and $\hat{R}_1^*$. It has a solution other than $\hat{R}_1 = 0 = \hat{R}_1^*$ if and only if

$$F_1(\sigma^2)F_2(\sigma^2) - F_3(\sigma^2)F_4(\sigma^2) = 0 \qquad (3.63)$$

where

$$F_1(\sigma^2) = \left[-\rho\sigma^2 K_0(kR_0) + \frac{\gamma}{R_0^3}[1 - k^2 R_0^2]kR_0 K_0'(kR_0) \right]$$

$$F_2(\sigma^2) = \left[\rho\sigma^2 I_0(kR_0^*) + \left[\frac{\gamma^*}{R_0^{*3}}[1 - k^2 R_0^{*2}] \right. \right.$$

$$\left. \left. -\rho^*\sigma^2 \frac{I_0(kR_0^*)}{kR_0^* I_0'(kR_0^*)} \right] kR_0^* I_0'(kR_0^*) \right]$$

$$F_3(\sigma^2) = \left[\rho\sigma^2 I_0(kR_0) - \frac{\gamma}{R_0^3}[1 - k^2 R_0^2]kR_0 I_0'(kR_0) \right]$$

and

$$F_4(\sigma^2) = \left[-\rho\sigma^2 K_0(kR_0^*) - \left[\frac{\gamma^*}{R_0^{*3}}[1 - k^2 R_0^{*2}] \right. \right.$$

$$\left. \left. -\rho^*\sigma^2 \frac{I_0(kR_0^*)}{kR_0^* I_0'(kR_0^*)} \right] kR_0^* K_0'(kR_0^*) \right]$$

This is a quadratic equation in σ^2 which determines two real values of σ^2 at each value of k^2. If $k^2 = 1/R_0^2$, one of the roots is zero, the other is positive; if $k^2 = 1/R_0^{*2}$, one of the roots is again zero but now the other is negative. The predictions of equation (3.63) are illustrated in Figure 3.3.

The two critical points, where σ^2 vanishes, were found earlier. But now it is certain that the stability of a two-fluid jet is determined by the radius of the inner fluid.

4

The Stability of a Liquid Bridge Lying Between Two Plane Walls

4.1 The End Conditions

It is our main aim in this essay to introduce a control variable. To do this, the liquid jet is required to lie between two parallel plane walls. Its equilibrium radius is denoted by R_0, as before, but its length is now finite and it is denoted by L_0. This is the control variable.

These new plane walls require new boundary conditions to be added to the problem and these new boundary conditions go a long way to determining the form of the solutions to the perturbation eigenvalue problem. However, only in special cases can the eigenfunctions be defined by their wavelengths. It is our second aim to see what needs to be done in the less special cases.

The nonlinear equations, the base equations and the perturbation equations written for the infinite jet in the third essay can be carried over, intact, to this essay. To these, equations must be added to account for the way in which the jet itself and its surface meet the solid walls.

To do this, two kinds of equations are introduced: the no-flow equations and the contact equations. The first specify that the solid walls must be impermeable to the jet; the second specify how its surface contacts these walls. Let the walls be solid planes lying at $z = 0$ and at $z = L_0$. Then no flow at these walls requires the fluid velocity to satisfy

$$\vec{i}_z \cdot \vec{v} = 0$$

and if the surface of the jet is denoted by

$$r = R(\theta, \; z, \; t, \; \epsilon)$$

the contact equations specify the conditions to be satisfied by R at these walls.

Two kinds of end conditions will be of interest to us, denoted as free and fixed. At a free end, the surface contacts the wall at right angles and R is required to satisfy

$$\frac{\partial R}{\partial z} = 0$$

whereas at a fixed end, R is specified and must then satisfy

$$R = R_0$$

where R_0 is assigned.

End conditions will be called free-free or free-fixed or fixed-fixed according to whether

$$\frac{\partial R}{\partial z}(z = 0) = 0 = \frac{\partial R}{\partial z}(z = L_0)$$

or

$$\frac{\partial R}{\partial z}(z = 0) = 0 \; , \; R(z = L_0) = R_0$$

or

$$R(z = 0) = R_0 = R(z = L_0)$$

is required to hold at the solid walls.

A photograph may help the readers fix their ideas. Figure 4.1 presents two photographs taken by Mason. It depicts a liquid bridge, first at a length just short of its critical length and then at a length just beyond its critical length, presumably on its way to breakup. The end conditions appear to be fixed-fixed and the corresponding critical length, as we will find later on, turns out to be

$$L_{0_{critical}} = 2\pi R_0 = \pi D_0$$

not

$$L_{0_{critical}} = \pi R_0$$

as it would be were the ends free instead of fixed.

Fixed ends would seem to be easy to establish in an experiment, by doing just as Mason has done, and his photographs confirm the predictions of calculations we aim to present.

To achieve neutral buoyancy and to eliminate the effect of gravity, Mason introduced a second, or ambient, fluid to take the place of free space. Of course, neither fluid is inviscid. The effect of viscosity, as well as the effect of the density of the second fluid, on the critical length is nil. The effect of

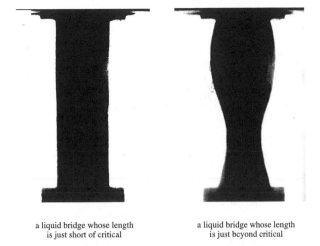

a liquid bridge whose length
is just short of critical

a liquid bridge whose length
is just beyond critical

Figure 4.1. *Two Photographs Taken by Mason Illustrating the Instability*
Reprinted from "An Experimental Determination of the Stable Length
of Cylindrical Liquid Bubbles" by [G. Mason], in Journal of Colloid
and Interface Science, Volume 32(1), 174, [1970], Elsevier Science (USA),
reproduced with permission from the publisher

density can already be seen in the first discussion note to the third essay;
the effect of viscosity will be taken up later on in this essay. It is worth
noticing, by looking at the bridge just short of its critical length, that
Mason was able to exercise good control over the shape of the surface in
the reference configuration.

The end conditions do not change the base solution to the nonlinear
equations. It remains as it was in the third essay, namely

$$\vec{v}_0 = \vec{0}, \; p_0 = \frac{\gamma}{R_0} \; \text{ and } \; R = R_0$$

where R_0 is a constant. The new conditions do make the perturbation
problem more interesting.

To write the perturbation eigenvalue problem needed in this essay, go
back to the perturbation equations in the third essay just after $\vec{v}_1$ has been
eliminated and substitute

$$p_1 = \hat{p}_1(r_0, \; z_0)e^{\sigma t_0}e^{im\theta_0}$$

and

$$R_1 = \hat{R}_1(z_0)e^{\sigma t_0}e^{im\theta_0}$$

into them, where m must take one of the values 0, ± 1, ± 2, $\cdots$ in order to
satisfy periodicity requirements.

Ordinarily, the z_0 dependence of p_1 and R_1 cannot be guessed, save in special cases, and so the factor e^{ikz_0} is not included in the normal mode expansion.

Then, $\hat{p}_1$ and $\hat{R}_1$ must satisfy

$$\frac{1}{r_0}\frac{\partial}{\partial r_0}\left[r_0\frac{\partial\hat{p}_1}{\partial r_0}\right] - \frac{m^2}{r_0^2}\hat{p}_1 + \frac{\partial^2\hat{p}_1}{\partial z_0^2} = 0, \quad 0 < r_0 < R_0, \quad 0 < z_0 < L_0 \quad (4.1)$$

and

$$\hat{p}_1 = -\frac{\gamma}{R_0^2}\left[\hat{R}_1 - m^2\hat{R}_1 + R_0^2\frac{d^2\hat{R}_1}{dz_0^2}\right] \quad (4.2)$$

$$\left.\begin{array}{c}\\ \\ \frac{\partial\hat{p}_1}{\partial r_0} = -\sigma^2\rho\hat{R}_1\end{array}\right\} \quad r_0 = R_0, \ 0 < z_0 < L_0$$

$$\frac{\partial\hat{p}_1}{\partial r_0} = -\sigma^2\rho\hat{R}_1 \quad (4.3)$$

as well as the no-flow and the contact conditions at the solid walls. These are

$$\frac{\partial\hat{p}_1}{\partial z_0} = 0 \quad \left\{\begin{array}{l} 0 < r_0 < R_0, \quad z_0 = 0 \\ \\ 0 < r_0 < R_0, \quad z_0 = L_0 \end{array}\right. \quad (4.4)$$

and either

$$\frac{d\hat{R}_1}{dz_0} = 0 \quad (4.5)$$

or

$$\hat{R}_1 = 0 \quad (4.6)$$

at each of the two ends $z_0 = 0$ and $z_0 = L_0$.

The problem presented by equations (4.1) through (4.6) is an eigenvalue problem, and unless the z_0 dependence of the eigenfunction can be guessed, it must remain a partial differential equation for $\hat{p}_1$ on the reference domain and an ordinary differential equation for $\hat{R}_1$ at its boundary.

Now, an odd thing about the work in the third essay is this: Nowhere did we introduce the condition that Rayleigh found to be indispensable, the condition that the volume of the jet remain fixed as its surface undergoes a displacement. To see what this amounts to, to first order, we write it in terms of the perturbation variables, and, then, as our work goes on, we try to see where it is important. Let this be our third goal.

4.2   The Fixed-Volume Requirement

Holding the volume of the jet fixed as its surface is displaced introduces a constraint on the solution to the nonlinear equations. It is

$$\pi R_0^2 L_0 = \int_0^{L_0} \int_0^{2\pi} \int_0^{R} r \; dr \; d\theta \; dz = \frac{1}{2} \int_0^{L_0} \int_0^{2\pi} R^2 \; d\theta \; dz$$

To find out what this tells us about R_1, substitute into it the expansion

$$R(\theta, \; z, \; t, \; \epsilon) = R_0 + \epsilon R_1 + \cdots$$

and obtain to first order in ϵ

$$\int_0^{L_0} \int_0^{2\pi} R_1 \; d\theta_0 \; dz_0 = 0$$

This is what the constant-volume requirement demands of the displacement.

By writing $R_1 = \hat{R}_1(z_0)e^{\sigma t_0}e^{im\theta_0}$, this demand can be seen to be satisfied whenever $m = \pm 1, \; \pm 2, \; \cdots$ and it plays no role in these cases. However, when $m = 0$, it reduces to

$$\int_0^{L_0} \hat{R}_1(z_0) \; dz_0 = 0 \tag{4.7}$$

and ordinarily this is an important part of the eigenvalue problem.

The only restriction that comes into play, then, is on axisymmetric displacements. However, if it is possible to write $\hat{R}_1(z_0) = \hat{\hat{R}}_1 e^{ikz_0}$, where $k = n\pi/L_0$, $n = 0, \; 1, \; 2, \; \cdots$, then the case $n = 0$ must be ruled out, but that is all; the fixed-volume requirement is satisfied, willy-nilly, for all other values of n. It does play an important part in our work in this essay, but not in the free-free case, nor in the case of an infinite jet, except, of course, to rule out the possibility of $k = 0$.

4.3   What Can Be Said Short of Solving the Perturbation Problem?

For all end conditions of interest, we can prove that σ^2 must be real, whence all critical points must correspond to $\sigma = 0$. This does not require the solution to the eigenvalue problem. To do this, multiply equation (4.1) by[1] $r_0 \bar{p}_1$ and integrate over r_0 and z_0. The result, on integrating by parts and

---

[1] An overbar denotes a complex conjugate.

using $\dfrac{\partial \hat{p}_1}{\partial z_0} = 0$ at $z_0 = 0$ and $z_0 = L_0$, is

$$0 = \int_0^{L_0} \left[R_0 \bar{\hat{p}}_1 (r_0 = R_0) \frac{d\hat{p}_1}{dr_0} (r_0 = R_0) \right] dz_0$$

$$- \int_0^{L_0} \int_0^{R_0} \left[r_0 \left| \frac{\partial \hat{p}_1}{\partial r_0} \right|^2 + \frac{m^2}{r_0} |\hat{p}_1|^2 \right] dr_0 \, dz_0$$

$$- \int_0^{L_0} \int_0^{R_0} r_0 \left| \frac{\partial \hat{p}_1}{\partial z_0} \right|^2 dr_0 \, dz_0 \qquad (4.8)$$

Then, eliminating $\hat{p}_1$ and $\dfrac{\partial \hat{p}_1}{\partial r_0}$ at $r_0 = R_0$ by using equations (4.2) and (4.3) and carrying out one more integration by parts, the first integral on the right-hand side comes down to

$$\frac{\rho \gamma}{R_0} \sigma^2 \int_0^{L_0} \left[[1 - m^2] |\hat{R}_1|^2 - R_0^2 \left| \frac{d\hat{R}_1}{dz_0} \right|^2 \right] dz_0$$

where $R_0^2 \bar{\hat{R}}_1 \dfrac{d\hat{R}_1}{dz_0}$ vanishes at $z_0 = 0$ and $z_0 = L_0$ on requiring free or fixed end conditions at either wall. Putting this into equation (4.8), there obtains

$$\frac{\rho \gamma}{R_0} \sigma^2 \int_0^{L_0} \left[[1 - m^2] |\hat{R}_1|^2 - R_0^2 \left| \frac{d\hat{R}_1}{dz_0} \right|^2 \right] dz_0$$

$$= \int_0^{L_0} \int_0^{R_0} \left[r_0 \left| \frac{\partial \hat{p}_1}{\partial r_0} \right|^2 + \frac{m^2}{r_0} |\hat{p}_1|^2 + r_0 \left| \frac{\partial \hat{p}_1}{\partial z_0} \right|^2 \right] dr_0 \, dz_0 \qquad (4.9)$$

which requires σ^2 to be real. It also tells us that σ^2 cannot be zero when m^2 is greater than zero unless $\hat{p}_1$ is zero and that σ^2 cannot be zero when m^2 equals zero unless $\hat{p}_1$ is constant. It also tells us that there must be a range of stable jets as long as L_0 is sufficiently small.

To see this, observe that[2]

$$\int_0^{L_0} \left| \frac{d\hat{R}_1}{dz_0} \right|^2 dz_0 \geq \lambda_1^2 \int_0^{L_0} |\hat{R}_1|^2 \, dz_0$$

whence the integral on the left-hand side of equation (4.9) must be less than

$$[1 - m^2 - R_0^2 \lambda_1^2] \int_0^{L_0} |\hat{R}_1|^2 \, dz$$

---

[2]See equation 36.11, p.164, Chapter 7 in H. F. Weinberger, *A First Course in Partial Differential Equations*. Wiley, New York, 1965.

where λ_1^2 is the smallest positive eigenvalue of the differential operator $-d^2/dz_0^2$ corresponding to the end conditions of interest. The right-hand side of equation (4.9) is always positive. Then, if λ_1^2 exceeds $1/R_0^2$, the integral on the left-hand side must be negative, hence σ^2 must be negative and the jet must be stable to any small displacement. As λ_1^2 must be a multiple of $1/L_0^2$, if the jet is sufficiently short, it must be stable to any small displacement; that is, if L_0 is less than some multiple of R_0, where the multiple will differ as the end conditions differ, the jet must be stable.

4.4   The Solution to the Perturbation Equations

Equations (4.1) and (4.4), together with the requirement that $\hat{p}_1$ remain bounded at $r_0 = 0$, can be satisfied by writing

$$\hat{p}_1 = A_{m0} r_0^{|m|} + \sum_{k>0} A_{mk} I_m(kr_0) \cos kz_0 \tag{4.10}$$

This equation is written for a fixed value of m, it satisfies all the demands on $\hat{p}_1$ save those at $r_0 = R_0$ [i.e., equations (4.2) and (4.3)], and it can be obtained by the method of separation of variables. The sum is over the admissible values of k and these are restricted by the end conditions on $\hat{p}_1$ to be

$$k = n\frac{\pi}{L_0}, \quad n = 0, \ 1, \ 2, \ \cdots$$

The constants A_{mk} remain to be determined.

The eigenvalue problem then reduces to equations (4.2) and (4.3) for $\hat{R}_1$ and the constants A_{mk}. It is

$$\left.\begin{aligned} \frac{d^2\hat{R}_1}{dz_0^2} + \frac{[1-m^2]}{R_0^2}\hat{R}_1 &= -\frac{1}{\gamma}\hat{p}_1(r_0 = R_0) \qquad (4.11)\\[2mm] \sigma^2\hat{R}_1 &= -\frac{1}{\rho}\frac{\partial\hat{p}_1}{\partial r_0}(r_0 = R_0) \qquad (4.12) \end{aligned}\right\} \quad 0 < z_0 < L_0$$

where $\hat{R}_1$ must satisfy homogeneous end conditions which depend on how the surface of the jet is required to contact the solid wall. Our job is to determine the values of σ^2 such that this problem has solutions other than $\hat{R}_1 = 0 = A_{mk}$.

The expectation is that each eigenvalue will correspond to an eigenfunction defined by an infinite sequence of nonzero coefficients A_{mk}. By this, the eigenfunctions need not be periodic and need not exhibit definite wavelengths. The exceptions are cases where only a finite number of the coefficients turn out to be other than zero. Then, the wavelength of a disturbance turns out to be a useful designation.

The case where both ends are free is such an exception; the other two cases are not. The free-free case is worked out first as though it were typical; that is, no short cuts are taken. Then, attention is drawn to other end conditions. These are more interesting. The critical conditions can be established, but the values of σ^2, when they are not zero, can only be approximated. The estimates of σ^2 are carried along just far enough to indicate the importance of the constant-volume condition.

4.4.1   Free-Free End Conditions

Let $\hat{R}_1$ satisfy[3] equation (4.5) at both ends, namely

$$\frac{d\hat{R}_1}{dz_0}(z_0 = 0) = 0 = \frac{d\hat{R}_1}{dz_0}(z_0 = L_0)$$

and notice that these are just the end conditions satisfied by $\hat{p}_1$. Then, like $\hat{p}_1$, $\hat{R}_1$ can be expanded in a cosine series and we write

$$\hat{R}_1 = B_{m0} + \sum_{k>0} B_{mk} \cos kz_0 \tag{4.13}$$

where, again, k is $n\pi/L_0$. This reduces the eigenvalue problem to an algebraic problem in σ^2, A_{mk} and B_{mk}.

The case $m = 0$ differs a little from the remaining cases and it is taken up first. In this case, $\hat{R}_1$ must satisfy the fixed-volume constraint, equation (4.7), and this requires $B_{00} = 0$ and so too $A_{00} = 0$, as can be seen on integrating

$$\frac{d^2\hat{R}_1}{dz_0^2} + \frac{1}{R_0^2}\hat{R}_1 = -\frac{1}{\gamma}\hat{p}_1(r_0 = R_0)$$

over $0 \leq z_0 \leq L_0$. Due to this, we can turn our attention to values of k greater than zero.

Then, substituting equations (4.10) and (4.13) into equations (4.11) and (4.12), there obtains

$$\sum_{k>0}\left[-k^2 + \frac{1}{R_0^2}\right]B_{0k}\cos kz_0 = -\frac{1}{\gamma}\sum_{k>0}A_{0k}I_0(kR_0)\cos kz_0$$

and

$$\sigma^2\sum_{k>0}B_{0k}\cos kz_0 = -\frac{1}{\rho}\sum_{k>0}A_{0k}kI_0'(kR_0)\cos kz_0$$

which is a set of two equations in the unknown coefficients A_{0k} and B_{0k}, where $k = n\pi/L_0$, $n = 1, 2, \cdots$. The values of σ^2 are those to which

---

[3]This case can be solved simply by putting $\hat{p}_1 = \hat{p}(r_0)\cos kz_0$ and $\hat{R}_1 = \hat{R}_1\cos kz_0$ directly into the eigenvalue problem. However, $\cos kz_0$, not e^{ikz_0}, must be used.

solutions other than $A_{0k} = 0 = B_{0k}$ can be found. What makes the two equations sufficient to the task at hand is that each must hold for all values of z_0.

These two equations can be solved term by term (i.e., at each value of k), whence they reduce to

$$[1 - R_0^2 k^2]B_{0k} = -\frac{R_0^2}{\gamma}I_0(kR_0)A_{0k}$$

and

$$\sigma^2 B_{0k} = -\frac{1}{\rho R_0}kR_0 I_0'(kR_0)A_{0k}$$

which are two, linear, homogeneous, algebraic equations in A_{0k} and B_{0k}. They determine solutions other than $A_{0k} = 0 = B_{0k}$ if and only if

$$\det\begin{pmatrix} \dfrac{R_0^2}{\gamma}I_0(kR_0) & [1 - R_0^2 k^2] \\[2ex] \dfrac{kR_0 I_0'(kR_0)}{\rho R_0} & \sigma^2 \end{pmatrix} = 0$$

This determines σ^2. It is

$$\sigma_{0k}^2 = \frac{\gamma}{\rho R_0^3}[1 - R_0^2 k^2]\frac{kR_0 I_0'(kR_0)}{I_0(kR_0)} \tag{4.14}$$

and it corresponds to the eigenfunction

$$\hat{p}_{1_{0k}} = A_{0k}I_0(kr_0)\cos kz_0$$

and

$$\hat{R}_{1_{0k}} = B_{0k}\cos kz_0$$

where $A_{0k}/B_{0k} = -\gamma[1 - R_0^2 k^2]/R_0^2 I_0(kR_0)$.

The eigenvalue σ_{0k}^2 determines whether or not the jet is stable to a small displacement in the shape of its eigenfunction. An arbitrary displacement, made up of a linear combination of many eigenfunctions, is then stable if and only if all of its constituent eigenfunctions are stable, each acting independently of the others. Each eigenfunction corresponds to one of the admissible values of k, each possesses a distinct wavelength and each is determined by the length of the jet.

Equation (4.14) is the dispersion formula for a finite jet in the case of free-free end conditions, subject to axisymmetric displacements. It looks just like the corresponding formula for an infinite jet, save only that the values of k are now restricted. The interpretation of the formula is then familiar. Indeed, the jet is stable to large wave number displacements and unstable to small wave number displacements, the most dangerous being the one corresponding to the smallest allowable wave number, π/L_0. If the

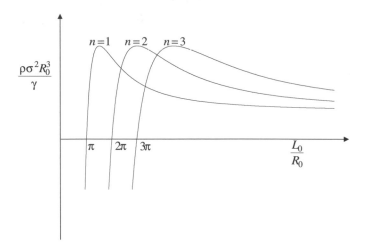

Figure 4.2. *The Effect of Bridge Length on σ^2 for Free-Free End Conditions at $m = 0$*

jet is stable to this displacement, it is stable to all displacements. The role played by the length of the jet is one of screening out very small wave number displacements and the stability condition is then a condition on how long a jet can be before its most unstable displacement reaches the critical condition corresponding to $\sigma^2 = 0$.

The length of the jet is the control variable. As L_0 increases, the jet reaches the critical condition when $L_0 = \pi R_0$. This is illustrated in Figure 4.2.

To go on to nonaxisymmetric displacements in order to see if any of the conclusions drawn thus far need to be modified, let m take any of the values ± 1, ± 2, $\cdots$. Then, the admissible values of k are increased by 1 to

$$k = n\frac{\pi}{L_0}, \quad n = 0, 1, 2, \cdots$$

and substituting the expansions of $\hat{p}_1$ and $\hat{R}_1$, equations (4.10) and (4.13), into equations (4.11) and (4.12), there now obtains

$$\frac{1 - m^2}{R_0^2} B_{m0} + \sum_{k>0}\left[-k^2 + \frac{[1 - m^2]}{R_0^2}\right] B_{mk} \cos kz_0$$

$$= -\frac{1}{\gamma} R_0^{|m|} A_{m0} - \frac{1}{\gamma}\sum_{k>0} A_{mk} I_m(kR_0) \cos kz_0$$

and

$$\sigma^2 B_{m0} + \sigma^2 \sum_{k>0} B_{mk} \cos kz$$

$$= -\frac{1}{\rho} A_{m0}|m|R_0^{|m|-1} - \frac{1}{\rho} \sum_{k>0} A_{mk} k I'_m(kR_0) \cos kz_0$$

The work carries on much as before. Again, the two equations can be solved term by term. The terms at $k = 0$ require

$$\frac{1-m^2}{R_0^2} B_{m0} = -\frac{1}{\gamma} R_0^{|m|} A_{m0}$$

and

$$\sigma^2 B_{m0} = -\frac{1}{\rho}|m|R_0^{|m|-1} A_{m0}$$

whereas the terms at any $k > 0$ require

$$[1 - m^2 - R_0^2 k^2]B_{mk} = -\frac{R_0^2}{\gamma} I_m(kR_0) A_{mk}$$

and

$$\sigma^2 B_{mk} = -\frac{1}{\rho R_0} k R_0 I'_m(kR_0) A_{mk}$$

Taking $k > 0$ first, the two equations in A_{mk} and B_{mk} determine a solution other than $A_{mk} = 0 = B_{mk}$ if and only if

$$\det \begin{pmatrix} \dfrac{R_0^2}{\gamma} I_m(kR_0) & [1 - m^2 - R_0^2 k^2] \\[2ex] \dfrac{kR_0 I'_m(kR_0)}{\rho R_0} & \sigma^2 \end{pmatrix} = 0$$

and this determines σ^2 as

$$\sigma^2_{mk} = \frac{\gamma}{\rho R_0^3}[1 - m^2 - R_0^2 k^2]\frac{kR_0 I'_m(kR_0)}{I_m(kR_0)}$$

This formula tells us that a jet of any length is stable to nonaxisymmetric disturbances so long as the wave number of the disturbance is not zero.

The case $k = 0$, wherein the jet is displaced uniformly in the z_0 direction, is a little more interesting. Taking k to be zero, the two equations in A_{m0} and B_{m0} determine a solution other than $A_{m0} = 0 = B_{m0}$ if and only if

$$\det \begin{pmatrix} \dfrac{1}{\gamma} R_0^{|m|} & \dfrac{1}{R_0^2} \dfrac{m^2}{} \\[2ex] \dfrac{1}{\rho}|m|R_0^{|m|-1} & \sigma^2 \end{pmatrix} = 0$$

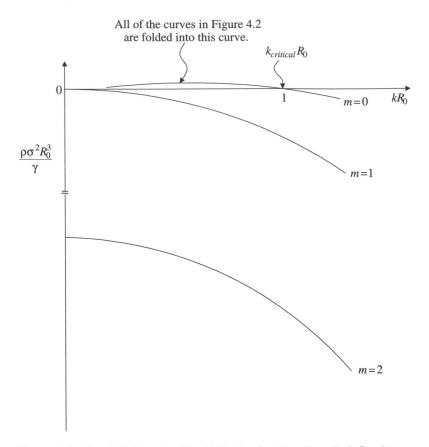

Figure 4.3. *The Stability of a Liquid Bridge for Free-Free End Conditions*

whence

$$\sigma_{m0}^2 = \frac{\gamma}{\rho R_0^3} \, |m|[1 - m^2]$$

and again all such displacements are stable, save a displacement corresponding to $m = \pm1$, which is neutral.

Displacements corresponding to $m = \pm1$ and $k = 0$ (i.e., to corrugated cylinders), require $\sigma^2 = 0$ for all values of L_0, but, of course, the jet is already unstable to axisymmetric displacements at all small finite values of k.

Our $m > 0$ results are illustrated in Figure 4.3, where $\rho\sigma^2 R_0^3/\gamma$ is plotted against kR_0, thereby folding together the variables n and L_0.

Before going on to other end conditions, observe, one last time, that in the case where the ends are free, each eigenfunction is a multiple[4] of $\cos kz_0$ for some value of k. By this, it is possible to describe an eigenfunction in terms of its wavelength, namely in terms of $\lambda = 2\pi/k = 2L_0/n$; this will be lost as we go on.

4.4.2 Other End Conditions

If the ends are constrained differently, solving the eigenvalue problem is not so easy. Again, all the conditions on $\hat{p}_1$ can be satisfied, save those at $r_0 = R_0$, by expanding $\hat{p}_1$ as in equation (4.10), namely

$$\hat{p}_1 = A_{m0}r_0^{|m|} + \sum_{k>0} A_{mk}I_m(kr_0)\cos kz_0$$

where the values of k are as they were before. However, now the form of $\hat{R}_1$ cannot be guessed and it must be determined in terms of the coefficients A_{mk} by solving equation (4.11), namely

$$\frac{d^2\hat{R}_1}{dz_0^2} + \frac{[1-m^2]}{R_0^2}\hat{R}_1 = -\frac{1}{\gamma}\hat{p}_1(r_0 = R_0)$$

In doing this, each value of m must be taken up in turn and $\hat{R}_1$ can be written

$$m = 0: \ \hat{R}_1 = A\cos\frac{z_0}{R_0} + B\sin\frac{z_0}{R_0}$$

$$-\frac{R_0^2}{\gamma}\left[A_{00} + \sum_{k>0}\frac{A_{0k}I_0(kR_0)}{[1-k^2R_0^2]}\cos kz_0\right] \qquad (4.15)$$

$$m = 1: \ \hat{R}_1 = A + B\frac{z_0}{R_0}$$

$$-\frac{R_0^2}{\gamma}\left[\frac{1}{2}A_{10}R_0\left[\frac{z_0}{R_0}\right]^2 + \sum_{k>0}\frac{A_{1k}I_1(kR_0)}{[-k^2R_0^2]}\cos kz_0\right] \qquad (4.16)$$

etc.

where, in each case, the constants A and B must be determined to satisfy the new end conditions; for example, in the case of fixed-fixed end

---

[4]They are not multiples of e^{ikz_0}; although $\sin kz_0$ can be retained when the jet is infinitely long, causing no technical problem, it must be omitted when the jet is finite, unless other end conditions are introduced, whereupon it does present a technical problem ending in the loss of eigenfunctions at definite wavelengths.

conditions

$$\hat{R}_1(z_0 = 0) = 0 = \hat{R}_1(z_0 = L_0)$$

must be satisfied.

This done, the result can be substituted into equation (4.12) to obtain

$$\sigma^2 \hat{R}_1 = -\frac{1}{\rho} \frac{\partial \hat{p}_1}{\partial r_0}(r_0 = R_0)$$

$$= -\frac{1}{\rho} \left[A_{m0}|m|R_0^{|m|-1} + \sum_{k>0} A_{mk}kI'_m(kR_0)\cos kz_0 \right] \quad (4.17)$$

which must hold on the entire interval $0 < z_0 < L_0$. This can be turned into an infinite set of linear, homogeneous, algebraic equations in the coefficients A_{mk} by multiplying by $\cos k'z_0$, $k' = n'\pi/L_0$, $n' = 0, 1, 2, \cdots$, and integrating over the interval $(0, L_0)$.

All cases where m is not zero work out as just explained, but the case $m = 0$ is special. In that case, upon multiplying equation (4.17) by 1, which corresponds to $n' = 0$, and carrying out the integration, the right-hand side vanishes. This would require σ^2 to be zero were it not for the constant-volume requirement, equation (4.7). This equation then replaces the $n' = 0$ equation in the course of determining the coefficients A_{0k}, but only if $m = 0$.

This is the first time the constant-volume requirement has come into the solution of the perturbation problem in an important way. We may observe that while it was important in using Rayleigh's work principle, it was not important in the perturbation problem when the ends were free, other than to rule out $k = 0$ when $m = 0$.

Now, our system of infinitely many equations will have a solution other than $A_{mk} = 0$, for all values of k, only if σ^2 takes on special values. These will be the eigenvalues and they will determine whether or not a jet will be stable.

Ordinarily, an eigenfunction will be composed of infinitely many values of k and the hope of identifying its wavelength will be lost. In fact, when the ends are not free, it is no longer possible to talk about a dispersion formula in the ordinary way, as the idea of wavelength no longer has any meaning. Still, the non zero eigenvalues are important and it is fair to say that many stability problems wind up like this: requiring the solution of infinitely many linear, homogeneous, algebraic equations.

It will be apparent, by now, that some sort of approximation must be introduced in order to obtain estimates of the eigenvalues in the non-free-free cases. We intend to go on and present some simple estimates in order to illustrate the method outlined above. But first we turn to the critical conditions and work out their dependence on the end conditions. This requires no approximation and makes the role of the constant-volume requirement in the calculation explicit.

The plan is to go back to the eigenvalue problem and look for solutions where $\sigma^2 = 0$. This will determine the critical length of the liquid bridge. That done, some estimates of σ^2 will be presented.

4.5   The Critical Length of a Bridge

To determine the critical value of L_0, put $\sigma^2 = 0$ directly into the eigenvalue problem and look for solutions other than $\hat{p}_1$ constant and $\hat{R}_1$ equal to zero.

Supposing the most unstable disturbance to be axisymmetric (i.e., requiring m to be zero), $\hat{p}_1$ can be expanded, as before, as

$$\hat{p}_1 = A_{00} + \sum_{k>0} A_{0k} I_0(kr_0) \cos kz_0$$

and requiring σ^2 to be zero places a simple restriction on this expansion. It comes from equation (4.12) and it is

$$\frac{\partial \hat{p}_1}{\partial r_0}(r_0 = R_0) = 0$$

which requires

$$\sum_{k>0} A_{0k} k I_0'(kR_0) \cos kz_0 = 0$$

To make this so for all z_0, $0 < z_0 < L_0$, all coefficients A_{0k}, $k > 0$, must be set equal to zero and the expansion of $\hat{p}_1$, at neutral conditions, reduces to

$$\hat{p}_1 = A_{00}$$

that is, under neutral conditions $\hat{p}_1$ must be constant.

Using this, equation (4.11) is simply

$$\frac{d^2 \hat{R}_1}{dz_0^2} + \frac{\hat{R}_1}{R_0^2} = C \tag{4.18}$$

where C is shorthand for $-A_{00}/\gamma$. This equation, together with homogeneous end conditions and the integral condition coming from the constant-volume requirement, is an eigenvalue problem in $\hat{R}_1$, C and L_0. Ordinarily, it has only the solution $\hat{R}_1 = 0 = C$, but it may have a solution other than this for certain values of L_0, which now plays the role of an eigenvalue.

Equation (4.18) has the solution

$$\hat{R}_1 = C R_0^2 + A \cos \frac{z_0}{R_0} + B \sin \frac{z_0}{R_0} \tag{4.19}$$

and using this solution, the eigenvalue problem reduces to an algebraic problem in C, A and B where C, A and B must be determined to satisfy the integral condition and the two end conditions. As these are homogeneous, there is always the solution where A, B and C all vanish. Our job is to see what is required to get a solution other than this. This is done case by case where the contact conditions of interest and the corresponding end conditions are as follows:

contact conditions	end conditions
free-free	$\dfrac{d\hat{R}_1}{dz_0}(z_0 = 0) = 0 = \dfrac{d\hat{R}_1}{dz_0}(z_0 = L_0)$
free-fixed	$\dfrac{d\hat{R}_1}{dz_0}(z_0 = 0) = 0 = \hat{R}_1(z_0 = L_0)$
fixed-fixed	$\hat{R}_1(z_0 = 0) = 0 = \hat{R}_1(z_0 = L_0)$

4.5.1   Free-Free End Conditions

In this case, equations (4.18) and (4.7) require C to be zero. Then, the free end condition at $z_0 = 0$ requires B to be zero. To satisfy the remaining free end condition in a way that does not make A zero, L_0 must take one of the values

$$\pi R_0, \; 2\pi R_0, \; \cdots$$

The critical value of L_0 is then

$$\pi R_0$$

as determined earlier.

Notice, by determining the critical condition in this way, the constant-volume condition appears to play a role, even in the case where the end conditions are free. It requires C to be zero and, by this, $\hat{p}_1$ to be zero.[5]

4.5.2   Free-Fixed End Conditions

Again, the free end condition at $z_0 = 0$ requires B to be zero. Hence, only

$$\hat{R}_1(z_0 = L_0) = 0$$

and

$$\int_0^{L_0} \hat{R}_1 \, dz_0 = 0$$

---

[5] Had the factor $\cos k z_0$ been put in at the outset, as in the third essay, the constant-volume condition would have been satisfied already and it would only serve to eliminate the possibility $k = 0$.

remain to be satisfied. These two equations determine A and C via

$$CR_0^2 + A\cos\frac{L_0}{R_0} = 0$$

and

$$CR_0^2 L_0 + AR_0\sin\frac{L_0}{R_0} = 0$$

which have a solution other than $C = 0 = A$ if and only if

$$\det\begin{pmatrix} R_0^2 & \cos\dfrac{L_0}{R_0} \\ \\ R_0^2 L_0 & R_0\sin\dfrac{L_0}{R_0} \end{pmatrix} = R_0^3\sin\frac{L_0}{R_0} - R_0^2 L_0\cos\frac{L_0}{R_0} = 0$$

or if and only if

$$\frac{L_0}{R_0} = \tan\frac{L_0}{R_0}$$

By this, the critical value of L_0 is nearly

$$\frac{3}{2}\,\pi R_0$$

4.5.3 Fixed-Fixed End Conditions

The fixed end condition[6] at $z_0 = 0$ can be used to eliminate A. Doing this leads to

$$A = -CR_0^2$$

The remaining two conditions then determine B and C. They are

$$\hat{R}_1(z_0 = L_0) = 0 \tag{4.20}$$

and

$$\int_0^{L_0} \hat{R}_1\,dz_0 = 0 \tag{4.21}$$

The first is satisfied by any values of B and C if L_0/R_0 takes the value 2π, and at this value of L_0/R_0, the second is satisfied by any value of B and by $C = 0$. This makes $2\pi R_0$ a candidate for the critical value of L_0.

Equation (4.20) can also be satisfied by any value of B and by $C = 0$ if L_0/R_0 takes the value π; but then, equation (4.21) cannot be satisfied

---

[6]By three different calculations and by many experiments, Plateau knew, in the case of fixed end conditions and as early as 1873, the critical length of a liquid bridge to be $2\pi R_0$.

unless $B = 0$. By this, πR_0 cannot be a critical value of L_0. It is the fixed-volume requirement that eliminates πR_0 as a critical value of L_0.

This takes care of the only two positive values of L_0/R_0 on the interval $[0,\ 2\pi]$ where $\sin(L_0/R_0) = 0$ and we can move on to the remaining values of L_0/R_0 in this interval. Then, using equation (4.20) to eliminate B, there obtains

$$\hat{R}_1 = CR_0^2\left[1 - \cos\frac{z_0}{R_0}\right] - CR_0^2\frac{\left[1 - \cos\dfrac{L_0}{R_0}\right]}{\sin\dfrac{L_0}{R_0}}\sin\frac{z_0}{R_0}$$

which, when substituted into equation (4.21), produces

$$CR_0^3\frac{\dfrac{L_0}{R_0}\sin\dfrac{L_0}{R_0} - 2 + 2\cos\dfrac{L_0}{R_0}}{\sin\dfrac{L_0}{R_0}} = 0$$

and this can be satisfied by a value of C other than zero if and only if

$$\frac{L_0}{R_0} = \frac{2\left[1 - \cos\dfrac{L_0}{R_0}\right]}{\sin\dfrac{L_0}{R_0}}$$

This equation determines the possible critical values of L_0/R_0 at which $\sin(L_0/R_0)$ is not zero and it needs to be solved only on the interval $0 < L_0/R_0 < 2\pi$, excluding $L_0/R_0 = \pi$. First, no solutions lie on the interval $\pi < L_0/R_0 < 2\pi$, as the right-hand side is negative there. Then, as L_0/R_0 approaches π from the left, the right-hand side approaches $+\infty$ and lies above the left-hand side. This holds true on the interval $\frac{1}{2}\pi < L_0/R_0 < \pi$, which means that if a solution can be found, it can only be on the interval $0 < L_0/R_0 < \frac{1}{2}\pi$. Near zero, both sides are asymptotic to L_0/R_0. To see which is larger, their derivatives can be compared. Doing this produces the result that no solutions can be found as long as

$$1 < \frac{2[1 - \cos\dfrac{L_0}{R_0}]}{1 - \cos^2\dfrac{L_0}{R_0}} = \frac{2}{1 + \cos\dfrac{L_0}{R_0}}$$

This being true on the interval $0 < L_0/R_0 < \frac{1}{2}\pi$, and indeed on the interval $0 < L_0/R_0 < \pi$, the critical value of L_0/R_0 cannot be less than 2π and hence it must be 2π. By this, the critical value of L_0 must be

$$2\pi R_0$$

These results, gathered in Table 4.1, illustrate the important role end conditions play in determining the stable lengths of a liquid bridge lying

Table 4.1. *The Critical Lengths for Various End Conditions*

end conditions	critical conditions
free-free	πR_0
free-fixed	nearly $\dfrac{3}{2}\pi R_0$
fixed-fixed	$2\pi R_0$

between two plane walls. It can be noticed that the more the ends are confined, the greater the critical length (i.e., the more stable the bridge).

Other than by the end conditions, the critical length is determined entirely by the radius of the liquid bridge. It does not depend on the density of the liquid or on the tension at its surface. Nor does it depend on the viscosity of the liquid making up the bridge, but viscosity remains to be taken into account. This will be done shortly.

4.6   Rough Estimates of σ^2

In the case of end conditions other than free-free, the eigenvalue problem is not easy to solve. Yet, we can illustrate how the calculation can be carried out and produce a few rough estimates. The case of fixed-free end conditions will be taken up as the work is then a little easier than it is in the fixed-fixed case and attention will be directed to the cases $m = 0$ and $m = 1$.

4.6.1   The Case $m = 0$

The case $m = 0$ is special in two ways. It is the only case in which the $n' = 0$ moment equation produces no information and it is the only case in which the fixed-volume condition is not automatically satisfied. In this case, then, equation (4.7) takes the place of the $n' = 0$ moment equation.

Now, when m is zero, $\hat{R}_1$ is given by equation (4.15), namely by

$$\hat{R}_1 = A\cos\frac{z_0}{R_0} + B\sin\frac{z_0}{R_0} - \frac{R_0^2}{\gamma}\left[A_{00} + \sum_{k>0}\frac{A_{0k}I_0(kR_0)}{[1 - k^2R_0^2]}\cos kz_0\right]$$

where A and B must be set in order to satisfy the assigned end conditions, here fixed-free. These conditions, equation (4.6) at $z_0 = 0$ and equation (4.5) at $z_0 = L_0$, require

$$0 = A - \frac{R_0^2}{\gamma}\left[A_{00} + \sum_{k>0}A_{0k}\frac{I_0(kR_0)}{1 - k^2R_0^2}\right] \tag{4.22}$$

and

$$0 = -A \sin \frac{L_0}{R_0} + B \cos \frac{L_0}{R_0} \tag{4.23}$$

while equation (4.7) requires

$$0 = A \sin \frac{L_0}{R_0} - B \left[\cos \frac{L_0}{R_0} - 1 \right] - \frac{R_0^2}{\gamma} A_{00} \frac{L_0}{R_0} \tag{4.24}$$

Now, equations (4.23) and (4.24) can be used to eliminate A and B in favor of A_{00}, and by doing this, there obtains

$$B = \frac{R_0^2}{\gamma} A_{00} x$$

and

$$A = \frac{R_0^2}{\gamma} A_{00} x \cot x$$

where $x = \dfrac{L_0}{R_0}$.

Then, substituting for A in equation (4.22) leads to

$$0 = \frac{R_0^2}{\gamma} A_{00} x \cot x - \frac{R_0^2}{\gamma} \left[A_{00} + \sum_{k>0} A_{0k} \frac{I_0(kR_0)}{1 - k^2 R_0^2} \right] \tag{4.25}$$

and this takes the place of the $n' = 0$ moment equation.

To derive the higher-moment equations, substitute the formulas for A and B into equation (4.15) to get

$$\hat{R}_1 = \frac{R_0^2}{\gamma} A_{00} x \cot x \, \cos \frac{z_0}{R_0} + \frac{R_0^2}{\gamma} A_{00} x \sin \frac{z_0}{R_0}$$

$$- \frac{R_0^2}{\gamma} \left[A_{00} + \sum_{k>0} A_{0k} \frac{I_0(kR_0)}{[1 - k^2 R_0^2]} \cos k z_0 \right]$$

and then put this into

$$\sigma^2 \hat{R}_1 = -\frac{1}{\rho} \frac{d\hat{p}_1}{dr_0} (r_0 = R_0) = -\frac{1}{\rho} \sum_{k>0} A_{0k} k I_0'(kR_0) \cos k z_0 \tag{4.26}$$

which is equation (4.12) with its right-hand side evaluated at $m = 0$ by using equation (4.10). The result is an equation in A_{00}, A_{01}, etc. which must be satisfied at all values of z_0 on the interval $0 < z_0 < L_0$. To turn this into a sequence of linear, algebraic equations in the unknowns, multiply it by $\cos k' z_0$ and integrate the product over $0 < z_0 < L_0$; notice that $k = n\pi/L_0$, $n = 1, 2, \cdots$, in the sum, and $k' = n'\pi/L_0$, $n' = 1, 2, \cdots$.

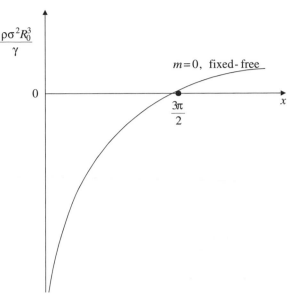

Figure 4.4. *The Effect of Bridge Length on σ^2 for Fixed-Free End Conditions at $m = 0$*

To get the second equation in the sequence, let $n' = 1$, multiply equation (4.26) by $\cos(\pi z_0/L_0)$, and integrate over $(0, L_0)$ to obtain

$$\sigma^2 \frac{R_0^3}{\gamma} A_{00} \frac{x}{1 - \dfrac{\pi^2}{x^2}} - \frac{\sigma^2 R_0^3}{\gamma} A_{01} \frac{I_0\left(\dfrac{\pi}{x}\right)}{1 - \dfrac{\pi^2}{x^2}} \frac{x}{2} = -\frac{1}{\rho} A_{01} \frac{1}{2}\pi \, I_0'\left(\frac{\pi}{x}\right)$$

This is an equation in A_{00} and A_{01}. Other equations obtain in the same way (viz., letting $n' = 2$ produces an equation in A_{00} and A_{02}, letting $n' = 3$ produces an equation in A_{00} and A_{03}, etc).

Now, only in equation (4.25), the first equation in our sequence, do all the unknown coefficients appear. To get an estimate of the eigenvalues, then, the series in this equation can be cut off after N terms and the result can be solved together with the N moment equations produced by setting $n' = 1, 2, \cdots, N$. By this, estimates of N of the eigenvalues can be obtained.

To illustrate how the calculation goes and at the same time obtain an interesting result, truncate the series to one term to get

$$0 = A_{00}[x \cot x - 1] - A_{01} \frac{I_0\left(\dfrac{\pi}{x}\right)}{1 - \dfrac{\pi^2}{x^2}}$$

and solve this together with the $n' = 1$ moment equation in A_{00} and A_{01}.

Two linear, homogeneous, algebraic equations in A_{00} and A_{01} result. These produce an estimate of one eigenvalue by requiring a solution other than $A_{00} = 0 = A_{01}$. It is

$$\sigma^2 \frac{R_0^3 \rho}{\gamma} = \frac{-\dfrac{\pi}{2} \dfrac{I_0'\left(\dfrac{\pi}{x}\right)}{I_0\left(\dfrac{\pi}{x}\right)} [x \cot x - 1]\left[1 - \dfrac{\pi^2}{x^2}\right]^2}{\left[x - \dfrac{x}{2} [x \cot x - 1]\left[1 - \dfrac{\pi^2}{x^2}\right]\right]} \tag{4.27}$$

and we observe that σ^2 vanishes at just those values of x where $1 - x \cot x$ vanishes. The smallest of these values is just short of $\frac{3}{2}\pi$ and this is the neutral point obtained earlier. By this, our simplest estimate produces the critical point entirely correctly. Notice also that σ^2 does not vanish at $x = \pi$; in fact, at $x = \pi$, $\sigma^2 \frac{R_0^3 \rho}{\gamma}$ takes the value $-\frac{4}{3}\frac{I_0'(1)}{I_0(1)}$. Equation (4.27) is illustrated in Figure 4.4 as a graph of $\sigma^2 R_0^3 \rho / \gamma$ versus L_0/R_0. As x goes to zero, $\sigma^2 R_0^3 \rho / \gamma$ goes to $-\infty$ as $-\pi^3/x^3$.

4.6.2   The Case $m = 1$

In the case $m = 1$, $\hat{R}_1$ is given by equation (4.16), namely by

$$\hat{R}_1 = A + B\frac{z_0}{R_0} - \frac{R_0^2}{\gamma}\left[\frac{1}{2} A_{10} R_0 \frac{z_0^2}{R_0^2} + \sum_{k>0} A_{1k} \frac{I_1(kR_0)}{-k^2 R_0^2} \cos kz_0\right]$$

where A and B must be set to satisfy the assigned end conditions. Again, the end conditions

$$\hat{R}_1(z_0 = 0) = 0 = \frac{d\hat{R}_1}{dz_0}(z_0 = L_0)$$

are of interest and these require

$$0 = A - \frac{R_0^2}{\gamma} \sum_{k>0} A_{1k} \frac{I_1(kR_0)}{-k^2 R_0^2} \tag{4.28}$$

and

$$0 = \frac{B}{R_0} - \frac{R_0^2}{\gamma} A_{10} \frac{L_0}{R_0}$$

To add to these two equations a sequence of equations in A, B, A_{10}, A_{11}, $\cdots$ substitute equation (4.16) into

$$\sigma^2 \hat{R}_1 = -\frac{1}{\rho}\frac{d\hat{p}_1}{dr_0}(r_0 = R_0) = -\frac{1}{\rho}\left[A_{10} + \sum_{k>0} A_{1k} kI_1'(kR_0) \cos kz_0\right]$$

which is equation (4.12) with its right-hand side evaluated at $m = 1$, multiply by $\cos k'z_0$, where $k' = n'\pi/L_0$, $n' = 0, 1, 2, \cdots$, and integrate the

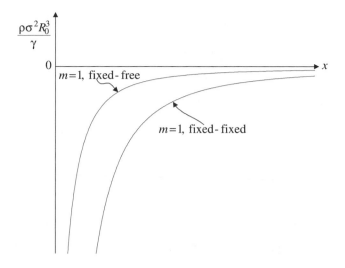

Figure 4.5. *The Effect of Bridge Length on σ^2 for Fixed-Free and Fixed-Fixed End Conditions at $m = 1$*

product over $0 < z_0 < L_0$. The result in case $n' = 0$ is

$$\sigma^2 \left[AL_0 + B\frac{L_0^2}{2R_0} - \frac{R_0^2}{\gamma}\left[\frac{1}{2} A_{10}R_0 \frac{L_0^3}{3R_0^2}\right]\right] = -\frac{1}{\rho} A_{10}L_0$$

Other equations corresponding to $n' = 1, 2, \cdots$ can be obtained.

Now, only equation (4.28), coming from the left-hand end condition, contains more than a few of the unknown constants. If it is truncated to A_{10}, that is, if $A_{11}, A_{12}, \cdots$ are all set to zero, a very simple result obtains. It is

$$A = 0$$

and

$$B = \frac{R_0^3}{\gamma} A_{10}\frac{L_0}{R_0}$$

whereupon the $n' = 0$ moment equation requires

$$\sigma^2 \frac{R_0^3 \rho}{\gamma} = -\frac{3}{x^2} \tag{4.29}$$

in order to make a solution, other than $B = 0 = A_{10}$, possible.

This is an estimate of only one eigenvalue and its accuracy is not known. Taking more equations into account, and, by this, more coefficients, not only produces estimates of more eigenvalues but, hopefully, better estimates of those already obtained.

It is of some interest to record the $m = 1$ result in the case of fixed-fixed end conditions. It is

$$\sigma^2 \frac{R_0^3 \rho}{\gamma} = -\frac{12}{x^2} \tag{4.30}$$

Equations (4.29) and (4.30) are illustrated in Figure 4.5.

Now, these approximations can be improved. Yet, instead of doing this, we turn to a question whose answer has been put off while we have been working out the effect of various end conditions on the stability of an inviscid liquid bridge.

4.7   What Can Be Said About Viscosity Short of Solving the Eigenvalue Problem?

Taking into account the viscosity of the fluid making up the bridge, and also the density and the viscosity of the ambient fluid used to eliminate the effect of gravity, introduces a technical difficulty. It is this: The two fluids must satisfy the no-slip condition where they come into contact with the solid walls of the vessel in which an experiment is being run. The two fluids must be divided by an interface. This interface must touch the solid walls in a line known as a contact line. Hence, if the end condition holding at this contact line is other than a fixed end condition, we would need to explain how the contact line is able to slide along the walls in apparent violation of the no-slip condition.

To avoid facing this difficulty, which, in any case, should not make viscosity more important than it otherwise might be, we confine our attention to the case of fixed end conditions whereupon the contact line is fixed and no-slip is required to hold on either side of this fixed line. Mason's photograph in Figure 4.1 is enough to indicate that a fixed end condition can be obtained in an experiment. In our idealization of his experiment we go on and require no flow to hold across the end surfaces in which contact lines lie.

Figure 4.6, then, illustrates the case of interest. Two fluids are bounded by solid walls upon which no-slip is required to hold. No more need be said about these walls. Their presence is required to make the problem finite in a simple way. Both fluids are viscous, their viscosities being denoted by μ and μ^*. The only surface that can be displaced is the interface dividing the two fluids, and where it contacts the side walls, it is required to remain fixed.

The equations that must hold on the domain are

$$\rho^* \frac{\partial \vec{v}^*}{\partial t} + \rho^* \vec{v}^* \cdot \nabla \vec{v}^* = -\nabla p^* + 2\mu^* \nabla \cdot \overset{\leftrightarrow}{D}^*$$

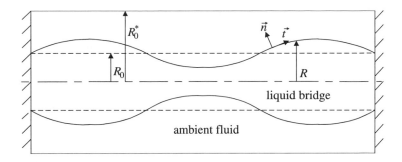

Figure 4.6. *A Liquid Bridge Satisfying Fixed End Conditions*

$$\rho\frac{\partial \vec{v}}{\partial t} + \rho\vec{v}\cdot\nabla\vec{v} = -\nabla p + 2\mu\nabla\cdot\vec{D}$$

and

$$\nabla\cdot\vec{v}^* = 0 = \nabla\cdot\vec{v}$$

The equations that must hold at the deflecting surface are

$$\vec{n}\cdot\vec{v}^* = u = \vec{n}\cdot\vec{v}$$

$$\vec{t}\cdot\vec{v}^* = \vec{t}\cdot\vec{v}$$

and

$$\vec{n}\cdot\vec{T}^* - \vec{n}\cdot\vec{T} = -\gamma\,2H\,\vec{n}$$

The last equation can be resolved into components normal and tangential to the surface. Using $\vec{T} = -p\vec{I} + 2\mu\vec{D}$, it can be written

$$-p^* + 2\mu^*\vec{n}\vec{n}:\vec{D}^* - \left[-p + 2\mu\vec{n}\vec{n}:\vec{D}\right] = -\gamma\,2H$$

and

$$2\mu^*\vec{t}\,\vec{n}:\vec{D}^* - 2\mu\vec{t}\,\vec{n}:\vec{D} = 0$$

The deflecting surface is denoted by

$$r = R(\theta,\,z,\,t,\,\epsilon)$$

and determining the function R is part of solving the nonlinear problem.

This problem has a simple solution, and it is denoted by the subscript zero to indicate that it is the base solution whose stability is to be determined. In this solution, $\vec{v}_0^*$, $\vec{v}_0$, p_0^*, p_0 and R_0 are all constant and, in fact, $\vec{v}_0^*$ and $\vec{v}_0$ must be zero.

It is important to observe that $\vec{D}_0^*$ and $\vec{D}_0$ are also both zero. The vanishing of $\vec{v}_0^*$ and $\vec{v}_0$ simplifies the perturbation equations on the reference domain; this and the vanishing of $\vec{\vec{D}}_0^*$ and $\vec{\vec{D}}_0$ simplify the perturbation equations at its boundary.

Turning then to the perturbation equations, on the reference domain,

$$\sigma \rho^* \vec{v}_1^* = -\nabla_0 p_1^* + 2\mu^* \nabla_0 \cdot \vec{\vec{D}}_1^* \tag{4.31}$$

$$\sigma \rho \vec{v}_1 = -\nabla_0 p_1 + 2\mu \nabla_0 \cdot \vec{\vec{D}}_1 \tag{4.32}$$

and

$$\nabla_0 \cdot \vec{v}_1^* = 0 = \nabla_0 \cdot \vec{v}_1 \tag{4.33}$$

must hold; at the reference interface between the two fluids

$$\vec{n}_0 \cdot \vec{v}_1^* = \sigma R_1 = \vec{n}_0 \cdot \vec{v}_1 \tag{4.34}$$

$$\vec{t}_0 \cdot \vec{v}_1^* = \vec{t}_0 \cdot \vec{v}_1 \tag{4.35}$$

$$-p_1^* + 2\mu^* \vec{n}_0 \vec{n}_0 : \vec{\vec{D}}_1^* - \left[-p_1 + 2\mu \vec{n}_0 \vec{n}_0 : \vec{\vec{D}}_1 \right]$$

$$= -\gamma \left[\frac{R_1}{R_0^2} + \frac{1}{R_0^2} \frac{\partial^2 R_1}{\partial \theta_0^2} + \frac{\partial^2 R_1}{\partial z_0^2} \right] \tag{4.36}$$

and

$$2\mu^* \vec{t}_0 \vec{n}_0 : \vec{\vec{D}}_1^* = 2\mu \vec{t}_0 \vec{n}_0 : \vec{\vec{D}}_1 \tag{4.37}$$

must hold, where the time dependence has been eliminated by introducing the factor $e^{\sigma t_0}$. By eliminating t_0 in favor of σ, the perturbation equations have been presented as an eigenvalue problem, where σ denotes the eigenvalue, and our job is to determine the eigenvalues as they depend on the parameters of the problem: the radii R_0 and R_0^*, the length L_0, the densities ρ and ρ^*, the viscosities μ and μ^*, etc.

Due to the simplicity of the base solution, its signature is nowhere to be seen in the perturbation equations and this is what will, in the end, lead us to the conclusion that the neutral point is independent of the viscosity of the jet as well as of the viscosity of the surrounding fluid.

Before going on, it is worth observing that the eigenvalue problem has the solution $\sigma = 0$ corresponding to $\vec{v}_1^*$ and $\vec{v}_1$ both zero, p_1^* and p_1 both constant, but R_1 not zero, where R_1 is determined by solving equation (4.36) in the form

$$\frac{p_1^* - p_1}{\gamma} = \left[\frac{R_1}{R_0^2} + \frac{1}{R_0^2} \frac{\partial^2 R_1}{\partial \theta_0^2} + \frac{\partial^2 R_1}{\partial z_0^2} \right]$$

while satisfying the required end conditions and the constant-volume condition. These are just the inviscid fluid formulas. As none of this depends in any way on the viscosity of either fluid, a neutral condition is obtained that is the same whatever values the viscosities of the two fluids may take. It is equally important to observe that this would not have been the case were $\vec{\bar{D}}_0^*$ and $\vec{\bar{D}}_0$ not zero. They would then make a contribution to the normal stress equation at the boundary of the reference configuration.

The plan, then, is to try to learn something about the solution to this eigenvalue problem, short of actually solving it. To do this, multiply equation (4.31) by $\vec{v}_1^* \cdot$ and equation (4.32) by $\vec{v}_1 \cdot$, add, and integrate the result over the reference domain. All terms evaluated at the solid walls vanish. Then, using equation (4.33) on the reference domain along with equation (4.34) at the interface between the two fluids, there obtains

$$
\sigma \left[\rho^* \int_{V_0^*} dV_0^* \, |\vec{v}_1^*|^2 + \rho \int_{V_0} dV_0 \, |\vec{v}_1|^2 \right]
$$

$$
= \int_{S_0} dA_0 \; \bar{\sigma} \bar{R}_1 [p_1^* - p_1] + \int_{S_0} dA_0 \left[- 2\mu^* \vec{n}_0 \cdot \vec{\bar{D}}_1^* \cdot \vec{v}_1^* + 2\mu \vec{n}_0 \cdot \vec{\bar{D}}_1 \cdot \vec{v}_1 \right]
$$

$$
- 2\mu^* \int_{V_0^*} dV_0^* \; \vec{\bar{D}}_1^* : \vec{\bar{D}}_1^* - 2\mu \int_{V_0} dV_0 \; \vec{\bar{D}}_1 : \vec{\bar{D}}_1 \tag{4.38}
$$

where S_0 denotes the deflecting interface in its reference configuration and where the signs on the right-hand side can be accounted for by the fact that $\vec{n}_0$ points out of V_0 into V_0^*.

Equation (4.38) can be simplified in an important way by writing, at the interface,

$$
\vec{v}_1^* = \vec{n}_0 [\vec{n}_0 \cdot \vec{v}_1^*] + \vec{t}_0 [\vec{t}_0 \cdot \vec{v}_1^*]
$$

and

$$
\vec{v}_1 = \vec{n}_0 [\vec{n}_0 \cdot \vec{v}_1] + \vec{t}_0 [\vec{t}_0 \cdot \vec{v}_1]
$$

Doing this and using the remaining equations at the interface [viz., equations (4.35), (4.36) and (4.37)], there obtains

$$\sigma \left[\rho^* \int_{V_0^*} dV_0^* \, |\vec{v}_1^*|^2 \, + \, \rho \int_{V_0} dV_0 \, |\vec{v}_1|^2 \right]$$

$$= \bar{\sigma}\gamma \int_{S_0} dA_0 \, \bar{R}_1 \left[\frac{R_1}{R_0^2} + \frac{1}{R_0^2} \frac{\partial^2 R_1}{\partial \theta_0^2} + \frac{\partial^2 R_1}{\partial z_0^2} \right] - 2\mu^* \int_{V_0^*} dV_0^* \, \overset{\Rightarrow}{D}_1^* : \overset{\Rightarrow}{D}_1^*$$

$$- 2\mu \int_{V_0} dV_0 \, \overset{\Rightarrow}{D}_1 : \overset{\Rightarrow}{D}_1 \tag{4.39}$$

and what turns out is that the tangential equations can be used to eliminate the viscosities of the fluids from the surface integral. The viscosities come into equation (4.39) only by way of the domain equations, just as they would if the interface were nondeflecting. This is the important legacy of the simple base solution where $\vec{v}_0^* = \vec{0} = \vec{v}_0$.

Although the integrals in equation (4.39) depend on the eigenfunctions, and these are not known, some important conclusions can still be drawn about the eigenvalues and, hence, about the stability of the bridge. To do this, it helps to introduce the shorthand notation

$$[KE] = \rho^* \int_{V_0^*} dV_0^* \, |\vec{v}_1^*|^2 + \rho \int_{V_0} dV_0 \, |\vec{v}_1|^2$$

and

$$[VISC] = 2\mu^* \int_{V_0^*} dV_0^* \, \overset{\Rightarrow}{D}_1^* : \overset{\Rightarrow}{D}_1^* + 2\mu \int_{V_0} dV_0 \, \overset{\Rightarrow}{D}_1 : \overset{\Rightarrow}{D}_1$$

where both are real and not negative. In fact, $[KE]$ vanishes if and only if $\vec{v}_1 = \vec{0} = \vec{v}_1^*$; likewise, so long as μ and μ^* are not zero, $[VISC]$ vanishes if and only if $\vec{v}_1 = \vec{0} = \vec{v}_1^*$. Due to the presence of the rigid walls, if either $\vec{v}_1$ or $\vec{v}_1^*$ is constant, the constant must be zero.

In terms of $[KE]$ and $[VISC]$, equation (4.39) can be written

$$\sigma[KE] = -\bar{\sigma}[PE] - [VISC] \tag{4.40}$$

where $[PE]$ is shorthand for

$$-\gamma \int_{S_0} dA_0 \left[\frac{1 - m^2}{R_0^2} |R_1|^2 - \left| \frac{dR_1}{dz_0} \right|^2 \right]$$

where the factor $e^{im\theta_0}$ has been introduced and where $[PE]$ is real. Now, R_1 depends only on z_0, its time and angle dependence having been removed.

To see what equation (4.40) tells us, put $\sigma = a + ib$ and resolve it into its real and imaginary parts. This turns equation (4.40) into

$$a[KE] = -a[PE] - [VISC] \tag{4.41}$$

and

$$b[KE] = b[PE] \tag{4.42}$$

from which some conclusions can be drawn.

First, let $\mu^* = 0 = \mu$, which is the inviscid case. Then, equations (4.41) and (4.42) reduce to

$$a[KE] = -a[PE]$$

and

$$b[KE] = b[PE]$$

whence a and b cannot both be non zero and, by this, σ must be either real or purely imaginary. There are two cases.

In the first case, let $b = 0$, $a \neq 0$ whereupon $[KE] = -[PE]$ and, hence, $[PE]$ must be non positive.

In the second case, let $b \neq 0$, $a = 0$ whereupon $[KE] = [PE]$ and, hence, $[PE]$ must be non negative.

Now, specification of an eigenfunction requires the specification of the two functions $\vec{v}_1$ and R_1, where R_1 cannot vanish. To any possible eigenfunction, we need to decide if the corresponding eigenvalue takes the form $\sigma = a \neq 0$ or the form $\sigma = ib \neq 0$.

To any eigenfunction where m^2 is greater than zero, $[PE]$ must be non negative[7] and the eigenvalue must then take the form $\sigma = ib$. By this, only axisymmetric eigenfunctions can lead to real values of σ.

Turn then to eigenfunctions where $m = 0$ and where $[PE]$, now given by

$$[PE] = -\gamma \int_{S_0} dA_0 \left[\frac{|R_1|^2}{R_0^2} - \left| \frac{dR_1}{dz_0} \right|^2 \right]$$

may be of either sign. If $[PE]$ is negative, the eigenvalue takes the form $\sigma = a$, where a may be of either sign; if $[PE]$ is positive, the eigenvalue takes the form $\sigma = ib$.

Now, observe that the inequality

$$\int_{S_0} dA_0 \left| \frac{dR_1}{dz_0} \right|^2 \geq \lambda_1^2 \int_{S_0} dA_0 |R_1|^2$$

must hold, where $dA_0 = 2\pi R_0 \, dz_0$ and $0 \leq z_0 \leq L_0$. Then, introduce the requirement

$$\frac{1}{R_0^2} < \lambda_1^2$$

---

[7]In case $m^2 = 1$, it might appear that an eigenfunction in which R_1 is a constant would be a possible solution, as then $[PE] = 0$. However, that requires $[KE] = 0$ whereupon $\vec{v}_1$ must be zero, but $\vec{n} \cdot \vec{v}_1 = \sigma R_1$ would then require σ to be zero and, so too, both a and b.

where λ_1^2 is the fundamental eigenvalue of $-d^2/dz_0^2$ corresponding to fixed end conditions. By this, there obtains

$$\gamma 2\pi R_0 \int_0^{L_0} dz_0 \left[\left| \frac{dR_1}{dz_0} \right|^2 - \frac{|R_1|^2}{R_0^2} \right] > \gamma 2\pi R_0 \int_0^{L_0} dz_0 \left[\lambda_1^2 - \frac{1}{R_0^2} \right] |R_1|^2 > 0$$

whence $[PE]$ must be positive and all eigenfunctions must correspond to eigenvalues such that $b \neq 0$, $a = 0$.

The requirement $1/R_0^2 < \lambda_1^2$ is a restriction on the geometry of the jet inasmuch as $\lambda_1^2 \propto 1/L_0^2$. Upon being satisfied, it asserts solutions to the eigenvalue problem of the form $b \neq 0$, $a = 0$, while ruling out solutions of the form $b = 0$, $a \neq 0$. By this, it defines a range of liquid bridges short enough to be stable to small displacements of any kind.

Now, turn to the case where μ and μ^* are not both zero. Then, either $[KE]$ and $[VISC]$ both vanish or neither vanish. Let $[VISC]$ be positive and require a and b to satisfy equations (4.41) and (4.42), namely

$$a[KE] = -a[PE] - [VISC]$$

and

$$b[KE] = b[PE]$$

Let b be other than zero, then a must be given by

$$a = \frac{-[VISC]}{2[KE]}$$

and a is not zero, as above, but it must be negative.

This leads to an important conclusion: So long as b is not zero, neither is a zero whereupon the real part of σ can be zero if and only if σ itself is zero. But $\sigma = 0$ is the only case where the eigenvalue problem is easy to solve, and earlier, we found the solution to be independent of the viscosities of the fluids.

The case where $[VISC]$ is zero is simple. Then, $\vec{v}_1^* = \vec{0} = \vec{v}_1$ obtains and this takes us back to the inviscid case where, to obtain a solution to the eigenvalue problem, R_1 must not be zero but $[PE]$ must vanish.

Short of solving the eigenvalue problem, we can conclude that in case $\mu^* = 0 = \mu$, to any solution where b is not zero, the corresponding value of a must be zero and the bridge exhibits sustained oscillations when displaced in this way. If a little viscosity is added, all the solutions where b is not zero now die out, as a must be negative. The requirement for this is $[PE] > 0$ and geometric conditions that the bridge must satisfy to make this hold have been presented above in the course of dealing with the inviscid case.

In the case where $[KE] = 0$, and therefore $[VISC] = 0$, the real part of σ (i.e., a), cannot be determined by the foregoing equations, but going back to the eigenvalue problem, this is just the case $\sigma = 0$.

The case where viscosity is dominant can be obtained by setting $\rho^* = 0 = \rho$. Then, $[KE] = 0$, but $[VISC] \geq 0$ and there obtains

$$0 = -a[PE] - [VISC]$$

and

$$0 = b[PE]$$

Unless $[PE] = 0$, there are only solutions where $a \neq 0$ and $b = 0$. Then, a is negative whenever $[PE]$ is positive whereas it is positive whenever $[PE]$ is negative. In both cases, b is zero. Again, the neutral case corresponds to $[PE] = 0$ and whether or not this is satisfied does not depend on the viscosities of the fluids.

A calculation like this will be carried out in the sixth essay. The conclusion will be the same there as it is here: The critical point does not depend on the viscosities of the fluids.

5

The Stability of a Spinning Jet in Rigid Rotation

5.1 The Physics of the Instability

In this essay, liquid bridges[1] remain of interest, but now, instead of lying at rest as earlier, they are set into rotation at a uniform angular speed.

To understand the physics of the problem, let us recall what we already know about a jet at rest and pay attention to axisymmetric displacements of the jet. These displacements introduce a longitudinal variation of the jet diameter along the length of the jet in the form of a sequence of crests and troughs. Such a displacement changes both the longitudinal curvature and the transverse curvature of the surface, the latter change being due to the fact that the diameter of the jet is increased under a crest, whereas it is decreased under a trough. Now, these curvature changes affect the pressure variation along the jet. On the one hand, the pressure increases under a crest, but it decreases under a trough, due to the longitudinal curvature variation. On the other hand, the transverse curvature variation decreases the pressure under a crest, whereas it increases it under a trough. Now, these two changes in the pressure, opposing one another, can come into balance when the displacement of the jet has just the right wavelength, for the transverse curvature is independent of the wavelength whereas the longitudinal curvature depends strongly on the wavelength. However, as

---

[1]The reader saw in the first essay how little it matter whether a liquid thread was a bridge or a jet. That is true here as well, as we limit ourselves to free-free end conditions where the surface of a bridge contacts a solid wall.

both changes in pressure are proportional to the surface tension, the surface tension plays no part in determining the critical wavelength.

Now, let the jet spin about its long axis at a uniform angular speed, and imagine spinning with it so that its motion is not noticed. Instead, what is noticed is a force on the fluid acting outward, causing the pressure of the fluid to increase outward. Again, a displacement of the surface of the jet can be introduced to see what is required in order that the pressure balance not be upset.[2] Doing this, we notice something new. Due to the centripetal acceleration, there is an increased pressure under a crest but a reduced pressure under a trough. This leads us to the conclusion that the balance discovered in the earlier essays ought to be shifted to shorter wavelengths when the jet is set to spinning. It also leads us to expect that the surface tension will now influence the value of the critical wavelength, as the surface tension comes into only one part of the balance. This balance is again a pressure balance at no flow; the critical wavelength should remain independent of the viscosity of the fluid making up the jet.

These ideas are going to be seen again in the next essay, where the centripetal acceleration will be replaced by a gravitational force.

Now, the endnote and discussion notes can be read or not at the readers pleasure, but the derivation of the base solution appears in the endnote. It establishes the facts of the problem as they would be if the fluid were confined by a rigid wall in place of its own surface tension. This is important, as spinning liquids introduce stability problems in their own right and these carry over, willy-nilly, to spinning jets. The endnote also introduces the idea of angular momentum stratification and the important function, denoted Φ, by which a stable stratification of angular momentum can be identified. A formula obtained there is used in the essay.

Our aim then is to establish the wave number of the critical displacement by solving the perturbation equations. This calculation is of some interest just due to the fact that the base state of a spinning jet is not quite so simple as that of a jet at rest, and due to this, this essay fits well between the earlier essays and the next essay.

5.2 The Nonlinear Equations

We direct our attention to a spinning liquid bridge held together by its surface tension, executing its motions in free space. This means that whatever lies outside the bridge itself is not subject to perturbation and the velocity and pressure there can be set to zero.

---

[2]As before, no motion is introduced by doing this; that is, in the laboratory frame, the rigid motion is maintained.

To define the problem, let an inviscid liquid bridge be confined by two parallel plane walls, one at $z = 0$ and the other at $z = L_0$. The walls and the liquid bridge between them are then set into motion at a uniform angular speed about the axis of the bridge. The end conditions where the surface of the bridge contacts the walls are taken to be free-free; that is, the surface contacts the walls at right angles.[3]

The present domain of the spinning bridge is denoted by

$$r < R(z, \theta, t)$$

where the function R must be determined, and the motion of the fluid making up the bridge must satisfy the nonlinear equations on this domain and at its boundary. These equations are

$$\rho \frac{\partial \vec{v}}{\partial t} + \rho \vec{v} \cdot \nabla \vec{v} = -\nabla p$$

and

$$\nabla \cdot \vec{v} = 0$$

on the domain and

$$\vec{n} \cdot \vec{v} = u$$

and

$$p = -\gamma \, 2H$$

at its lateral surface.

There is a simple solution to these equations and, as it is the base solution whose stability we seek to determine, the subscripts zero are introduced in writing it. It is

$$v_{r_0} = 0 = v_{z_0}$$

$$v_{\theta_0} = r_0 \Omega_0$$

and

$$\frac{dp_0}{dr_0} = \rho r_0 \Omega_0^2$$

where

$$p_0(r_0 = R_0) = \frac{\gamma}{R_0}$$

The variable R_0 denotes the equilibrium radius of the spinning bridge and Ω_0 denotes its angular speed. Now, Ω_0 can be an arbitrary function of r_0,

---

[3] Again the fluid is taken to be inviscid. Again the critical point does not depend on the viscosity of the fluid making up the jet. But now there appears to be some doubt about this. That doubt is brought to light at the end of the essay.

but our interest is in the special case where Ω_0 is constant, corresponding to a uniform angular speed, and to emphasize this, Ω_0 is written henceforth as A.

This, then, is the base solution whose stability is in question. This base flow enjoys a stable stratification of its angular momentum. In fact, as Ω_0 is constant, Rayleigh's function Φ_0 is just $4\Omega_0^2$ whence it is positive everywhere in the base flow. It remains, then, to discover how a fluid bounded by a flexible wall acts differently in the way it responds to a disturbance, than does a fluid bounded by a rigid wall.

The difference is due to the fact that the surface of a liquid bridge can deflect whenever a disturbance is imposed on the bridge and the variable used to account for this in the perturbation problem is denoted by R_1. It comes into the perturbation problem via the expansion of R that is

$$R(z,\ t,\ \epsilon) = R_0 + \epsilon R_1(z_0,\ t_0) + \cdots$$

where R_1 is part of the solution to the perturbation problem. It is zero if the fluid is bounded by a rigid wall.

This expansion and all other formulas in this essay take only axisymmetric displacements into account.

5.3   The Perturbation Equations

The perturbation equations on the reference domain and on its boundary obtain in the usual way as explained in the second essay. However, the base solution is a little more interesting than it was in the earlier essays. Terms appear due to both v_{θ_0} and $\dfrac{dp_0}{dr_0}$ not being zero.

The reference domain is the region

$$0 < r_0 < R_0,\ 0 < z_0 < L_0$$

and the perturbation equations thereon can be taken from the endnote. They are

$$\left.\begin{array}{c}
\dfrac{\partial v_{r_1}}{\partial t_0} - 2Av_{\theta_1} = -\dfrac{1}{\rho}\dfrac{\partial p_1}{\partial r_0} \\[3mm]
\dfrac{\partial v_{\theta_1}}{\partial t_0} + 2Av_{r_1} = 0 \\[3mm]
\dfrac{\partial v_{z_1}}{\partial t_0} - \dfrac{1}{\rho}\dfrac{\partial p_1}{\partial z_0} \\[3mm]
\dfrac{\partial v_{r_1}}{\partial r_0} + \dfrac{v_{r_1}}{r_0} + \dfrac{\partial v_{z_1}}{\partial z_0} = 0
\end{array}\right\} \quad 0 < r_0 < R_0,\ 0 < z_0 < L_0$$

Again, Ω_0 is written as A to emphasize that it is constant. The terms involving A distinguish the spinning jet equations from those corresponding to a jet at rest.

To these equations must be added the perturbation equations at the lateral surface and at the ends of the reference jet. At the lateral surface there obtains

$$v_{r_1} = \frac{\partial R_1}{\partial t_0}, \quad r_0 = R_0, \quad 0 < z_0 < L_0$$

and

$$p_1 + R_1 \frac{dp_0}{dr_0} = -\frac{\gamma}{R_0^2}\left[R_1 + R_0^2 \frac{\partial^2 R_1}{\partial z_0^2}\right], \quad r_0 = R_0, \quad 0 < z_0 < L_0$$

The first equation is familiar, but that is due to the requirement of axial symmetry. In the case of a nonaxisymmetric disturbance, it would include a new term not seen in the earlier essays. The second equation is new, the new term being due to the pressure variation in the base solution.

The remaining equations come from the requirements imposed on the liquid bridge where it is in contact with the solid walls, namely from

$$v_z = 0, \quad z = 0, \quad z = L_0$$

and

$$\frac{\partial R}{\partial z} = 0, \quad z = 0, \quad z = L_0$$

and they turn out to be

$$v_{z_1} = 0, \quad z_0 = 0, \quad z_0 = L_0$$

and

$$\frac{\partial R_1}{\partial z_0} = 0, \quad z_0 = 0, \quad z_0 = L_0$$

The perturbation problem, then, is an initial-value problem, and to solve it, turn it into an eigenvalue problem in the usual way by writing

$$v_{r_1} = \hat{v}_{r_1}(r_0)e^{\sigma t_0} \cos k z_0$$

$$v_{\theta_1} = \hat{v}_{\theta_1}(r_0)e^{\sigma t_0} \cos k z_0$$

$$v_{z_1} = \hat{v}_{z_1}(r_0)e^{\sigma t_0} \sin k z_0$$

$$p_1 = \hat{p}_1(r_0)e^{\sigma t_0} \cos k z_0$$

and

$$R_1 = \hat{R}_1 e^{\sigma t_0} \cos k z_0$$

Then, by limiting k to the discrete values[4] $n\pi/L_0$, $n = 1, 2, \cdots$, the end conditions on v_{z_1} and R_1 can be seen to be satisfied.

In terms of $\hat{v}_{r_1}$, etc., the perturbation eigenvalue problem obtains. It is

$$\sigma \hat{v}_{r_1} - 2A\hat{v}_{\theta_1} = -\frac{1}{\rho}\frac{d\hat{p}_1}{dr_0} \tag{5.1}$$

$$\sigma \hat{v}_{\theta_1} + 2A\hat{v}_{r_1} = 0 \tag{5.2}$$

$$\sigma \hat{v}_{z_1} = \frac{1}{\rho}k\hat{p}_1 \tag{5.3}$$

and

$$\frac{d\hat{v}_{r_1}}{dr_0} + \frac{\hat{v}_{r_1}}{r_0} + k\hat{v}_{z_1} = 0 \tag{5.4}$$

on $0 < r_0 < R_0$ and

$$\hat{v}_{r_1} = \sigma \hat{R}_1 \tag{5.5}$$

and

$$\hat{p}_1 + \rho R_0 A^2 \hat{R}_1 = -\frac{\gamma}{R_0^2}[1 - k^2 R_0^2]\hat{R}_1 \tag{5.6}$$

at $r_0 = R_0$. In addition, $\hat{v}_{r_1}$ and $\hat{p}_1$ must be bounded at $r_0 = 0$.

Notice that this is just the rigid-wall problem of the endnote in case $\hat{R}_1$ is set to zero and the last equation is dropped. It is also the axisymmetric case of the problem solved in the fourth essay in the case where A is set to zero.

5.4   The Critical Condition

To determine the critical wavelength, we look for a solution to the eigenvalue problem where σ is zero. To see if there is such a solution, other than $\hat{v}_{r_1}$, $\hat{v}_{\theta_1}$, $\hat{v}_{z_1}$, $\hat{p}_1$ and $\hat{R}_1$ all zero, set σ to zero and observe that if A is not zero, the only solution to equations (5.1), (5.2), (5.3) and (5.4) is $\hat{v}_{1_1}$, $\hat{v}_{\theta_1}$, $\hat{v}_{z_1}$ and $\hat{p}_1$ all zero. However, $\hat{R}_1$ may or may not vanish. The necessary and sufficient condition that $\hat{R}_1$ not vanish, and by this the necessary and sufficient condition that $\sigma = 0$ be a solution to the eigenvalue problem, is given by equation (5.6) as

$$1 - k^2 R_0^2 + \frac{\rho}{\gamma}R_0^3 A^2 = 0$$

---

[4]Again, the possibility $k = 0$ is ruled out by requiring the volume of the jet to remain fixed on displacement.

This result, which determines the critical value of k^2, was also found in the first discussion note by a Rayleigh work calculation.

In the case of an infinitely long jet, this formula tells us the value of the critical wavelength and it indicates that the critical displacement shifts to shorter wavelengths the higher the value of A. In the case of a liquid bridge, where $k = \pi n/L_0$, the formula tells us the lengths at which a neutral disturbance can be found, once A is set. By this, a spinning jet is more like a jet at rest than like a spinning liquid confined by rigid walls. In the case of rigid walls, by setting σ to zero there obtains $\hat{v}_{r_1}$, $\hat{v}_{\theta_1}$, $\hat{v}_{z_1}$ and $\hat{p}_1$ all zero, but $\hat{R}_1$ cannot be other than zero, whereupon σ cannot be zero as long as A is not zero. By this, $\sigma = 0$ cannot be an eigenvalue in the rigid-wall problem.

Turning to the jet, there is an important difference between spinning jets and jets at rest. If $\sigma = 0$ is an eigenvalue in case A is not zero, then $\hat{v}_{r_1}$, $\hat{v}_{\theta_1}$, $\hat{v}_{z_1}$ and $\hat{p}_1$ must all vanish, but $\hat{R}_1$ need not, so long as $1 - k^2 R_0^2 + \frac{\rho}{\gamma} R_0^3 A^2 = 0$. If $\sigma = 0$ is an eigenvalue in the case A is zero, then $\hat{p}_1$ must vanish, but $\hat{R}_1$ need not, so long as $1 - k^2 R_0^2 = 0$. In this case, $\hat{v}_{r_1}$, $\hat{v}_{\theta_1}$ and $\hat{v}_{z_1}$ cannot be determined and need not vanish, whether or not $\hat{R}_1$ vanishes.

If the shortest neutral length is also the critical length, as it is for a jet at rest, then the critical length is

$$\left[\frac{\pi^2 R_0^2}{1 + \dfrac{\rho}{\gamma} R_0^3 A^2} \right]^{\frac{1}{2}}$$

This result is illustrated later on in Figures 5.1, 5.2 and 5.3.

5.5   An Integral Formula for σ^2

To solve the eigenvalue problem, eliminate $\hat{v}_{\theta_1}$, $\hat{v}_{z_1}$ and $\hat{R}_1$ in favor of $\hat{v}_{r_1}$ and $\hat{p}_1$. To do this, use equations (5.2) and (5.3) to eliminate $\hat{v}_{\theta_1}$ and $\hat{v}_{z_1}$ in equations (5.1) and (5.4) and then use equation (5.5) to eliminate $\hat{R}_1$ in equation (5.6) to obtain

$$[\sigma^2 + 4A^2]\hat{v}_{r_1} = -\frac{\sigma}{\rho} \frac{d\hat{p}_1}{dr_0}$$

and

$$\sigma \left[\frac{d}{dr_0} + \frac{1}{r_0} \right] \hat{v}_{r_1} + \frac{k^2}{\rho} \hat{p}_1 = 0$$

on $0 < r_0 < R_0$ and

$$\sigma \hat{p}_1 = -\frac{\gamma}{R_0^2} \left[1 - k^2 R_0^2 + \frac{\rho}{\gamma} R_0^3 A^2 \right] \hat{v}_{r_1}$$

at $r_0 = R_0$, where both $\hat{v}_{r_1}$ and $\hat{p}_1$ must remain bounded at $r_0 = 0$. Then, either eliminate $\hat{p}_1$ in favor of $\hat{v}_{r_1}$ to get

$$\sigma^2 \frac{d}{dr_0} \left[\frac{d}{dr_0} + \frac{1}{r_0} \right] \hat{v}_{r_1} - k^2 [\sigma^2 + 4A^2] \hat{v}_{r_1} = 0, \quad 0 < r_0 < R_0 \qquad (5.7)$$

and

$$\sigma^2 \left[\frac{d}{dr_0} + \frac{1}{r_0} \right] \hat{v}_{r_1} = \frac{\gamma}{\rho} \frac{k^2}{R_0^2} \left[1 - k^2 R_0^2 + \frac{\rho}{\gamma} R_0^3 A^2 \right] \hat{v}_{r_1}, \quad r_0 = R_0 \qquad (5.8)$$

or eliminate $\hat{v}_{r_1}$ in favor of $\hat{p}_1$ to get

$$\sigma^2 \left[\frac{d}{dr_0} + \frac{1}{r_0} \right] \frac{d}{dr_0} \hat{p}_1 - k^2 [\sigma^2 + 4A^2] \hat{p}_1 = 0, \quad 0 < r_0 < R_0$$

and

$$\sigma [\sigma^2 + 4A^2] \hat{p}_1 = \frac{\gamma \sigma}{\rho R_0^2} \left[1 - k^2 R_0^2 + \frac{\rho}{\gamma} R_0^3 A^2 \right] \frac{d\hat{p}_1}{dr_0}, \quad r_0 = R_0$$

Our work will be in terms of $\hat{v}_{r_1}$; the reader should work the problem in terms of $\hat{p}_1$.

Now, notice that once again, as in the earlier essays, σ^2, not σ, is seen to be the eigenvalue. Then, notice that in the case A^2 is zero, the problem is unlike any problem where A^2 is not zero. In the first case, σ^2 does not appear in the domain equation whereupon just one value of σ^2 corresponds to each value of k^2. In the second case, many values of σ^2 correspond to each value of k^2. In fact, in the second case, the domain equation tells us that $\hat{v}_{r_1}$ must be a multiple of

$$J_1(\lambda r_0)$$

where $\sigma^2 \lambda^2 = -k^2 [\sigma^2 + 4A^2]$. Then, substituting this into the equation at $r_0 = R_0$ produces an equation in analytic functions of σ^2 whose infinitely many isolated roots determine infinitely many values of σ^2.

Now, it is possible to draw some useful conclusions about σ^2 short of solving equations (5.7) and (5.8). To do this, multiply equation (5.7) by $r_0 \bar{\hat{v}}_{r_1}$, integrate the product over $0 < r_0 < R_0$ and use integration by parts one time. The result of doing this can be found in the endnote as equation (5.11), save that now $\Phi_0 = 4A^2$. It is

$$\sigma^2 \left[\bar{\hat{v}}_{r_1} r_0 \frac{d\hat{v}_{r_1}}{dr_0} \right]_{r_0=0}^{r_0=R_0} - \sigma^2 \int_0^{R_0} \left[\left| \frac{d\hat{v}_{r_1}}{dr_0} \right|^2 + \frac{|\hat{v}_{r_1}|^2}{r_0^2} + k^2 |\hat{v}_{r_1}|^2 \right] r_0 \, dr_0$$

$$= k^2 4A^2 \int_0^{R_0} |\hat{v}_{r_1}|^2 r_0 \, dr_0$$

and it only remains to evaluate the first term at $r_0 = 0$ and at $r_0 = R_0$, where at $r_0 = R_0$, by equation (5.8), we have

$$\sigma^2 R_0 \frac{d\hat{v}_{r_1}}{dr_0} = -\sigma^2 \hat{v}_{r_1} + \frac{\gamma}{\rho} \frac{k^2}{R_0} \left[1 - k^2 R_0^2 + \frac{\rho}{\gamma} R_0^3 A^2\right] \hat{v}_{r_1}$$

Doing this, there obtains

$$-\sigma^2 \left[\left|\hat{v}_{r_1}(r_0 = R_0)\right|^2 + \int_0^{R_0} \left[\left|\frac{d\hat{v}_{r_1}}{dr_0}\right|^2 + \frac{|\hat{v}_{r_1}|^2}{r_0^2} + k^2 |\hat{v}_{r_1}|^2\right] r_0 \, dr_0\right]$$

$$= k^2 4 A^2 \int_0^{R_0} |\hat{v}_{r_1}|^2 r_0 \, dr_0$$

$$- \frac{\gamma}{\rho} \frac{k^2}{R_0} \left[1 - k^2 R_0^2 + \frac{\rho}{\gamma} R_0^3 A^2\right] \left|\hat{v}_{r_1}(r_0 = R_0)\right|^2 \tag{5.9}$$

and this equation produces two easy conclusions. The first is that σ^2 must be real; the second is that σ^2 must be negative if k^2 is large enough to make $\left[1 - k^2 R_0^2 + \frac{\rho}{\gamma} R_0^3 A^2\right]$ negative. The bridge, then, is stable to all short wavelength displacements (i.e., displacements corresponding to large values of k^2). Again, it is the long wavelength displacements that may be dangerous.

Now, equation (5.9) is indeterminate in the case where $\hat{v}_{r_1}$ is equal to zero. Hence, this case requires special attention, but it presents no problem. This is due to the fact that $\hat{v}_{r_1} = 0$ if and only if $\sigma^2 = 0$. To see this, go back to equations (5.1) through (5.6) and set $\hat{v}_{r_1}$ to zero. Then, $\hat{v}_{\theta_1}$, $\hat{v}_{z_1}$ and $\hat{p}_1$ must also be zero, and by this $\hat{R}_1$, must not be zero. To have $\hat{R}_1$ other than zero requires σ^2, as well as $[1 - k^2 R_0^2 + \frac{\rho}{\gamma} R_0^3 A^2]$, to be zero. If, on the other hand, σ^2 is set to zero, then $\hat{v}_{r_1}$, $\hat{v}_{\theta_1}$, $\hat{v}_{z_1}$ and $\hat{p}_1$ must all be zero and so too $[1 - k^2 R_0^2 + \frac{\rho}{\gamma} R_0^3 A^2]$ in order that $\hat{R}_1$ not also be zero. From this, we draw an important conclusion: The value of k^2 at which $[1 - k^2 R_0^2 + \frac{\rho}{\gamma} R_0^3 A^2]$ is zero is the only value of k^2 at which $\sigma^2 = 0$ can satisfy the eigenvalue problem, and in that case, and only in that case, can $\hat{v}_{r_1}$ be zero. Outside this important special case, equation (5.9) can be used.

Now, let k^2 be such that $[1 - k^2 R_0^2 + \frac{\rho}{\gamma} R_0^3 A^2]$ vanishes, but let σ^2 be an eigenvalue other than zero. Then, equation (5.9) tells us that any such eigenvalue must satisfy

$$-\sigma^2 = 4 A^2 \frac{k^2 \int_0^{R_0} |\hat{v}_{r_1}|^2 r_0 \, dr_0}{k^2 \int_0^{R_0} |\hat{v}_{r_1}|^2 r_0 \, dr_0 + P}$$

where

$$P = |\hat{v}_{r_1}(r_0 = R_0)|^2 + \int_0^{R_0} \left[\left| \frac{d\hat{v}_{r_1}}{dr_0} \right|^2 + \frac{|\hat{v}_{r_1}|^2}{r_0^2} \right] r_0 \, dr_0$$

and where P must be positive. By this, the eigenvalues, other than zero, at the critical value of k^2, must be negative. In fact, they must lie greater than $-4A^2$, less than zero. This is our third conclusion. The fourth is this: If any eigenvalue is positive, then $[1 - k^2 R_0^2 + \frac{\rho}{\gamma} R_0^3 A^2]$ must be positive.

The readers can draw these same four conclusions from the eigenvalue problem stated in terms of $\hat{p}_1$, should they choose to do so.

If $\hat{v}_{r_1}$ were to vanish at $r_0 = R_0$, but not vanish on the domain, equation (5.9) would tell us that σ^2 must be negative. By this, $\hat{R}_1$ must be zero as well, whereupon the problem would reduce to the one worked out in the endnote where the jet is held together by a rigid wall. Due to this, we require, henceforth, that $\hat{v}_{r_1}$ not vanish at $r_0 = R_0$ unless it vanishes everywhere.

This being so, by equation (5.9) the sign of σ^2 must be the sign of

$$-4A^2 k^2 \int_0^{R_0} |\hat{v}_{r_1}|^2 r_0 \, dr_0 + \frac{\gamma k^2}{\rho R_0} \left[1 - k^2 R_0^2 + \frac{\rho}{\gamma} R_0^3 A^2 \right] |\hat{v}_{r_1}(r_0 = R_0)|^2$$

where the first term is always negative and always stabilizing. This is due to the fact that angular momentum is stably stratified in the base flow. The sign of σ^2 will then depend on the second term, and for large enough values of k^2, it too is negative and stabilizing.

Now, the length of the bridge determines the admissible values of k^2 to be $\frac{\pi^2}{L_0^2} n^2$, $n = 1, 2, \cdots$. To any value of L_0, the factor

$$\left[1 - \frac{\pi^2}{L_0^2} n^2 R_0^2 + \frac{\rho}{\gamma} R_0^3 A^2 \right]$$

denoted $F(n, L_0)$, will be negative if n is made large enough and will take its largest value if $n = 1$. Hence, all else being held fixed, L_0 can always be made small enough so that F will be negative when $n = 1$ and even more negative when $n > 1$. A liquid bridge of that length, or any shorter length, will then be stable to all axisymmetric displacements.

As L_0 increases, this will continue to be the case until L_0 just reaches its critical value where

$$\left[1 - \frac{\pi^2}{L_0^2} R_0^2 + \frac{\rho}{\gamma} R_0^3 A^2 \right] = 0$$

(i.e., where the factor F just reaches zero at $n = 1$, but remains negative at $n = 2, 3, \cdots$). The critical value of L_0 so determined and denoted $L_{0_{critical}}$

is

$$\frac{\pi R_0}{\sqrt{1 + \frac{\rho}{\gamma} R_0^3 A^2}}$$

and it now, for the first time, depends on the value of the surface tension.

Before L_0 reaches its critical value, σ^2 is negative for all values of n and $\hat{v}_{r_1}$ is not zero. Yet, when L_0 does reach its critical value, there is at least one eigenfunction at $n = 1$ such that $\sigma^2 = 0$. It is $\hat{v}_{r_1}$, $\hat{v}_{\theta_1}$, $\hat{v}_{z_1}$ and $\hat{p}_1$ all zero but $\hat{R}_1$ not zero. All eigenvalues corresponding to $n = 2$, 3, $\cdots$ remain negative as do any other eigenvalues corresponding to $n = 1$. To the other eigenvalues at $n = 1$, while $F(n = 1, L_0)$ is zero, $\hat{v}_{r_1}$ is not zero, and due to this, they must be less than zero. Certainly all the eigenvalues corresponding to $n = 2$, 3, $\cdots$ must be less than zero, as $F(n, L_0)$ must be negative.

Now, if L_0 is increased just beyond its critical value, we can only imagine that the zero eigenvalue at $n = 1$ turns positive while all the other eigenvalues corresponding to $n = 1$ and all the eigenvalues corresponding to $n = 2$, 3, $\cdots$ remain negative. But equation (5.9) is not precise enough to establish the truth of this.

To illustrate the critical condition and the other neutral points, all determined by

$$\frac{1}{R_0^2} \left[1 - \frac{\pi^2}{L_0^2} n^2 R_0^2 + \frac{\rho}{\gamma} R_0^3 A^2 \right] = 0$$

a graph of $\left[\frac{1}{R_0^2} + \frac{\rho}{\gamma} R_0 A^2 \right]$ versus R_0 can be drawn and this can be done at a fixed value of L_0 and at $n = 1$ for several values of A^2 (Figure 5.1), at a fixed value of A^2 and at $n = 1$ for several values of L_0 (Figure 5.2), and at fixed values of L_0 and A^2 for $n = 1$ and $n = 2$ (Figure 5.3). It helps to notice that if F is zero at $n = 1$, then it is surely negative at $n = 2$, 3, $\cdots$. If F is zero at $n = 2$, then it is surely positive at $n = 1$.

Figure 5.1 illustrates the destabilizing effect of spinning a liquid bridge. In this figure, L_0 is fixed and so too the disturbance. Long wavelength disturbances are the most dangerous and $n = 1$ corresponds to the longest wavelength disturbance possible. The skinny bridges on the left are unstable, and they are unstable for the usual reason that bridges at rest would be unstable; namely, they cannot stand long wavelength disturbances. However, skinny bridges would soon become stable as their radii are increased, which increase their tolerance to long wavelength disturbances, were it not for the destabilizing effect of rotation. The fat bridges on the right become unstable due to their inability to withstand the pressure required to hold the fluid on its circular course. If A^2 is small enough, there is a stable range of bridge radii; the window of stability is lost as A^2 increases and all bridges

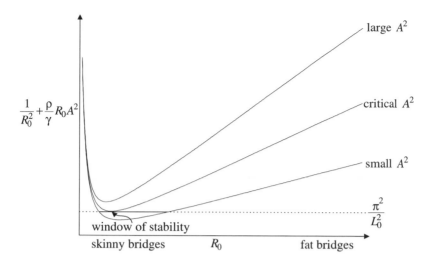

Figure 5.1. *Stability of a Spinning Bridge: L_0 Fixed, $n = 1$*

are unstable if A^2 is large enough. The dependence on γ is in accord with the bridge's ability to contain the buildup of pressure.

Figure 5.2 illustrates the critical value of L_0, not defined as above at an assigned value of R_0, but now defined so that only if L_0 lies below its critical value, does a stable range of R_0 open up.

Figure 5.3 illustrates the fact that the range of bridge radii stable to $n = 2$ displacements contains the range stable to $n = 1$ displacements. It restates the fact that liquid jets are more stable to short wavelength displacements than they are to long wavelength displacements.

Now, observe again that the sign of σ^2 is the sign of

$$-4A^2 k^2 \int_0^{R_0} |\hat{v}_{r_1}|^2 r_0 \, dr_0 + \frac{\gamma}{\rho} \frac{k^2}{R_0} \left[1 - k^2 R_0^2 + \frac{\rho}{\gamma} R_0^3 A^2 \right] |\hat{v}_{r_1}(r_0 = R_0)|^2$$

where $k^2 = \dfrac{\pi^2}{L_0^2} n^2$, $n = 1, 2, \cdots$. The first term by itself is in accord with Rayleigh's criterion. It tells us that fluids in rotation at constant angular speed must be stable, as in this case Φ_0 is just $4A^2$, a positive number. This is independent of L_0, the length of the liquid column. However, the first term stands alone only if the fluid is confined by a rigid wall (i.e., only if $\hat{v}_{r_1}$ vanishes at $r_0 = R_0$). The second term takes into account the deflection of the lateral surface that is possible when the rigid wall is removed and the liquid is held together by its surface tension. It adds something new. Now, no matter how strongly the positive stratification of angular momentum acts to stabilize the flow on the domain, there will come a value of L_0,

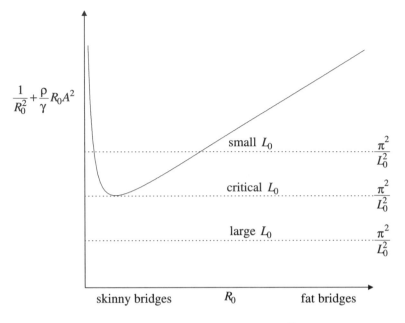

Figure 5.2. *Stability of a Spinning Bridge: A^2 Fixed, $n = 1$*

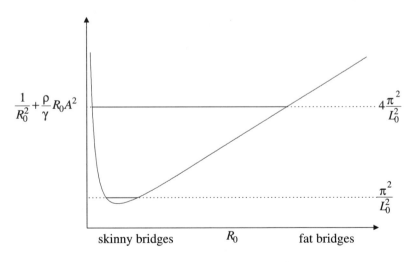

Figure 5.3. *Stability of a Spinning Bridge: L_0 and A^2 Fixed, $n = 1$ and $n = 2$*

where stability is lost. This value of L_0 will arrive the sooner, the stronger the stabilization of the flow on the domain.

It turns out that equation (5.9) holds even if Ω_0 is not constant and Φ_0 is not $4\Omega_0^2$ and not constant. In this case, the sign of σ^2 is the sign of

$$-k^2 \int_0^{R_0} \Phi_0 |\hat{v}_{r_1}|^2 r_0 \, dr_0$$

$$+\frac{\gamma}{\rho} \frac{k^2}{R_0} \left[1 - k^2 R_0^2 + \frac{\rho}{\gamma} R_0^3 \Omega_0^2 (r_0 = R_0) \right] |\hat{v}_{r_1}(r_0 = R_0)|^2$$

and in this more general case, something interesting appears. Even if the stratification of angular momentum on the domain is destabilizing (i.e., even if Φ_0 is negative), there are stable displacements of the liquid bridge and these correspond to large values of k or large values of n. Although this is not possible when the liquid is confined by a rigid wall, when the wall is removed, short enough wavelength displacements can withstand an adverse angular momentum stratification, and if the liquid bridge were short enough to filter out all the dangerous long wavelength displacements, it would be stable under adverse conditions.

It remains only to go on and solve the eigenvalue problem to determine the values of σ^2 and hence the growth rates or oscillation frequencies of a displacement of wave number k. Although this is not difficult, neither is it instructive. Instead, we turn to another problem, very much like this one, gravity now taking the place of the centrifugal force in establishing the base pressure field.

The effect of viscosity remains as before. It does not change the critical point. A calculation like that in the fourth essay can be carried out to establish this.[5]

---

[5]The effect of viscosity is explained by Hocking and by Gillis (cf. L. M. Hocking. The Stability of a Rigidly Rotating Column of Liquid. *Mathematika*, 7:1, 1960 and J. Gillis. Stability of a Column of Rotating Viscous Liquid. *Proc. Cambridge Phil. Soc.*, 57:152, 1967).

The readers could profitably work through these papers. If they do, they will find, as expected, that taking viscosity into account does not alter the critical point at $m = 0$. Then, they will find that viscosity appears to alter the critical point at $m \neq 0$ and in an odd way. The second discussion note is presented to throw some light on this mystery. It uncovers inviscid critical points not appearing in the above papers.

5.6   ENDNOTE

5.6.1   The Stability of Circular Couette Flow
Confined by Two Rigid Walls

Let a liquid lying between two concentric circular cylinders be set into circular motion. The walls of the cylinders are rigid and their lengths are not important; they can be taken to be infinite.

Let the axis Oz lie along the common axis of the two cylinders. Then, the nonlinear equations describing the motion of the fluid, in the case of an axisymmetric flow, can be written

$$\frac{\partial v_r}{\partial t} + \vec{v} \cdot \nabla\, v_r - \frac{v_\theta^2}{r} = -\frac{1}{\rho}\frac{\partial p}{\partial r}$$

$$\frac{\partial v_\theta}{\partial t} + \vec{v} \cdot \nabla\, v_\theta + \frac{v_r v_\theta}{r} = 0$$

$$\frac{\partial v_z}{\partial t} + \vec{v} \cdot \nabla\, v_z = -\frac{1}{\rho}\frac{\partial p}{\partial z}$$

and

$$\frac{\partial v_r}{\partial r} + \frac{v_r}{r} + \frac{\partial v_z}{\partial z} = 0$$

where

$$\vec{v} \cdot \nabla = v_r \frac{\partial}{\partial r} + v_z \frac{\partial}{\partial z}$$

The viscous terms have been omitted.

These equations can be satisfied by

$$v_r = 0 = v_z$$

$$v_\theta = V(r)$$

and

$$p = \int \rho \frac{V^2}{r}\, dr$$

where the function V is arbitrary. This is the base flow whose stability is in question.

It is worth noticing that the function V would be restricted to

$$V = Ar + \frac{B}{r}$$

were a little viscosity to be added to the fluid. The first term corresponds to a rigid rotation at uniform angular speed, A, while the second term is irrotational. The base flow in the essay is restricted to be a rigid rotation.

Rayleigh's work principle can be used to pick out the functions V defining stable base flows. This calculation can be found in Chandrasekhar's book and it leads to Rayleigh's criterion for the stability of a circular flow, namely

$$\frac{d}{dr}[rV]^2 > 0$$

for all r in the flow domain. We shall use it in the form

$$\Phi(r) > 0$$

where, in terms of the local angular speed, denoted by Ω, we have

$$\Phi = \frac{1}{r^3}\frac{d}{dr}[r^2\Omega]^2 = 2\Omega\left[\Omega + \frac{d}{dr}[r\Omega]\right]$$

where $V = r\Omega$ and where Ω need not be constant.

Rayleigh's criterion is ordinarily stated in physical terms like this: A circular flow of an inviscid fluid is stable to axisymmetric displacements if and only if the magnitude of its angular momentum increases outward. A flow for which this is so is said to exhibit a stable stratification of angular momentum.

To see how Rayleigh's criterion comes out of the solution to the perturbation problem, let the base flow of an inviscid fluid in circular motion be denoted by the subscript zero and be specified by

$$v_{\theta_0} = r\Omega_0(r)$$

$$v_{r_0} = 0 = v_{z_0}$$

and

$$p_0 = \rho\int r\Omega_0^2\, dr$$

Then, introduce Φ_0 via

$$\Phi_0 = 2\Omega_0\left[\Omega_0 + \frac{d}{dr}[r\Omega_0]\right]$$

where Ω_0 denotes the local angular speed of the base flow.

In this calculation, the flow is confined by two rigid cylinders: one at $r = R^*$, the other at $r = R$. This being the case (i.e., no surface is being displaced), the domain is fixed in advance and no reference domain need be introduced, whereupon r, z, and t can be written where we would ordinarily write r_0, z_0 and t_0.

To determine whether or not this base flow is stable to a small displacement, the perturbation equations must be solved and, again requiring axial symmetry, these equations take the form

$$\frac{\partial v_{r_1}}{\partial t} - 2\Omega_0 v_{\theta_1} = -\frac{1}{\rho}\frac{\partial p_1}{\partial r}$$

$$\frac{\partial v_{\theta_1}}{\partial t} + \left[\Omega_0 + \frac{d}{dr}[r\Omega_0]\right] v_{r_1} = 0$$

$$\frac{\partial v_{z_1}}{\partial t} = -\frac{1}{\rho}\frac{\partial p_1}{\partial z}$$

and

$$\frac{\partial v_{r_1}}{\partial r} + \frac{v_{r_1}}{r} + \frac{\partial v_{z_1}}{\partial z} = 0$$

To determine a solution, substitute into these equations

$$v_{r_1} = \hat{v}_{r_1} e^{\sigma t} e^{ikz}$$

$$v_{\theta_1} = \hat{v}_{\theta_1} e^{\sigma t} e^{ikz}$$

$$v_{z_1} = \hat{v}_{z_1} e^{\sigma t} e^{ikz}$$

and

$$p_1 = \hat{p}_1 e^{\sigma t} e^{ikz}$$

which turns the initial-value problem into an eigenvalue problem, namely

$$\sigma \hat{v}_{r_1} - 2\Omega_0 \hat{v}_{\theta_1} = -\frac{1}{\rho}\frac{d\hat{p}_1}{dr}$$

$$\sigma \hat{v}_{\theta_1} + \left[\Omega_0 + \frac{d}{dr}[r\Omega_0]\right]\hat{v}_{r_1} = 0$$

$$\sigma \hat{v}_{z_1} = -\frac{1}{\rho}ik\hat{p}_1$$

and

$$\frac{d\hat{v}_{r_1}}{dr} + \frac{\hat{v}_{r_1}}{r} + ik\hat{v}_{z_1} = 0$$

where k denotes the wave number of the disturbance and σ, the eigenvalue, determines its growth rate.

To solve this problem, eliminate $\hat{v}_{\theta_1}$ from the first two equations and $\hat{v}_{z_1}$ from the last two to obtain

$$[\sigma^2 + \Phi_0]\hat{v}_{r_1} = -\frac{\sigma}{\rho}\frac{d\hat{p}_1}{dr}$$

and

$$\sigma\left[\frac{d\hat{v}_{r_1}}{dr} + \frac{\hat{v}_{r_1}}{r}\right] + \frac{k^2}{\rho}\hat{p}_1 = 0$$

Then, as Φ_0 is not ordinarily constant, the eigenvalue problem is more easily reduced to one equation by eliminating $\hat{p}_1$ in favor of $\hat{v}_{r_1}$. The result is

$$\sigma^2 \frac{d}{dr}\left[\frac{d}{dr} + \frac{1}{r}\right]\hat{v}_{r_1} - k^2[\sigma^2 + \Phi_0]\hat{v}_{r_1} = 0 \tag{5.10}$$

where $\hat{v}_{r_1}$ must vanish at $r = R^*$ and $r = R$. This is where the rigid walls play their role.

Again, it is σ^2, not σ, that turns out to be the eigenvalue and σ^2 cannot be zero, unless Φ_0 is zero everywhere.

Now, this is a Sturm–Liouville problem and although Sturm–Liouville theory can tell us the signs of the eigenvalues in terms of the variation of the function Φ_0, we turn to a simple calculation, which is at the heart of our work in the essay, to produce Rayleigh's criterion.

To see how σ^2 depends on Φ_0, multiply equation (5.10) by $r\bar{\hat{v}}_{r_1}$ and integrate the result over $R^* < r < R$ to get

$$\sigma^2 \int_{R^*}^{R} \bar{\hat{v}}_{r_1} \frac{d}{dr}\left[r\frac{d\hat{v}_{r_1}}{dr}\right] dr - \sigma^2 \int_{R^*}^{R} \frac{|\hat{v}_{r_1}|^2}{r}\, dr - k^2\sigma^2 \int_{R^*}^{R} |\hat{v}_{r_1}|^2 r\, dr$$

$$= k^2 \int_{R^*}^{R} \Phi_0|\hat{v}_{r_1}|^2 r\, dr$$

Then, integrate the first term by parts to get

$$\sigma^2 \left[\bar{\hat{v}}_{r_1} r \frac{d\hat{v}_{r_1}}{dr}\right]_{r=R^*}^{r=R} - \sigma^2 \int_{R^*}^{R} \left[\left|\frac{d\hat{v}_{r_1}}{dr}\right|^2 + \frac{|\hat{v}_{r_1}|^2}{r^2} + k^2|\hat{v}_{r_1}|^2\right] r\, dr$$

$$= k^2 \int_{R^*}^{R} \Phi_0|\hat{v}_{r_1}|^2 r\, dr \tag{5.11}$$

which is a useful formula, both here and in the essay.

In this formula, the difference between a liquid confined by rigid walls and one confined by its surface tension appears only in the first term. Here, the rigid walls require

$$\hat{v}_{r_1} = 0 \text{ at } r = R^* \text{ and } r = R$$

whereupon the first term vanishes and we get

$$\sigma^2 = -\frac{k^2 \displaystyle\int_{R^*}^{R} \Phi_0|\hat{v}_{r_1}|^2 r\, dr}{\displaystyle\int_{R^*}^{R} \left[\left|\frac{d\hat{v}_{r_1}}{dr}\right|^2 + \frac{|\hat{v}_{r_1}|^2}{r^2} + k^2|\hat{v}_{r_1}|^2\right] r\, dr}$$

where $\hat{v}_{r_1}$ cannot be zero so long as Φ_0 is not zero.

This formula tells us that the eigenvalues must be real and it tells us that they must be negative if Φ_0 is everywhere positive. This is just Rayleigh's criterion for the stability of a circular flow confined by rigid walls, deduced here from the perturbation equations.

5.7 DISCUSSION NOTES

5.7.1 Rayleigh's Work Principle

In this discussion note, Rayleigh's work principle is used to determine whether or not a spinning jet is stable to a displacement of its surface.

The calculation is carried out by an observer who sees no motion but who sees, instead, an outward force acting on the fluid making up the jet.

Again, the surface of the jet undergoes a displacement. If the total potential energy of the jet is increased then the jet is stable to the displacement. Otherwise the jet is unstable.

The total potential energy is made up of two parts: the surface potential energy, which is the product of the surface area and the surface tension, and the potential energy of the fluid due to the fact that it is spinning.

Let R_0 denote the radius of the undisplaced jet; then let the surface of the displaced jet be denoted by

$$r = R + \epsilon \cos kz$$

where ϵ is small, where R remains to be determined to hold the volume of the jet fixed and where the wavelength of the displacement, denoted by λ, is $2\pi/k$. Then, over one wavelength, the area of the displaced surface, denoted by A, is

$$A = \int_0^\lambda 2\pi r \frac{ds}{dz}\, dz$$

where, to sufficient accuracy,

$$\frac{ds}{dz} = \frac{1}{2}\, \epsilon^2 k^2 \sin^2 kz + 1$$

whereupon

$$A = 2\pi R\lambda + 2\pi R \frac{1}{2}\, \epsilon^2 k^2 \frac{1}{2}\, \lambda$$

By this, then, the increase in the surface area of the jet, due to a displacement of its surface, is

$$2\pi R\lambda - 2\pi R_0\lambda + \frac{\pi}{2}\, R\epsilon^2 k^2 \lambda$$

To determine the value of R, let the volume of the jet, again over one wavelength, be denoted by V. Then, V is given by

$$V = \int_0^\lambda \pi[R + \epsilon \cos kz]^2\, dz \;=\; \pi R^2\lambda + \pi\epsilon^2 \frac{1}{2}\, \lambda$$

and as V must remain at the value $\pi R_0^2\lambda$, R^2 turns out to be

$$R^2 = R_0^2 - \frac{1}{2}\, \epsilon^2$$

whence, to sufficient accuracy, R and R^4 are given by

$$R = R_0 \left[1 - \frac{\epsilon^2}{4R_0^2} \right]$$

and

$$R^4 = R_0^4 \left[1 - \frac{\epsilon^2}{R_0^2} \right]$$

It remains to determine the centrifugal potential energy of the fluid making up the jet. This is a calculation such as an observer, participating in the motion of the jet, would make.

Let Ω denote the constant angular speed at which a laboratory observer would see the jet spinning. Then, to the moving observer, the centrifugal force per unit mass, appears to be

$$r\Omega^2 \vec{i}_r$$

which introduces the centrifugal potential energy per unit mass,

$$-\frac{1}{2} r^2 \Omega^2$$

By this, the increase in the centrifugal potential energy of the jet, over one wavelength, due to the displacement $r = R + \epsilon \cos kz$ is

$$\int_0^\lambda \int_0^{R+\epsilon \cos kz} -\rho \frac{1}{2} r^2 \Omega^2 2\pi r \, dr \, dz \; - \; \int_0^\lambda \int_0^{R_0} -\rho \frac{1}{2} r^2 \Omega^2 2\pi r \, dr \, dz$$

$$= -\rho\Omega^2 \frac{\pi}{4} \int_0^\lambda [R + \epsilon \cos kz]^4 \, dz \; + \; \rho\Omega^2 \frac{\pi}{4} \int_0^\lambda R_0^4 \, dz$$

which, to sufficient accuracy, is

$$-\rho\Omega^2 \frac{\pi}{4} \left[R^4 \lambda + 6R^2 \epsilon^2 \frac{1}{2} \lambda \right] + \rho\Omega^2 \frac{\pi}{4} R_0^4 \lambda$$

The critical displacement is the one whose wavelength is such that the total potential energy of the jet neither increases nor decreases. To determine its wave number, set the sum of the two potential energy increases to zero, namely

$$-\rho\Omega^2 \frac{\pi}{4} \lambda \left[R^4 + 3R^2\epsilon^2 - R_0^4 \right] + \gamma 2\pi\lambda \left[R - R_0 + \frac{1}{4} R\epsilon^2 k^2 \right] = 0$$

and then use the formulas for R, R^2 and R^4 in terms of R_0 and ϵ^2 to get

$$-\rho\Omega^2 \frac{\pi}{4} \lambda \left[-R_0^2\epsilon^2 + 3R_0^2\epsilon^2 \right] + \gamma 2\pi\lambda \left[-\frac{\epsilon^2}{4R_0} + \frac{1}{4} R_0\epsilon^2 k^2 \right] = 0$$

which determines the critical value of k^2. It is given by

$$k_{critical}^2 R_0^2 = 1 + \frac{\rho R_0^3 \Omega^2}{\gamma}$$

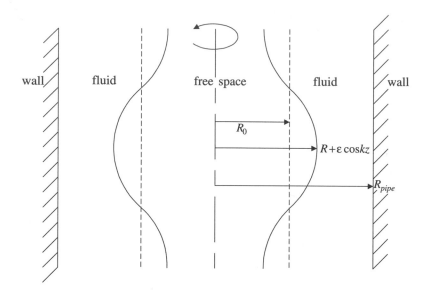

Figure 5.4. *A Fluid Layer Pressed Against a Spinning Wall*

This formula, which also comes out of the solution to the perturbation equations, tells us that the critical value of k^2 not only depends upon γ but that it also increases as Ω^2 increases. This much is in accord with our expectations.

It is worth carrying our calculations on to another case. Often, liquids in the form of annular layers can be found pressed against the inside walls of spinning pipes. This is illustrated in Figure 5.4 where some notation is indicated.

In this case, Rayleigh's work principle tells us that spinning the pipe stabilizes the layer of fluid pressed against its wall. The calculation is much as it was above. Let V denote the volume of the annular layer, taken over one wavelength, when its surface is displaced as indicated in the figure, and let A denote the area of its displaced inner surface, again taken over one wavelength. Then, there obtains

$$V = \int_0^\lambda \int_{R+\epsilon \cos kz}^{R_{pipe}} 2\pi r \ dr \ dz = \pi R_{pipe}^2 \lambda - \int_0^\lambda \pi \left[R + \epsilon \cos kz\right]^2 \ dz$$

$$= \pi R_{pipe}^2 \lambda - \pi R^2 \lambda - \pi \epsilon^2 \frac{1}{2} \lambda$$

which determines R^2 in terms of R_0^2 and ϵ^2 upon requiring V to remain at its undisplaced value, $\pi \left[R_{pipe}^2 - R_0^2\right] \lambda$. The result is, again,

$$R^2 = R_0^2 - \frac{1}{2} \epsilon^2$$

and again, to sufficient accuracy, there obtains

$$R = R_0 \left[1 - \frac{\epsilon^2}{4R_0^2} \right]$$

and

$$R^4 = R_0^4 \left[1 - \frac{\epsilon^2}{R_0^2} \right]$$

Likewise, the formula for A is just as before, namely

$$A = \int_0^\lambda 2\pi [R + \epsilon \cos kz] \left[1 + \frac{1}{2} \epsilon^2 k^2 \sin^2 kz \right] dz = 2\pi R\lambda + \pi R\epsilon^2 k^2 \frac{1}{2} \lambda$$

whence the surface potential energy increase due to the displacement of the surface is just

$$\gamma \left[2\pi [R - R_0]\lambda + \pi R\epsilon^2 k^2 \frac{1}{2} \lambda \right]$$

and, to sufficient accuracy, this is

$$\gamma \left[2\pi \frac{-\epsilon^2}{4R_0} \lambda + \pi R_0 \epsilon^2 k^2 \frac{1}{2} \lambda \right] = \frac{\gamma \frac{1}{2} \pi \epsilon^2 \lambda}{R_0} \left[-1 + R_0^2 k^2 \right]$$

In this case, the centrifugal potential energy of the fluid making up the layer, taken over one wavelength, is given by

$$\int_0^\lambda \int_{R+\epsilon \cos kz}^{R_{pipe}} -\rho \frac{1}{2} r^2 \Omega^2 2\pi r \, dr \, dz$$

$$= -\frac{1}{4} \pi \rho \Omega^2 \int_0^\lambda \left[R_{pipe}^4 - [R + \epsilon \cos kz]^4 \right] \, dz$$

and this, to order ϵ^2, is

$$-\frac{1}{4} \pi \rho \Omega^2 \left[R_{pipe}^4 \lambda - R^4 \lambda - 6R^2 \epsilon^2 \frac{1}{2} \lambda \right]$$

Its increase, namely

$$-\frac{1}{4} \pi \rho \Omega^2 \left[R_0^4 \lambda - R^4 \lambda - 3R^2 \epsilon^2 \lambda \right]$$

is then, to sufficient accuracy,

$$-\frac{1}{4} \pi \rho \Omega^2 \lambda \left[\epsilon^2 R_0^2 - 3R_0^2 \epsilon^2 \right] = \frac{1}{2} \pi \rho \Omega^2 \lambda \epsilon^2 R_0^2$$

The wave number of the critical displacement can now be determined by requiring the increase in the total potential energy to vanish, namely by requiring

$$\frac{\gamma\frac{1}{2}\pi\epsilon^2\lambda}{R_0}\left[-1+R_0^2k^2\right]+\frac{1}{2}\pi\rho\Omega^2\lambda\epsilon^2R_0^2=0$$

This determines the critical value of k. It is given by[6]

$$k_{critical}^2R_0^2=1-\frac{\rho R_0^3\Omega^2}{\gamma}$$

This tells us that spinning the pipe stabilizes the layer on its wall. It decreases the critical value of k^2, turning unstable displacements of the layer at rest into stable displacements of the layer in rotation.

5.7.2   The Case Where m Is Not Zero

In a footnote at the end of the essay, two papers, one by Hocking, one by Gillis, were introduced. In these papers the following claim is made: Upon taking into account the viscosity of the liquid, the value of the surface tension needed to maintain the critical condition increases over what it would have been had the fluid been inviscid. This means that viscosity is destabilizing, an unusual, but not unheard of, result. However, this unusual claim may not be true, it may be due to an oversight, that is, some inviscid critical points may have been missed.

To see if this is so, we take up the inviscid case and let k and m be unrestricted. Notice, first, that the base solution is given by

$$v_{\theta_0}=Ar_0,\quad v_{z_0}=0=v_{r_0},\quad\frac{\partial p_0}{\partial r_0}=\rho A^2r_0$$

whence the perturbation equations, written as an eigenvalue problem, are then given by

$$[\sigma+imA]\hat{v}_{r_1}-2A\hat{v}_{\theta_1}=-\frac{1}{\rho}\frac{d\hat{p}_1}{dr_0}$$

$$[\sigma+imA]\hat{v}_{\theta_1}+2A\hat{v}_{r_1}=-\frac{1}{\rho}\frac{im}{r_0}\hat{p}_1$$

$$[\sigma+imA]\hat{v}_{z_1}=-ik\hat{p}_1$$

---

[6]The reader can go on and fill the central free space in Figure 5.4 with a fluid of density ρ^* and then learn by a Rayleigh work calculation that

$$[\rho^*-\rho]R_0^3\Omega^2+\gamma[1-k^2R_0^2]=0$$

This reproduces both of the results in this endnote; the first on setting ρ to zero and the second on setting ρ^* to zero.

and

$$\frac{d\hat{v}_{r_1}}{dr_0} + \frac{\hat{v}_{r_1}}{r_0} + \frac{im}{r_0}\hat{v}_{\theta_1} + ik\hat{v}_{z_1} = 0$$

on $0 < r_0 < R_0$ and

$$\hat{v}_{r_1} = [\sigma + imA]\hat{R}_1$$

and

$$\hat{p}_1 = -\frac{\gamma}{R_0^2}\left[1 - m^2 - k^2 R_0^2 + \frac{\rho R_0^3 A^2}{\gamma}\right]\hat{R}_1$$

at $r_0 = R_0$, where all variables must remain finite at $r_0 = 0$.

Now, so long as A is not zero, these equations have a simple solution where $\hat{v}_{r_1}, \hat{v}_{\theta_1}, \hat{v}_{z_1}, \hat{p}_1$ and $\hat{R}_1$ are not all zero. It corresponds to $\sigma + imA = 0$, whence to $\sigma = -imA$, and the eigenfunction is given by $\hat{v}_{r_1}, \hat{v}_{\theta_1}, \hat{v}_{z_1}$ and $\hat{p}_1$ all zero but $\hat{R}_1 \neq 0$.[7] And it obtains if and only if

$$\left[1 - m^2 - k^2 R_0^2 + \frac{\rho R_0^3 A^2}{\gamma}\right] = 0 \tag{5.12}$$

is satisfied.

As the real part of σ vanishes, equation (5.12) defines the critical value of any one of the variables A, γ, R_0, etc. in terms of the values of the other variables. Introducing L as $\dfrac{\gamma}{\rho R_0^3 A^2}$, the critical value of L is given by

$$\frac{1}{k^2 R_0^2 + m^2 - 1}$$

This is the result presented by both Hocking and Gillis in the case $k \neq 0$, $m = 0$. But it is not their result in the case $k = 0$, $m \neq 0$. Yet, in this second case,[8] it is their result if the viscosity of the fluid is taken into account. Hence their claim.

So what is going on? Can there be other critical points? To see how other critical points might be found let us look briefly at the case $k = 0$ while maintaining m and A other than zero. On setting k to zero the eigenvalue problem simplifies to

$$[\sigma + imA]\hat{v}_{r_1} - 2A\hat{v}_{\theta_1} = -\frac{1}{\rho}\frac{d\hat{p}_1}{dr_0} \tag{5.13}$$

$$[\sigma + imA]\hat{v}_{\theta_1} + 2A\hat{v}_{r_1} = -\frac{1}{\rho}\frac{im}{r_0}\hat{p}_1 \tag{5.14}$$

---

[7]To understand this solution in physical terms, i.e., to understand how $\vec{v}_1$ can be zero while R_1 is given by $\hat{R}_1 e^{ikz} e^{im\theta} e^{-imAt}$, the reader needs to notice that to an observer in rotation at angular speed A, the surface, while deflected, appears to be at rest.

[8]In the case where k is zero, both Hocking and Gillis explain why $m = 0$ and $m = \pm 1$ are of no interest.

$$[\sigma + imA]\hat{v}_{z_1} = 0 \tag{5.15}$$

and

$$\frac{d\hat{v}_{r_1}}{dr_0} + \frac{\hat{v}_{r_1}}{r_0} + \frac{im}{r_0}\hat{v}_{\theta_1} = 0 \tag{5.16}$$

on $0 < r_0 < R_0$ and

$$\hat{v}_{r_1} = [\sigma + imA]\hat{R}_1 \tag{5.17}$$

and

$$\hat{p}_1 = -\frac{\gamma}{R_0^2}\left[1 - m^2 + \frac{\rho R_0^3 A^2}{\gamma}\right]\hat{R}_1 \tag{5.18}$$

at $r_0 = R_0$, where, as above, all variables must remain finite at $r_0 = 0$.

Again, the solutions found above where $\sigma + imA$ is zero remain solutions at $k = 0$, now subject to

$$\left[1 - m^2 - \frac{\rho R_0^3 A^2}{\gamma}\right] = 0$$

being satisfied. And it may be observed that $\hat{v}_{z_1}$ cannot be determined, but this is physically reasonable.

There may be other solutions where $\sigma + imA = 0$ and there may also be solutions where $\sigma + imA = i\lambda$, $\lambda \neq 0$. These latter solutions would lead to critical conditions and they can be obtained by solving equations (5.13) and (5.14) for $\hat{v}_{r_1}$ and $\hat{v}_{\theta_1}$ in terms of $\hat{p}_1$ and $\dfrac{d\hat{p}_1}{dr_0}$ and then substituting the results into equation (5.16) and into the result of eliminating $\hat{R}_1$ between equations (5.17) and (5.18). Any real values of λ which admit solutions other than $\hat{p}_1 = 0$ would lead to critical conditions and such solutions appear to us to be the source of the inviscid critical conditions obtained by both Hocking and Gillis.

The readers may wish to join us in trying to decide if we really like the idea of finding more than one critical value of L at each value of m.

6

The Stability of a Heavy Fluid Lying over a Light Fluid: The Rayleigh–Taylor Instability

Let two fluids of different densities lie one over the other. Then, the instability of the plane interface separating the two fluids, if it occurs, is called the Rayleigh–Taylor instability.

The fluids are incompressible; their densities are uniform and the densities are denoted by ρ and ρ^*. The surface dividing the two liquids can be displaced and determining its stability is the goal of our work.

Figure 6.1 presents three photographs taken by us and a sketch. They indicate what the problem is. In the photographs, the interface separates a heavy fluid and a light fluid, the heavy fluid lying above the light fluid. The diameter of the glass tube containing the heavy fluid is less than the critical diameter at which the interface becomes unstable. These photographs make two additional points.

First, the interface can be made to be as nearly planar as one might like, by controlling the pressure above the heavy fluid. Second, fixed end conditions, where the interface comes into contact with the solid walls, can be achieved in an experiment, as the interface seems to attach itself to the sharp edge.

This second point will be important later in the essay when we introduce the viscosity of the two fluids, for fixed end conditions do not require us to take into account the motion of a three-phase contact line at the same time that the no-slip condition is being imposed along the solid walls.

To begin, however, we take the fluids to be inviscid, whereupon neither free end conditions, nor any other end conditions, pose contact line difficulties. Then, our experiment can be idealized as Figure 6.2, where the two fluids lie inside a rigid container. This figure draws attention to the interface

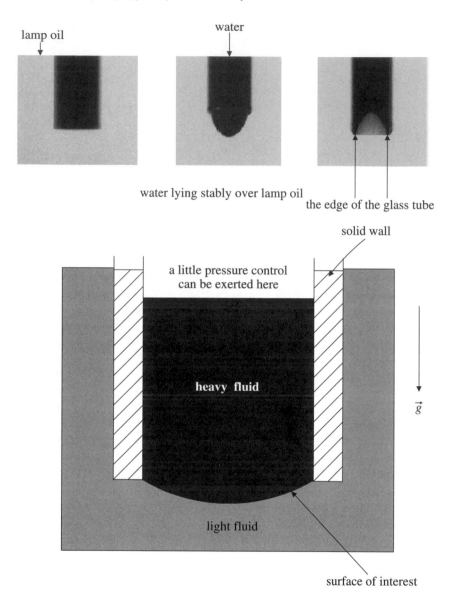

Figure 6.1. *Photographs Illustrating Fixed End Conditions and a Sketch Defining the Problem*

separating the two fluids. The problem is reduced from three dimensions to two, at no great loss, and to fix the coordinate system, the axis Oz is taken to point upward, opposite to the direction in which gravity acts. The axis Ox lies in the plane of the equilibrium interface. The idealization presented

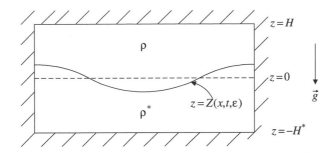

Figure 6.2. *An Idealization of an Experiment: Free-Free End Conditions*

in Figure 6.2 is not great due to the fact that at critical conditions there is no flow.

The Rayleigh–Taylor instability is taken up in this essay because in some ways it is like the jet instability, but in other ways it is not.

In both cases, the change in the area of a surface, when it is displaced, determines what is going to happen. In the case of the jet, the equilibrium surface is a cylinder and, when it is displaced, its area may increase or it may decrease and this tells us whether or not the jet is stable to the displacement. As there are two offsetting ways in which the surface area of a jet can change, the critical point is established when these changes just balance one another. By this, the critical point is independent of the surface tension, as the surface tension is simply a factor multiplying every kind of change.

In the new problem, the equilibrium surface is a plane, its area always increases on displacement, and the resulting potential energy increase depends on the magnitude of the surface tension. If $\rho > \rho^*$, this stands over and against the gravitational potential energy decrease that the fluid itself undergoes when its surface is displaced. The critical point, then, is determined by requiring these two changes to be just offsetting and this leads to a critical point which now depends on the value of the surface tension.

The equations satisfied by the perturbation variables at the reference surface are now a little more interesting due to the fact that the base state itself is a little more interesting. It is this contribution of the base state to the perturbed pressure at the boundary that is at the heart of the problem. The base state pressure gradient also contributed to the perturbed pressure at the boundary in the essay just before this, where a jet was set to spinning. But there, the pressure variation due to spinning was an add on; here, the pressure variation due to gravity is everything.

Before going on, it is worthwhile to think back to the jet problem and remember that in the use of Rayleigh's work principle, it was the area of the displaced surface that came into the calculation whereas in the perturbation calculation, it was its curvature. The area and the curvature must be telling us the same thing, but we did not work out how area changes depend on curvature. The calculation is a little easier to do in this case and it is done in the first discussion note. There too, it is seen how the gravitational potential energy can be determined.

Now, the critical point, either a critical wavelength or a critical width, does not depend on the viscosities of the two fluids and the reasons for this are much the same as in the case of the jet (viz., the critical state is a state of rest). To begin, then, the viscosities of the two fluids are set to zero before being reintroduced in the fourth discussion note, where a calculation taking viscosity into account is presented. This calculation is just like the corresponding calculation in the fourth essay.

We turn first to Rayleigh's work principle, but the reader who remembers how important the constant-volume condition was in the jet problem will be struck by the minor role that it plays in this problem, at least until the third discussion note is read.[1]

6.1   Rayleigh's Work Principle

Let two fluids lie at rest, the heavier above the lighter, as illustrated in Figure 6.2. The initial state corresponds to a planar interface at $z = 0$. At first, the two liquids can be thought to lie in infinitely wide horizontal layers of thicknesses H_0 and H_0^*. The interface is then displaced and our job is to determine the change in the total potential energy of the two liquids, the total being the sum of the gravitational and the surface potential energies.

Let the displacement[2] be $z = Z(x) = \epsilon \cos kx$ and restrict the calculation to one wavelength, λ, in the horizontal direction, where $\lambda = 2\pi/k$. Then, per unit depth, the potential energy increase due to the surface area[3] increase is given by

$$\gamma \int_0^\lambda dx \ \sqrt{1 + Z_x^2} \ - \gamma \int_0^\lambda dx$$

while the gravitational potential energy increase is given by

---

[1] It might be fairer to say that the constant-volume condition plays about the same role in both problems when it comes to solving the perturbation equations. It is in the use of Rayleigh's work principle that the importance of its role differs.

[2] The possibility $k = 0$ is excluded by requiring the volumes of the layers to remain fixed as the displacement is carried out.

[3] The system is now two dimensional. So, by volume and area is meant volume and area per unit depth into the page of Figure 6.2.

$$\int_0^\lambda \int_Z^{H_0} \rho g z \; dz \; dx \; + \; \int_0^\lambda \int_{-H_0^*}^{Z} \rho^* g z \; dz \; dx \; - \; \int_0^\lambda \int_0^{H_0} \rho g z \; dz \; dx$$

$$- \int_0^\lambda \int_{-H_0^*}^{0} \rho^* g z \; dz \; dx$$

The first, to order ϵ^2, is

$$\gamma \frac{1}{4} \epsilon^2 k^2 \lambda$$

while the second, requiring no approximation, is

$$-\frac{1}{4} g [\rho - \rho^*] \epsilon^2 \lambda$$

The gravitational contribution, per unit width, then, is independent, not only of k but also of H_0 and H_0^*. The fact that H_0 and H_0^* do not appear in this result means that their values play no role in determining the critical condition.[4]

The total potential energy increase, per unit width, due to the displacement $z = \epsilon \cos kx$ is obtained by adding the two contributions and dividing by λ. It is, to order ϵ^2,

$$\frac{1}{4} \epsilon^2 \left[\gamma k^2 - g[\rho - \rho^*] \right]$$

Now, according to Rayleigh's work principle, a system is stable to a displacement if the displacement increases the potential energy of the system; otherwise the system is unstable. Surface tension, then, is always stabilizing and there are only stable displacements if ρ is less than ρ^*, whereas there are both stable and unstable displacements if ρ is greater than ρ^*.

Let $\rho - \rho^*$ be positive and let the surface be displaced into the form $z = \epsilon \cos kx$. Then, gravity always supplies work, whereas surface tension always requires work. The stability question turns on whether gravity supplies too little, just enough or more than enough work to carry out the displacement. Now, the total potential energy increase per unit width is just the net work required per unit width and, of the two terms making up the total potential energy increase, the surface tension term depends on k, but the gravity term is free of k. At large values of k, the surface tension requires a large amount of work, more than gravity can supply, and additional work must be put in to carry out the displacement. This is the recipe for stability. At small values of k, gravity can supply all the work required by the surface tension, with some left over. This is the recipe for instability. Thus, the interface is stable to short wavelength displacements, unstable to long wavelength displacements.

---

[4]Nor did R_{pipe} play a role in determining the critical condition in the first discussion note to the fifth essay.

At neutral conditions, the work required to displace the surface is just balanced by the work gravity is prepared to do and this determines the critical value of k^2. It is[5]

$$g \frac{[\rho - \rho^*]}{\gamma}$$

The value of γ now determines the critical wavelength at which the balance is struck. This was not the case in the jet problem.

If the system were finite in horizontal extent, instead of infinite, then the permissible values of k would not be continuous but would, instead, be discrete, though infinite in number. Denoting the width by L_0, the admissible values of k would be restricted to $\pi n / L_0$, $n = 1, 2, \cdots$, if free-free end conditions must be satisfied at the points where the interface comes into contact with the side walls. Then, the increase in potential energy would be

$$\frac{1}{4} \epsilon^2 \left[\gamma \frac{\pi^2}{L_0^2} n^2 - g[\rho - \rho^*] \right]$$

and again, if $\rho - \rho^*$ is positive, this can be either positive or negative depending on the values of γ, g, etc. If it is positive for $n = 1$, then it is positive for all values of n and, hence, for all displacements. By setting the increase in the potential energy to zero, then, and putting n equal to one, the critical horizontal width can be determined in terms of γ, g and $[\rho - \rho^*]$. It is given by

$$\pi \sqrt{\frac{\gamma}{g[\rho - \rho^*]}}$$

and for all widths less than this, heavy over light is stable to all small displacements. If the width is greater than its critical value, at least one displacement, and maybe more, but not more than a finite number, is unstable. The width of the layer then is the control variable.

6.2   The Nonlinear Equations

We would like to establish these conclusions, drawn from Rayleigh's work principle, by solving the perturbation equations to learn the fate of a small displacement imposed on the horizontal surface dividing two fluids. To do

---

[5]It is possible to determine the surface area increase to higher order in ϵ, and on doing this, the surface potential energy increase, per unit width, turns out to be

$$\frac{1}{4} \epsilon^2 \left[\gamma k^2 - \frac{3}{16} \gamma k^4 \epsilon^2 + \cdots - g[\rho - \rho^*] \right]$$

It is tempting to think that a nonlinear correction to the critical value of k can be obtained by setting this to zero, but, of course, this is nonsense.

this, the first step is to write the nonlinear equations for the motion of the fluids. In the configuration indicated in Figure 6.2, they are, first on the domain,

$$\left. \begin{array}{c} \rho \dfrac{\partial \vec{v}}{\partial t} + \rho \vec{v} \cdot \nabla \vec{v} = -\nabla p + \rho \vec{g} \\[2em] \nabla \cdot \vec{v} = 0 \end{array} \right\} \qquad 0 < x < L_0, \ Z < z < H_0, \ t > 0$$

and

$$\left. \begin{array}{c} \rho^* \dfrac{\partial \vec{v}^*}{\partial t} + \rho^* \vec{v}^* \cdot \nabla \vec{v}^* \\[1em] = -\nabla p^* + \rho^* \vec{g} \\[1em] \nabla \cdot \vec{v}^* = 0 \end{array} \right\} \qquad 0 < x < L_0, \ -H_0^* < z < Z, \ t > 0$$

where the fluids are taken to be inviscid and where $z = Z(x, \ t, \ \epsilon)$ specifies the interface lying between the two fluids. The function Z is part of the solution to the problem. The value of ϵ identifies the members of a class of problems, possibly by the amplitude of an initial displacement.

Then, at the solid walls which form the boundary of the domain, no flow is required, whereupon

$$\vec{n} \cdot \vec{v} = 0$$

must hold at the top wall and along part of the side walls, while

$$\vec{n} \cdot \vec{v}^* = 0$$

must hold along the remainder of the side walls and at the bottom wall.

There remains the interface separating the two fluids and its motion is our main interest. Along this deflecting surface two conditions must hold, namely

$$\vec{n} \cdot \vec{v} = u = \vec{n} \cdot \vec{v}^*$$

and

$$p - p^* = \gamma \, 2H$$

where $\vec{n}$ denotes the normal to this surface, u denotes the normal speed and $2H$ denotes the mean curvature, taken to be negative at a crest where the lower fluid has been displaced upward. Letting $\vec{n}$ point out of the lower fluid into the upper fluid, $\vec{n}$, u and $2H$ can be expressed in terms of Z via

$$\vec{n} = \frac{-Z_x \vec{i} + \vec{k}}{\sqrt{Z_x^2 + 1}}$$

$$u = \frac{Z_t}{\sqrt{Z_x^2 + 1}}$$

and

$$2H = \frac{Z_{xx}}{[Z_x^2 + 1]^{\frac{3}{2}}}$$

The first condition at the deflecting surface expresses the fact that neither fluid may cross their common interface. The second specifies the pressure difference across a curved surface as a multiple of its mean curvature.

The nonlinear equations have at least one simple solution. It is

$$\vec{v}_0 = \vec{0}, \quad p_0 = -\rho g z_0$$

$$\vec{v}_0^* = \vec{0}, \quad p_0^* = -\rho^* g z_0$$

and

$$Z_0 = 0$$

and it is denoted by the subscript zero to indicate that it is the base solution whose stability is to be determined by solving the perturbation equations.

By this, the two fluids are divided in the reference configuration by the horizontal plane $z_0 = Z_0 = 0$; in any other configuration, they are divided by the surface $z = Z(x, t, \epsilon)$.

6.3   The Perturbation Equations

To derive the perturbation equations, a mapping of the reference domain into the current domain must be introduced and the variables defined on the present domain must be expanded along this mapping in terms of variables defined on the reference domain. At the interface between the two fluids, the mapping can be expanded in powers of ϵ as

$$z = Z(x, t, \epsilon) = Z_0(x_0, t_0) + \epsilon Z_1(x_0, t_0) + \frac{1}{2}\epsilon^2 Z_2(x_0, t_0) + \cdots$$

where $z_0 = Z_0(x_0, t_0)$ denotes the interface in the reference configuration and, while $Z_0(x_0, t_0)$ must be zero, $Z_1(x_0, t_0)$, $Z_2(x_0, t_0)$, etc. must be determined in order to specify the position of this interface and, by this, the configuration of the two fluids, in the present domain.

The solid boundaries do not require the introduction of special notation. These walls are not displaced.

The perturbation equations then obtain on carrying out the calculations as explained in the second essay. They are, first on the reference domain,

$$\rho \frac{\partial \vec{v}_1}{\partial t_0} = -\nabla_0 p_1$$

$$\nabla_0 \cdot \vec{v}_1 = 0$$

$\left. \right\}$ $\quad 0 < x_0 < L_0, \quad 0 < z_0 < H_0, \quad t_0 > 0$

and

$$\rho^* \frac{\partial \vec{v}_1^*}{\partial t_0} = -\nabla_0 p_1^*$$

$$\nabla_0 \cdot \vec{v}_1^* = 0$$

$\left. \right\}$ $\quad 0 < x_0 < L_0, \quad -H_0^* < z_0 < 0, \quad t_0 > 0$

Then, at the solid walls making up the boundary of the reference domain,

$$\vec{n}_0 \cdot \vec{v}_1 = 0$$

must hold along the top and the upper part of the side walls, while

$$\vec{n}_0 \cdot \vec{v}_1^* = 0$$

must hold along the bottom and the lower part of the side walls.

Turning to the interface separating the two fluids in the reference configuration, there are conditions holding across the interface and conditions holding at its ends. Across the interface,

$$\vec{i}_{z_0} \cdot \vec{v}_1 = \frac{\partial Z_1}{\partial t_0} = \vec{i}_{z_0} \cdot \vec{v}_1^*, \quad 0 < x_0 < L_0, \quad z_0 = 0, \quad t_0 > 0$$

and

$$\left[p_1 + Z_1 \frac{\partial p_0}{\partial z_0} \right] - \left[p_1^* + Z_1 \frac{\partial p_0^*}{\partial z_0} \right] = \gamma \frac{\partial^2 Z_1}{\partial x_0^2}, \quad 0 < x_0 < L_0, \quad z_0 = 0, \quad t_0 > 0$$

must hold. The second of these two equations is the perturbed pressure difference equation. It has terms coming from the base solution, as it did in the fifth essay but did not in the third and fourth essays. The end conditions to be satisfied by Z_1 at $x_0 = 0$ and at $x_0 = L_0$ depend on how the surface meets the side walls. The free case requires

$$\frac{\partial Z_1}{\partial x_0} = 0$$

whereas the fixed case requires

$$Z_1 = 0$$

The free-free case is taken up in the essay, at least until viscosity is taken into account. Upon introducing viscosity, fixed end conditions make things a little easier. Other end conditions are introduced in the second and third discussion notes.

There remains one last condition. It is the requirement that the displacement be carried out at constant volume and it places the demand that the solution to the perturbation equations satisfy

$$\int_0^{L_0} Z_1 \, dx_0 \; = \; 0$$

The signature of the base solution can be seen in the perturbation problem by looking at the pressure difference equation at the interface. The perturbed pressure difference has two sources, both originating in the displacement of the interface. On the one hand, the perturbed surface curvature causes a perturbed pressure difference; on the other hand, the prevailing pressure gradient, established in the reference state, induces a perturbed pressure difference as the reference interface is displaced and invades new ground. The terms $Z_1 \dfrac{dp_0}{dz_0}$ and $Z_1 \dfrac{dp_0^*}{dz_0}$, then, arise in just the same way as does the term $R_1 \dfrac{dp_0}{dr_0}$ in the previous essay.

6.4   The Solution to the Perturbation Equations

The perturbation problem is linear and homogeneous, determining $\vec{v}_1$, p_1, $\vec{v}_1^*$, p_1^* and Z_1 once their initial values have been specified. To solve this problem, we turn it into an eigenvalue problem by writing its solution as a product, introducing the factor $e^{\sigma t_0}$. Then, the eigenvalue σ determines the growth rate of a displacement in the form of the corresponding eigenfunction. Eliminating $\vec{v}_1$ and $\vec{v}_1^*$ in favor of p_1 and p_1^* in the perturbation problem, as we did in the fourth essay, and then substituting

$$p_1 = \hat{p}_1(x_0, \ z_0)e^{\sigma t_0}$$

$$p_1^* = \hat{p}_1^*(x_0, \ z_0)e^{\sigma t_0}$$

and

$$Z_1 = \hat{Z}_1(x_0)e^{\sigma t_0}$$

produces the eigenvalue problem. It is

$$\nabla_0^2 \hat{p}_1 = 0 \tag{6.1}$$

on $0 < x_0 < L_0, \ \ 0 < z_0 < H_0$

$$\nabla_0^2 \hat{p}_1^* = 0 \tag{6.2}$$

on $0 < x_0 < L_0, \ \ -H_0^* < z_0 < 0$

$$\vec{n}_0 \cdot \nabla_0 \hat{p}_1 = 0 \tag{6.3}$$

or

$$\vec{n}_0 \cdot \nabla_0 \hat{p}_1^* = 0 \tag{6.4}$$

at all solid walls and

$$\frac{\partial \hat{p}_1}{\partial z_0} = -\sigma^2 \rho \hat{Z}_1 \tag{6.5}$$

$$\frac{\partial \hat{p}_1^*}{\partial z_0} = -\sigma^2 \rho^* \hat{Z}_1 \tag{6.6}$$

and

$$\left[\hat{p}_1 - \rho g \hat{Z}_1\right] - \left[\hat{p}_1^* - \rho^* g \hat{Z}_1\right] = \gamma \frac{d^2 \hat{Z}_1}{dx_0^2} \tag{6.7}$$

at the interface between the two fluids (viz., at $z_0 = 0$, $0 < x_0 < L_0$). To these equations must be added the constant-volume requirement, namely

$$\int_0^{L_0} \hat{Z}_1 \, dx_0 = 0 \tag{6.8}$$

The end conditions remain to be specified.

Notice that the eigenvalue, σ^2, is present only in the equations at the boundary; it is eliminated from the domain equations upon elimination of $\vec{v}_1$ and $\vec{v}_1^*$.

This is a system of linear, homogeneous equations for $\hat{p}_1$, $\hat{p}_1^*$ and $\hat{Z}_1$. For ordinary values of σ^2, it has only the solution

$$\hat{p}_1 = \text{constant} = \hat{p}_1^*$$

and

$$\hat{Z}_1 = 0$$

This corresponds to no displacement.[6] Our interest is in the special values of σ^2 to which solutions other than this can be found. These values of σ^2 then determine the stability of the base solution.

Let the end conditions be free-free. Then, $\hat{Z}_1$ must satisfy

$$\frac{d\hat{Z}_1}{dx_0} = 0, \quad x_0 = 0 \text{ and } x_0 = L_0$$

and these end conditions together with the side wall conditions

$$\frac{\partial \hat{p}_1}{\partial x_0} = 0, \quad x_0 = 0 \text{ and } x_0 = L_0, \quad 0 < z_0 < H_0$$

and

$$\frac{\partial \hat{p}_1^*}{\partial x_0} = 0, \quad x_0 = 0 \text{ and } x_0 = L_0, \quad -H_0^* < z_0 < 0$$

make it possible to obtain product solutions to the eigenvalue problem.

---

[6] Notice that the perturbation problem remains the same if the same constant is added to $\hat{p}_1$ and $\hat{p}_1^*$.

In this case, we can write

$$\hat{p}_1 = \hat{\hat{p}}_1(z_0) \cos k x_0$$

$$\hat{p}_1^* = \hat{\hat{p}}_1^*(z_0) \cos k x_0$$

and

$$\hat{Z}_1 = \hat{\hat{Z}}_1 \cos k x_0$$

where k must be one of the values[7] $\pi n/L_0$, $n = 1, 2, \cdots$. The end and side wall conditions are now satisfied. Then, substitute these products into equations (6.1) through (6.7) to obtain, on the domain,

$$\frac{d^2 \hat{\hat{p}}_1}{dz_0^2} - k^2 \hat{\hat{p}}_1 = 0, \quad 0 < z_0 < H_0 \tag{6.9}$$

and

$$\frac{d^2 \hat{\hat{p}}_1^*}{dz_0^2} - k^2 \hat{\hat{p}}_1^* = 0, \quad -H_0^* < z_0 < 0 \tag{6.10}$$

at the boundary of the domain (i.e., at the top and bottom walls),

$$\frac{d\hat{\hat{p}}_1}{dz_0} = 0, \quad z_0 = H_0 \tag{6.11}$$

and

$$\frac{d\hat{\hat{p}}_1^*}{dz_0} = 0, \quad z_0 = -H_0^* \tag{6.12}$$

and, at the reference interface,

$$\frac{d\hat{\hat{p}}_1}{dz_0} = -\sigma^2 \rho \hat{\hat{Z}}_1 \tag{6.13}$$

$$\left. \begin{array}{c} \\ \\ \end{array} \right\} \quad z_0 = 0$$

$$\frac{d\hat{\hat{p}}_1^*}{dz_0} = -\sigma^2 \rho^* \hat{\hat{Z}}_1 \tag{6.14}$$

and

$$\hat{\hat{p}}_1 - \hat{\hat{p}}_1^* = \left[g[\rho - \rho^*] - \gamma k^2 \right] \hat{\hat{Z}}_1, \quad z_0 = 0 \tag{6.15}$$

The set of equations (6.9) through (6.15) is an eigenvalue problem. For a displacement at any admissible wave number, it determines the growth constant.

---

[7] The possibility $n = 0$ is ruled out by equation (6.8), but this is the only role played by fixed volume in the free-free case.

6.4.1   The Critical Condition

The eigenvalue problem determines σ^2 as a function of k^2 and it is easy to see that σ^2 must be real. By this, stable displacements correspond to negative values of σ^2 or to imaginary values of σ. We begin by looking for the critical values of k^2. To do this, we set $\sigma^2 = 0$ and determine the special values of k^2 to which solutions[8] other than

$$\hat{\hat{p}}_1 = 0 = \hat{\hat{p}}_1^*, \quad \hat{\hat{Z}}_1 = 0$$

can be found.

Now, whether σ^2 is zero or not,

$$\hat{\hat{p}}_1 = A \cosh k[z_0 - H_0]$$

and

$$\hat{\hat{p}}_1^* = A^* \cosh k[z_0 + H_0^*]$$

satisfy equations (6.9), (6.10), (6.11) and (6.12) (i.e., all the equations save those at $z_0 = 0$). Then, at $z_0 = 0$, equations (6.13) and (6.14), with $\sigma^2 = 0$, require

$$A = 0 = A^*$$

and, hence,

$$\hat{\hat{p}}_1 = 0 = \hat{\hat{p}}_1^*$$

Putting this into equation (6.15) produces

$$0 = \left[g[\rho - \rho^*] - \gamma k^2 \right] \hat{\hat{Z}}_1$$

which determines the value of k^2 at which the eigenvalue problem, obtained by setting σ^2 to zero, has a solution other than $\hat{\hat{p}}_1$, $\hat{\hat{p}}_1^*$ and $\hat{\hat{Z}}_1$ all zero. It is

$$\frac{g[\rho - \rho^*]}{\gamma}$$

This is the critical value of k^2; it is the only value of k^2 at which σ^2 can be zero. Now, k^2 must be one of the values $\frac{\pi^2}{L_0^2} n^2$, $n = 1, 2, \cdots$; hence, to each value of n there corresponds a critical width and the smallest critical width corresponds to $n = 1$. It is

$$\pi \sqrt{\frac{\gamma}{g[\rho - \rho^*]}}$$

This is the result obtained earlier by use of Rayleigh's work principle and, again, it indicates that the thickness of the fluid layers plays no role in determining the critical point. This seems a little surprising. After all, common

---

[8]The constant that was arbitrary before must now be zero.

sense would seem to tell us that the width of a layer of water that can be maintained over air ought to depend on its thickness.

6.4.2   The Dispersion Formula: σ^2 versus k^2

To determine σ^2 as it depends on k^2, substitute

$$\hat{p}_1 = A \cosh k[z_0 - H_0]$$

and

$$\hat{p}_1^* = A^* \cosh k[z_0 + H_0^*]$$

into equations (6.13), (6.14) and (6.15) to obtain a set of three equations for the constants A, A^* and $\hat{Z}_1$, namely

$$-Ak \sinh kH_0 = -\sigma^2 \rho \hat{Z}_1$$

$$A^* k \sinh kH_0^* = -\sigma^2 \rho^* \hat{Z}_1$$

and

$$A \cosh kH_0 - A^* \cosh kH_0^* = \left[g[\rho - \rho^*] - \gamma k^2 \right] \hat{Z}_1$$

These equations determine σ^2 as a function of k^2. The result is

$$\sigma^2 = \frac{k}{\rho \coth kH_0 + \rho^* \coth kH_0^*} \left[g[\rho - \rho^*] - \gamma k^2 \right] \qquad (6.16)$$

This formula indicates the influence of the thickness of the layers on the oscillation frequency of stable displacements and on the growth constant of unstable displacements. If the thicknesses are large and k is taken to be positive, equation (6.16) reduces to

$$\sigma^2 = \frac{k}{\rho + \rho^*} \left[g[\rho - \rho^*] - \gamma k^2 \right] \qquad (6.17)$$

while if ρ^* is small, it reduces to[9]

$$\sigma^2 = \frac{k \tanh kH_0}{\rho} \left[g\rho - \gamma k^2 \right] \qquad (6.18)$$

   A graph of equation (6.18) is presented as Figure 6.3. It looks like Figure 3.1 of the third essay. Again, σ^2 turns and approaches zero as k^2 approaches zero. In this problem, as the wavelength of a displacement increases, more and more of the motion is horizontal, less and less vertical, yet only the vertical motion releases gravitational potential energy.

---

   [9]It is of some interest to notice that equation (6.18) cannot be obtained simply by deciding not to perturb the pressure of the lower fluid. There are no solutions $\hat{p}_1^* = 0$ where $\hat{p}_1$ is not zero unless $\rho^* = 0$.

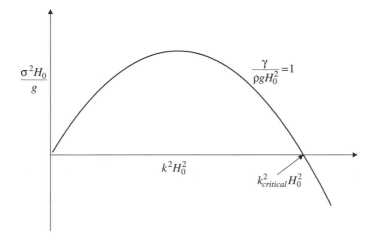

Figure 6.3. *A Graph of Equation (6.18)*

In these formulas, k must be replaced by $\pi n/L_0$, $n = 1,\ 2,\ \cdots$. If $\rho - \rho^*$ is positive and L_0 is large enough, the interface is unstable to some displacements, yet it is stable to many others, and this is according to the sign of $\left[g[\rho - \rho^*] - \gamma \dfrac{\pi^2}{L_0^2} n^2 \right]$. It is stable to all displacements if and only if L_0 is less than the critical width determined earlier.

6.5   An Observation Made by G.I. Taylor

The instability worked out in this essay is called the Rayleigh–Taylor instability. In simplest terms, it tells us this: If two fluids are separated by a horizontal plane interface, then the interface is unstable if $\vec{g}$ points out of the heavy fluid into the light fluid; otherwise it is stable. In this, and hereafter, the interface is taken to be of infinite horizontal extent.

Now, setting gravity aside, let two fluids, divided by a plane interface, be accelerated perpendicular to their interface; then, the interface is unstable if the acceleration points out of the light fluid into the heavy fluid; otherwise it is stable. This observation is due to G. I. Taylor.[10]

To see why it is so, let a laboratory observer O carry out an experiment in which two fluids are being accelerated at a uniform rate, denoted by $\vec{A}$, perpendicular to their interface.

---

[10]G.I. Taylor. The Instability of Liquid Surfaces When Accelerated in a Direction Perpendicular to Their Planes. I. *Proc. Roy. Soc. Lond. A*, 201:192, 1950.

Then, to an observer, O', displaced from O by a vector $\vec{R}$, where

$$\frac{d^2\vec{R}}{dt^2} = \vec{A}$$

the acceleration, $\vec{a}\,'$, of a particle of fluid is

$$\vec{a}\,' = \vec{a} - \vec{A}$$

where $\vec{a}$ is the acceleration of that same particle determined by O.

To this new observer, the equations describing the motion of the two fluids remain as they were earlier in the essay but with $\rho\vec{g}$ deleted on the right-hand side and $\rho\vec{A}$ inserted on the left-hand side. The calculation is now just as it was before but $-\vec{A}$ is in place of $\vec{g}$ and, by this, Taylor's conclusion is then obtained.

This tells us that a gravitationally unstable arrangement, heavy above light, can be stabilized by accelerating it downward, whereas a gravitationally stable arrangement, light above heavy, can be destabilized by again accelerating it downward.

6.6   The Stability of an Advancing Interface: The Saffman–Taylor Problem

Now, G. I. Taylor's observation has led to some experimental work.[11] In the unstable case, where the acceleration is directed out of the light fluid into the heavy fluid, a small displacement of the interface grows into what look like fingers of the light fluid penetrating into the heavy fluid. Something like this is also seen when one fluid is used to displace another fluid from a porous solid. Figure 6.4, which reproduces a photograph taken by Paterson, illustrates this instability. It is called the Saffman–Taylor instability.

To get some idea what this instability is all about, first let a horizontal interface be moving perpendicular to itself in the laboratory frame at a fixed velocity, denoted by $V\vec{k}$. Let it divide two fluids of differing densities and viscosities. The interface is not accelerating, and so, to an observer moving at the velocity $V\vec{k}$ vis-à-vis the laboratory observer, the problem is just Rayleigh's problem and the stability of the interface is again determined by whether $\vec{g}$ points out of the heavy fluid into the light fluid or the reverse. This is independent of the speed of the interface. This is what the Navier–Stokes description of fluid motion would tell us.

However, in a porous solid, things do not work according to Navier–Stokes. Darcy's law must be used and new results turn up due to the fact

---

[11]D.J. Lewis. The Instability of Liquid Surfaces When Accelerated in a Direction Perpendicular to Their Planes.II. *Proc. Roy. Soc. Lond. A*, 202:81, 1950.

Figure 6.4. *A Photograph Taken by Paterson Illustrating the Instability*
Reprinted from "Radial Fingering in a Hele-Shaw Cell" by L. Paterson, in
Journal of Fluid Mechanics, Volume 113, 513-529, (1981), Cambridge University
Press

that pressure gradients and their perturbations now feed directly into velocities and their perturbations. The viscosity of the fluid then determines
the factor by which velocities must be multiplied to get pressure differences. Darcy's law puts velocities in place of accelerations and then makes
predictions that velocities, instead of accelerations, can be used to stabilize
or destabilize plane interfaces.

To present a picture illustrating the cause of the instability let two fluids,
one very viscous, the other nearly inviscid, lie as indicated in Figure 6.5.
The pressure is fixed at some distance to both the left and the right of the
interface. The high pressure lies to the left; the pressure gradient appears
overwhelmingly in the very viscous fluid. At the top, the less viscous fluid
lies to the left; at the bottom, it lies to the right. In both cases, the fluid to
the left displaces the one to the right and the interface moves to the right. A
small disturbance of the interface is illustrated: a local region moving ahead
of the displacement front, its speed being greater than that of the front.
This alters the pressure gradients as indicated: the strengthened gradient
above tending to support the disturbance, the weakened gradient below

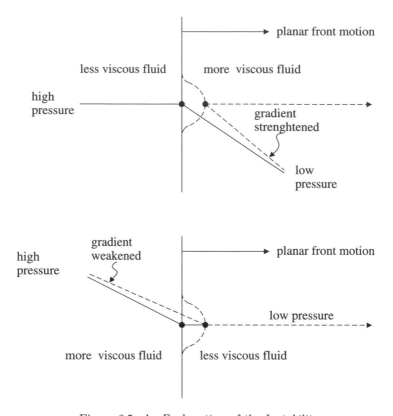

Figure 6.5. *An Explanation of the Instability*

definitely opposing the disturbance. The prediction of the pictures,[12] then, is that the displacement of a more viscous fluid by a less viscous fluid might be unstable, but the reverse is surely stable.

This is the conclusion of Saffman and Taylor and it is the aim of our work to reproduce their result. To do this, let a horizontal plane interface separate a fluid of density ρ and viscosity μ, above, from a fluid of a density ρ^* and viscosity μ^*, below. The direction of gravity is set in advance and

---

[12]This way of illustrating the cause of an instability will serve us well in the next three essays. In those essays, gradients of temperature and composition determine the stability of solidification and other fronts in just the same way that gradients of pressure determine the stability of fluid displacement fronts here. Diffusion emerges as an important theme here and henceforth.

It is left to the reader to make a prediction by taking account of a displacement to the left: a local region falling behind the front, its speed being less than that of the front.

Then, the effect of surface tension can be taken into account. In Figure 6.5, the pressure in the less viscous fluid is fixed whereupon surface tension weakens the strengthened gradient illustrated above and weakens even more the weakened gradient illustrated below. Hence, surface tension is predicted to have a stabilizing influence.

the interface is taken to be moving at a speed V in the $\vec{k}$ direction, upward if V is positive or downward if it is negative.

This motion takes place in a porous solid which is macroscopically uniform, having the same permeability, denoted κ, to both fluids. It is described by Darcy's law, namely by

$$\vec{v} = -\frac{\kappa}{\mu} \left[\nabla p - \rho \vec{g} \right]$$

and

$$\vec{v}^* = -\frac{\kappa}{\mu^*} \left[\nabla p^* - \rho^* \vec{g} \right]$$

where the interface is supposed to be smooth and these equations are supposed to hold up to the interface.

Before doing the calculation, let $\vec{v} - V\vec{k} = \vec{v}^*$; then, Darcy's law would tell us that

$$\frac{dp}{dz} = -\rho g \; - \; \frac{\mu V}{\kappa}$$

and

$$\frac{dp^*}{dz} = -\rho^* g \; - \; \frac{\mu^* V}{\kappa}$$

Earlier in this essay it was learnt that the pressure gradient in the base solution determines the stability of the base state. Carrying that idea over, and using these formulas, leads to some predictions. If V were zero but g were not, we would guess that the interface is or is not stable according to whether ρ is or is not less than ρ^*. Instead, if V were positive and g were zero, we would guess that the interface is or is not stable according to whether μ is or is not less than μ^*, and vice versa if V were negative. In fact, we might go on and guess that the interface is or is not stable according to whether

$$\rho g + \frac{\mu}{\kappa} V \quad \text{is or is not less than} \quad \rho^* g + \frac{\mu^*}{\kappa} V$$

Omitting the stabilizing effect of surface tension, this is the correct result. It makes some predictions that can be tested; for example, if ρ is greater than ρ^*, a gravitationally unstable configuration can be stabilized if μ is less than μ^* provided that V is positive and large enough.

To carry out the stability calculation, introduce an observer O' moving at the velocity $V\vec{k}$ as seen by a laboratory observer. Then, write all the equations as they would be seen by this observer but do not mark the variables with primes.

To begin, let the interface be two dimensional and denote its position by

$$z = Z(x, \, t, \, \epsilon)$$

Then, in terms of Z, the normal and the normal speed can be determined by

$$\vec{n} = \frac{-Z_x \, \vec{i} + \vec{k}}{\sqrt{Z_x^2 + 1}}$$

and

$$u = \frac{Z_t}{\sqrt{Z_x^2 + 1}}$$

Now, Darcy's law must hold above and below the interface. But Darcy's law is not form invariant under a change to an observer moving at constant speed. Taking this into account, above and below the interface,

$$\vec{v} + V\vec{k} = -\frac{\kappa}{\mu}\nabla[p + \rho g z], \quad z > Z(x, \, t, \, \epsilon)$$

and

$$\vec{v}^* + V\vec{k} = -\frac{\kappa}{\mu^*}\nabla[p^* + \rho^* g z], \quad z < Z(x, \, t, \, \epsilon)$$

must hold. In addition, requiring

$$\nabla \cdot \vec{v} = 0, \quad z > Z(x, \, t, \, \epsilon)$$

and

$$\nabla \cdot \vec{v}^* = 0, \quad z < Z(x, \, t, \, \epsilon)$$

produces

$$\nabla^2 p = 0, \quad z > Z(x, \, t, \, \epsilon)$$

and

$$\nabla^2 p^* = 0, \quad z < Z(x, \, t, \, \epsilon)$$

At the interface, two conditions must be met. The first is that neither fluid cross the interface. To satisfy this,

$$\vec{n} \cdot \vec{v} = u = \vec{n} \cdot \vec{v}^*$$

must hold, and in terms of Z, this is

$$-Z_x v_x + v_z = Z_t = -Z_x v_x^* + v_z^*$$

The second condition is that the pressure difference must be determined by the curvature of the surface, namely

$$p - p^* = \gamma \frac{Z_{xx}}{[1 + Z_x^2]^{3/2}}$$

must hold.

All of the time dependence in this problem comes out of whatever time dependence there might be in the surface motion. This transmits time dependence to the velocity by way of the no-flow condition and this, in turn, transmits time dependence to the pressure by way of Darcy's law.

These are the nonlinear equations and they admit a simple solution, denoted by the subscript zero and called the base solution. It is

$$Z_0 = 0$$

$$\vec{v}_0 = \vec{0} = \vec{v}_0^*$$

$$\nabla_0 p_0 = -\left[\frac{\mu}{\kappa}V + \rho g\right]\vec{k}$$

and

$$\nabla_0 p_0^* = -\left[\frac{\mu^*}{\kappa}V + \rho^* g\right]\vec{k}$$

The reference domain is then defined by the reference interface and it is

$$-\infty < x_0 < \infty, \quad z_0 = Z_0 = 0$$

To determine whether or not the base solution is stable to a displacement of its interface, the perturbation equations must be solved. To write the perturbation equations, the variables in the problem must be expanded along a mapping of the reference domain into the present domain. It is enough to write

$$Z(x, t, \epsilon) = Z_0(x_0, t_0) + \epsilon Z_1(x_0, t) + \cdots$$

where $Z_0 = 0$ and $Z_1(x_0, t_0)$ remains to be determined.

Substituting the expansions of the variables $\vec{v}$, p, etc. into the nonlinear equations produces, first, the base equations at zeroth order and then the perturbation equations at first order, all on the reference domain. At first order, there obtains

$$\left.\begin{array}{c}\vec{v}_1 = -\dfrac{\kappa}{\mu}\nabla_0 p_1 \\[2mm] \nabla_0 \cdot \vec{v}_1 = 0\end{array}\right\} \quad z_0 > 0$$

and

$$\left.\begin{array}{c}\vec{v}_1^* = -\dfrac{\kappa}{\mu^*}\nabla_0 p_1^* \\[2mm] \nabla_0 \cdot \vec{v}_1^* = 0\end{array}\right\} \quad z_0 < 0$$

which, on eliminating $\vec{v}_1$ and $\vec{v}_1^*$, are just

$$\nabla_0^2 p_1 = 0, \quad z_0 > 0$$

and

$$\nabla_0^2 p_1^* = 0, \quad z_0 < 0$$

These are the domain equations. At the interface, $z_0 = 0$, there obtains

$$\vec{n}_0 \cdot \vec{v}_1 = \frac{\partial Z_1}{\partial t_0} = \vec{n}_0 \cdot \vec{v}_1^*$$

and

$$\left[p_1 + Z_1 \frac{\partial p_0}{\partial z_0}\right] - \left[p_1^* + Z_1 \frac{\partial p_0^*}{\partial z_0}\right] = \gamma \frac{\partial^2 Z_1}{\partial x_0^2}$$

where the first reduces, on using Darcy's law, to

$$-\frac{\kappa}{\mu} \frac{\partial p_1}{\partial z_0} = \frac{\partial Z_1}{\partial t_0} = -\frac{\kappa}{\mu^*} \frac{\partial p_1^*}{\partial z_0}$$

The perturbation problem is then

$$\nabla_0^2 p_1 = 0, \quad 0 < z_0 < \infty$$

$$p_1 \to 0, \quad z_0 \to +\infty$$

$$\nabla_0^2 p_1^* = 0, \quad -\infty < z_0 < 0$$

$$p_1^* \to 0, \quad z_0 \to -\infty$$

where, at $z_0 = 0$,

$$-\frac{\kappa}{\mu} \frac{d\hat{p}_1}{dz_0} = \frac{\partial Z_1}{\partial t_0} = -\frac{\kappa}{\mu^*} \frac{d\hat{p}_1^*}{dz_0}$$

and

$$\left[p_1 + Z_1 \frac{dp_0}{dz_0}\right] - \left[p_1^* + Z_1 \frac{dp_0^*}{dz_0}\right] = \gamma \frac{\partial^2 Z_1}{\partial x_0^2}$$

must be satisfied. To solve it, turn it into an eigenvalue problem by writing

$$p_1 = \hat{p}_1(z_0) e^{\sigma t_0} e^{ikx_0}$$

$$p_1^* = \hat{p}_1^*(z_0) e^{\sigma t_0} e^{ikx_0}$$

and

$$Z_1 = \hat{Z}_1 e^{\sigma t_0} e^{ikx_0}$$

Then, the eigenvalue problem is

$$\frac{d^2 \hat{p}_1}{dz_0^2} - k^2 \hat{p}_1 = 0, \quad 0 < z_0 < \infty$$

$$\hat{p}_1 \to 0, \quad z_0 \to +\infty$$

$$\frac{d^2 \hat{p}_1^*}{dz_0^2} - k^2 \hat{p}_1^* = 0, \quad -\infty < z_0 < 0$$

$$\hat{p}_1^* \to 0, \quad z_0 \to -\infty$$

where, at $z_0 = 0$,

$$-\frac{\kappa}{\mu}\frac{d\hat{p}_1}{dz_0} = \sigma\hat{Z}_1 = -\frac{\kappa}{\mu^*}\frac{d\hat{p}_1^*}{dz_0}$$

and

$$\left[\hat{p}_1 + \hat{Z}_1\frac{dp_0}{dz_0}\right] - \left[\hat{p}_1^* + \hat{Z}_1\frac{dp_0^*}{dz_0}\right] = -k^2\gamma\hat{Z}_1$$

must hold.

The domain equations, as well as the far-field conditions, can be satisfied by writing

$$\hat{p}_1 = C\,e^{-|k|z_0}$$

and

$$\hat{p}_1^* = C^*\,e^{|k|z_0}$$

Then, by substituting these formulas into the equations holding at $z_0 = 0$, there obtain

$$\frac{\kappa}{\mu}|k|C = \sigma\hat{Z}_1$$

$$-\frac{\kappa}{\mu^*}|k|C^* = \sigma\hat{Z}_1$$

and

$$C - \left[\frac{\mu}{\kappa}V + \rho g\right]\hat{Z}_1 - C^* + \left[\frac{\mu^*}{\kappa}V + \rho^* g\right]\hat{Z}_1 = -\gamma k^2\hat{Z}_1$$

This is a system of three, homogeneous, linear, algebraic equations in the unknown constants C, C^* and $\hat{Z}_1$. It has solutions other than C, C^* and $\hat{Z}_1$ all zero only for special values of σ and these are the eigenvalues. Their dependence on k can be determined by requiring

$$\det\begin{pmatrix} \dfrac{\kappa}{\mu}|k| & 0 & -\sigma \\[2mm] 0 & -\dfrac{\kappa}{\mu^*}|k| & -\sigma \\[2mm] 1 & -1 & \left[-\dfrac{\mu}{\kappa} + \dfrac{\mu^*}{\kappa}\right]V + [-\rho + \rho^*]g + \gamma k^2 \end{pmatrix} = 0$$

to hold, which produces the result

$$\frac{\sigma}{|k|}\left[\frac{\mu}{\kappa} + \frac{\mu^*}{\kappa}\right] = \left[\frac{\mu}{\kappa} - \frac{\mu^*}{\kappa}\right]V + [\rho - \rho^*]g - k^2\gamma$$

This formula tells us the growth constant of a small displacement of wavelength $2\pi/k$. It indicates that surface tension is stabilizing and that a planar surface is stable to displacements of short enough wavelength, no matter the values of ρ, μ, ρ^*, μ^* and V. By making the lateral extent of an experiment small enough to filter out the dangerous long wavelengths, a stable invasion of one fluid into another can always be achieved.

If the surface tension vanishes, the result is just the one predicted before the perturbation calculation was carried out. But now the value of σ is known and it depends on $|k|$, where $|k|$ determines the rate at which p_1 and p_1^* die out as $|z_0| \longrightarrow \infty$. This hints at what underlies this instability.

Now, in case the only effect is the direction in which the front moves (i.e., in case γ and $\rho - \rho^*$ both vanish), the interface is stable if the front moves in one direction, but it is unstable if the motion is reversed. This dependence on the direction of front motion will be seen again in the essays on solidification, precipitation and electrodeposition.

6.7   DISCUSSION NOTES

6.7.1   The Energy Associated with the Motion of a Fluid

Let a fluid be confined by rigid side and bottom walls as indicated in Figure 6.6 where free space lies above its deflecting top surface. It is our aim in this discussion note to draw attention to the various forms of energy that can be associated with the motion of an inviscid fluid in this configuration. In so doing, we can see how the mean curvature of the surface comes into the determination of the surface energy.

The nonlinear equations describing the motion of the fluid can be written

$$\rho \frac{\partial \vec{v}}{\partial t} + \rho \vec{v} \cdot \nabla \vec{v} = -\nabla p + \rho \vec{g} \tag{6.19}$$

and

$$\nabla \cdot \vec{v} = 0 \tag{6.20}$$

The domain is indicated in Figure 6.6 and it remains to be determined.

The force per unit mass due to gravity, denoted $\vec{g}$, can be written in terms of the gravitational potential energy per unit mass, denoted ϕ, as

$$\vec{g} = -g\vec{k} = -\nabla \phi$$

where

$$\phi = gz$$

the axis Oz being vertically upward.

To determine how the several forms of energy associated with the motion of a fluid fit together, multiply equation (6.19) by $\vec{v}\cdot$ and integrate the product over the present domain of the fluid, denoted $V(t)$. The result, taking equation (6.20) into account, is

$$\iiint\limits_{V(t)} dV \ \left[\frac{\partial}{\partial t} \left[\frac{1}{2} \rho v^2 \right] + \nabla \cdot \left[\vec{v} \frac{1}{2} \rho v^2 \right] + \nabla \cdot \left[p\vec{v} \right] + \nabla \cdot \left[\rho \phi \vec{v} \right] \right] = 0$$

which, on using $\vec{n} \cdot \vec{v} = 0$ at the solid walls, comes down to

$$\iiint\limits_{V(t)} dV \ \frac{\partial}{\partial t} \left[\frac{1}{2} \rho v^2 \right] + \iint\limits_{S(t)} dA \ \vec{n} \cdot \left[\vec{v} \frac{1}{2} \rho v^2 + \vec{v} p + \vec{v} \rho \phi \right] = 0 \tag{6.21}$$

where $S(t)$ denotes the deflecting surface. Now, Leibnitz' rule can be used to take $\dfrac{\partial}{\partial t}$ outside the integral over $V(t)$ and, at the same time, $\vec{n} \cdot \vec{v}$ can be replaced by u or $\vec{n} \cdot \vec{u}$ at the deflecting surface. Doing this, equation (6.21)

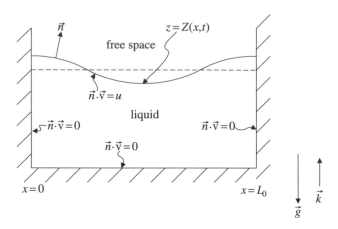

Figure 6.6. *A Fluid Confined by Side and Bottom Walls Whose Top Surface Is Free to Deflect*

can be written as

$$\frac{d}{dt} \iiint\limits_{V(t)} dV\, \frac{1}{2}\rho v^2 \;+\; \iint\limits_{S(t)} dA\, [\vec{n}\cdot\vec{u}p + \vec{n}\cdot\vec{u}\rho\phi] = 0 \qquad (6.22)$$

where $\dfrac{d}{dt}$ is an ordinary time derivative. Taking into account that ρ and ϕ do not depend on t and again using Leibnitz' rule, equation (6.22) can be turned into

$$\frac{d}{dt} \iiint\limits_{V(t)} dV\, \left[\frac{1}{2}\rho v^2 + \rho\phi\right] \;+\; \iint\limits_{S(t)} dA\, \vec{n}\cdot\vec{u}p = 0 \qquad (6.23)$$

Now, let $V(t)$ be two dimensional; then, the deflecting surface $S(t)$ is just the curve

$$z = Z(x,\ t)$$

and equation (6.23) takes the simple form

$$\frac{d}{dt} \iint\limits_{V(t)} dV\, \left[\frac{1}{2}\rho v^2 + \rho\phi\right] \;+\; \int\limits_{z=Z(x,t)} ds\, \vec{n}\cdot\vec{u}p = 0 \qquad (6.24)$$

where s denotes the arc length along the curve.[13] Then, by use of

$$p = -\gamma \frac{Z_{xx}}{[1 + Z_x^2]^{3/2}}$$

---

[13]The volume and the surface area are now per unit depth into the page.

$$\vec{n} \cdot \vec{u} = \frac{Z_t}{\sqrt{1 + Z_x^2}}$$

and

$$ds = \sqrt{1 + Z_x^2} \, dx$$

the second term on the left-hand side of equation (6.24) can be written

$$\int_{z=Z(x,\,t)} ds \, \vec{n} \cdot \vec{u} p = -\gamma \int_0^{L_0} dx \, \frac{Z_t Z_{xx}}{[1 + Z_x^2]^{3/2}} \tag{6.25}$$

where the integral over x is over the fixed width of the container, L_0, say, and where this integral can be written in terms of the surface area, namely

$$-\gamma \int_0^{L_0} dx \, \frac{Z_t Z_{xx}}{[1 + Z_x^2]^{3/2}} = \gamma \frac{d}{dt} \int ds \tag{6.26}$$

Then, substituting equations (6.25) and (6.26) into equation (6.24), there obtains

$$\frac{d}{dt} \left[\iint_{V(t)} dV \left[\frac{1}{2} \rho v^2 + \rho \phi \right] + \gamma \int_{z=Z(x,\,t)} ds \right] = 0$$

which tells us that the sum of the kinetic, gravitational and surface energy of the fluid remains fixed in time as a liquid confined by its surface tension undergoes its motion.

The term $\gamma \int ds$ is the surface energy as we have been using it. Equation (6.26), namely

$$\frac{d}{dt} \int ds = -\int_0^{L_0} dx \, Z_t \, \frac{Z_{xx}}{[1 + Z_x^2]^{3/2}} = \int ds \, \vec{n} \cdot \vec{u} \, [-2H]$$

tells us how the local mean curvature acts to increase the surface area. It is the factor by which the local surface speed must be multiplied to obtain the rate at which the local surface area increases. Indeed, wherever the mean curvature vanishes, the motion of the surface does not contribute to an increase in its area.

It only remains to derive equation (6.26). To do this, write

$$\frac{d}{dt} \int ds = \frac{d}{dt} \int_0^{L_0} dx \sqrt{1 + Z_x^2} = \int_0^{L_0} dx \frac{Z_x}{\sqrt{1 + Z_x^2}} \frac{\partial}{\partial t} Z_x$$

Then, write $\dfrac{\partial}{\partial t} Z_x = \dfrac{\partial}{\partial x} Z_t$ and integrate by parts. Requiring Z or Z_x to vanish at the ends (i.e., requiring fixed or free end conditions), produces[14] the required result:

$$\frac{d}{dt} \int ds = -\int_0^{L_0} dx \; Z_t \frac{\partial}{\partial x}\left[\frac{Z_x}{\sqrt{1+Z_x^2}}\right] = -\int_0^{L_0} dx \; Z_t \frac{Z_{xx}}{[1+Z_x^2]^{3/2}}$$

As a side result of carrying out this calculation, we learn how to determine the gravitational potential energy of a fluid when its surface is displaced. To be definite, let the origin of z lie in the horizontal equilibrium plane whence the bottom wall is at $z = -H_0$. The potential energy is then

$$\int_0^{L_0}\int_{-H_0}^{Z(x,t)} \rho\phi \; dz \; dx = \int_0^{L_0}\int_{-H_0}^{Z(x,t)} \rho g z \; dz \; dx = \rho g \int_0^{L_0} \frac{1}{2}[Z^2 - H_0^2] \; dx$$

whence any deflection of the surface away from its equilibrium position, $z = 0$, increases the gravitational potential energy of the fluid.

6.7.2   Other End Conditions: Rough Estimates of σ^2

Only free end conditions, end conditions where the interface separating the two fluids meets the side walls at right angles, were introduced in the essay. In this discussion note, other end conditions are introduced.

Just as the critical length of a liquid bridge was found to depend on the end conditions, so too, the critical width of two liquid layers, one above the other, ought to depend on the end conditions. Again, we might guess that it should increase as free end conditions are replaced by fixed end conditions, and we might also guess that the fixed-volume requirement ought to play a more important role as the end conditions move away from free-free.

To take other end conditions into account, go back to the eigenvalue problem in the essay just before free-free end conditions were imposed. Then, it is, at least in part,

$$\nabla_0^2 \hat{p}_1 = 0, \quad 0 < x_0 < L_0, \quad 0 < z_0 < H_0$$

and

$$\nabla_0^2 \hat{p}_1^* = 0, \quad 0 < x_0 < L_0, \quad -H_0^* < z_0 < 0$$

---

[14]This need not have been made so special. Using the theorem

$$\frac{d}{dt} \iint_A dA = \iint_A dA \; u \; [-2H]$$

would save some work and restore some generality. However, what has been presented fits in better with the essay and can be understood in elementary terms.

on the domain,

$$\frac{\partial \hat{p}_1}{\partial z_0} = 0, \quad 0 < x_0 < L_0, \quad z_0 = H_0$$

on the top wall,

$$\frac{\partial \hat{p}_1}{\partial x_0} = 0, \quad x_0 = 0 \text{ and } x_0 = L_0, \quad 0 < z_0 < H_0$$

and

$$\frac{\partial \hat{p}_1^*}{\partial x_0} = 0, \quad x_0 = 0 \text{ and } x_0 = L_0, \quad -H_0^* < z_0 < 0$$

on the side walls, and

$$\frac{\partial \hat{p}_1^*}{\partial z_0} = 0, \quad 0 < x_0 < L_0, \quad z_0 = -H_0^*$$

on the bottom wall. The equations at the solid walls express the no-flow condition.

All of these equations are satisfied by writing

$$\hat{p}_1 = \sum_k A_k \cosh k[z_0 - H_0] \cos kx_0$$

and

$$\hat{p}_1^* = \sum_k A_k^* \cosh k[z_0 + H_0^*] \cos kx_0$$

where $k = \dfrac{\pi}{L_0} n$, $n = 0, 1, 2, \cdots$.

This takes $\hat{p}_1$ and $\hat{p}_1^*$ out of the problem, putting in their place the constants A_0, A_1, etc. and A_0^*, A_1^*, etc. These remain to be determined, as does the function $\hat{Z}_1$. The equations holding at $z_0 = 0$ presented in the essay can be used to do this.

They are, first,

$$\gamma \frac{d^2 \hat{Z}_1}{dx_0^2} + g[\rho - \rho^*]\hat{Z}_1 = \sum_k [A_k \cosh kH_0 - A_k^* \cosh kH_0^*] \cos kx_0 \quad (6.27)$$

where $\hat{Z}_1$ must also satisfy the assigned end conditions at $x_0 = 0$ and at $x_0 = L_0$.

This determines $\hat{Z}_1$ in terms of A_0, A_1, $\cdots$, A_0^*, A_1^*, $\cdots$ whereupon these constants must be determined so that the remaining equations at $z_0 = 0$, the no-flow equations, are satisfied for all values of x_0, namely so that

$$\sum_{k>0} -A_k\, k\, \sinh kH_0\, \cos kx_0 = -\sigma^2 \rho \hat{Z}_1 \quad (6.28)$$

and

$$\sum_{k>0} A_k^* \, k \, \sinh k H_0^* \, \cos kx_0 \; = \; -\sigma^2 \rho^* \hat{Z}_1 \qquad (6.29)$$

It only remains, then, to multiply each of these equations by $\cos k'x_0$ and integrate the result over $0 < x_0 < L_0$, where $k' = \pi n'/L_0$, $n' = 0, 1, 2, \cdots$. By this, there obtain two sets of infinitely many homogeneous linear equations in the constants $A_0, A_1, \cdots, A_0^*, A_1^*, \cdots$. The eigenvalues are the special values of σ^2 to which solutions can be found, other than $A_0, A_1, \cdots, A_0^*, A_1^*, \cdots$ all zero.

Now, if either equation (6.28) or equation (6.29) is integrated over $0 < x_0 < L_0$, there obtains $\sigma^2 = 0$ unless $\int_0^{L_0} \hat{Z}_1 \, dx_0$ vanishes. This condition maintains the volume of both fluid layers fixed as their dividing surface is displaced. It takes the place of the leading moment equation in both infinite sequences of moment equations and it turns out to be sufficient, as A_0 and A_0^* appear only as their difference, $A_0 - A_0^*$.

To illustrate the way in which estimates of the eigenvalues can be obtained, we put $H_0^* = H_0$ and carry out a few steps of the calculation. Let the end conditions be fixed-free. Then, if the heavy fluid lies above the light fluid, the solution to equation (6.27), namely to

$$\frac{d^2 \hat{Z}_1}{dx_0^2} + \lambda^2 \hat{Z}_1 = \frac{A_0 - A_0^*}{\gamma} + \sum_{k>0} \frac{A_k - A_k^*}{\gamma} \cosh k H_0 \, \cos kx_0$$

where $\lambda^2 = \dfrac{\rho - \rho^*}{\gamma} g > 0$, can be written

$$\hat{Z}_1 = C \cos \lambda x_0 + D \sin \lambda x_0$$

$$+ \frac{A_0 - A_0^*}{\gamma \lambda^2} + \sum_{k>0} \frac{A_k - A_k^*}{[\lambda^2 - k^2]\gamma} \cosh k H_0 \cos kx_0$$

and it must be made to satisfy the two end conditions and the integral condition, namely

$$\hat{Z}_1(x_0 = 0) = 0 = \frac{d\hat{Z}_1}{dx_0}(x_0 = L_0) \quad \text{and} \quad \int_0^{L_0} \hat{Z}_1 \, dx_0 = 0$$

By this, there obtains

$$0 = C + \frac{A_0 - A_0^*}{\gamma \lambda^2} + \sum_{k>0} \frac{A_k - A_k^*}{[\lambda^2 - k^2]\gamma} \cosh k H_0 \qquad (6.30)$$

$$0 = -\lambda C \sin \lambda L_0 + \lambda D \cos \lambda L_0 \qquad (6.31)$$

and

$$0 = \frac{C}{\lambda} \sin \lambda L_0 - \frac{D}{\lambda} [\cos \lambda L_0 - 1] + \frac{A_0 - A_0^*}{\gamma \lambda^2} L_0 \qquad (6.32)$$

where equations (6.31) and (6.32) can be used to eliminate C and D in favor of $A_0 - A_0^*$. The result is

$$C = -\frac{A_0 - A_0^*}{\gamma \lambda^2} \lambda L_0 \cot \lambda L_0 \qquad (6.33)$$

and

$$D = -\frac{A_0 - A_0^*}{\gamma \lambda^2} \lambda L_0 \qquad (6.34)$$

It remains to satisfy equation (6.30) and to demand that equations (6.28) and (6.29) at the interface, namely

$$-\sigma^2 \rho \hat{Z}_1 = \sum_{k>0} -k A_k \sinh k H_0 \; \cos k x_0$$

and

$$-\sigma^2 \rho^* \hat{Z}_1 = \sum_{k>0} k A_k^* \sinh k H_0 \; \cos k x_0$$

be satisfied at each point on the interval $0 < x_0 < L_0$. It is enough to take the first of these two equations into account, for if it is satisfied by a set of coefficients A_k, $k = 1, \ 2, \ \cdots$, then the second equation can be satisfied by setting $A_k^* = -\dfrac{\rho^*}{\rho} A_k$, $k = 1, \ 2, \ \cdots$.

To begin our calculations, let us first determine the critical width of the liquid layers (viz., the smallest width at which $\sigma^2 = 0$ is a solution to the perturbation eigenvalue problem). To do this, set $\sigma^2 = 0$ and observe that equations (6.28) and (6.29) can be satisfied if and only if $A_k = 0 = A_k^*$ for all $k > 0$. Then, equation (6.30) reduces to

$$0 = C + \frac{A_0 - A_0^*}{\gamma \lambda^2}$$

and eliminating C between this and equation (6.33), there obtains

$$-\lambda L_0 \cot \lambda L_0 + 1 = 0$$

which determines the critical value of λL_0. It is just short of $\frac{3}{2}\pi$, whence

$$\frac{L_{0_{critical}}}{\sqrt{\dfrac{\gamma}{g[\rho - \rho^*]}}} \simeq \frac{3}{2} \pi$$

This exceeds the free-free critical width and the sequence of critical values recapitulates the sequence found in the case of the liquid bridge, as the same sequence of end conditions is entertained. The only difference is the length scale. Here, it is $\sqrt{\dfrac{\gamma}{g[\rho - \rho^*]}}$; there, it was R_0.

To estimate the eigenvalues other than zero, the moments of equation (6.28) must be determined. The simplest estimate obtains using only one moment and, by this, an estimate of only one eigenvalue results. To get it, multiply equation (6.28) by $\cos(\pi x_0/L_0)$ and integrate the product over the interval $0 < x_0 < L_0$ whereupon there obtains

$$-\rho\sigma^2 \left[C \left[\frac{-\lambda \sin \lambda L_0}{\lambda^2 - \frac{\pi^2}{L_0^2}} \right] + D \left[\frac{\lambda[\cos \lambda L_0 + 1]}{\lambda^2 - \frac{\pi^2}{L_0^2}} \right] \right]$$

$$+ \frac{A_1 - A_1^*}{\gamma[\lambda^2 - \frac{\pi^2}{L_0^2}]} \cosh\left(\frac{\pi H_0}{L_0}\right) \frac{L_0}{2} = -\frac{\pi}{L_0} A_1 \sinh\left(\frac{\pi H_0}{L_0}\right) \frac{L_0}{2} \qquad (6.35)$$

where C and D can be eliminated in favor of $A_0 - A_0^*$ via equations (6.33) and (6.34). It is worth noticing that all of the higher-moment equations would look just like this; for example, taking the next moment would produce an equation in which only C, D, $A_2 - A_2^*$ and A_2 would appear, etc.

Now, all of the coefficients are present only in equation (6.30), namely only in

$$\hat{Z}_1(x_0 = 0) = 0 = C + \frac{A_0 - A_0^*}{\gamma\lambda^2} + \sum_{k>0} \frac{A_k - A_k^*}{[\lambda^2 - k^2]\gamma} \cosh kH_0$$

but if the series in this equation is truncated to one term, it can be used together with the first moment equation, equation (6.35), to estimate σ^2. Truncating equation (6.30) to one term and eliminating C via equation (6.33) then produces

$$\frac{[A_1 - A_1^*]}{\left[\lambda^2 - \frac{\pi^2}{L_0^2}\right]\gamma} \cosh\frac{\pi H_0}{L_0} = [\lambda L_0 \cot \lambda L_0 - 1]\frac{A_0 - A_0^*}{\gamma\lambda^2} \qquad (6.36)$$

which leads, using $A_1^* = -\frac{\rho^*}{\rho}A_1$, to

$$A_1 = \frac{[\lambda L_0 \cot \lambda L_0 - 1]\gamma\left[\lambda^2 - \frac{\pi^2}{L_0^2}\right]}{\left[1 + \frac{\rho^*}{\rho}\right]\cosh\frac{\pi H_0}{L_0}} \frac{A_0 - A_0^*}{\gamma\lambda^2} \qquad (6.37)$$

Then, using equations (6.33), (6.34), (6.36) and (6.37) to substitute for C, D, $\frac{[A_1 - A_1^*]}{\gamma[\lambda^2 - \frac{\pi^2}{L_0^2}]} \cosh\frac{\pi H_0}{L_0}$ and A_1 in equation (6.35) and dividing out

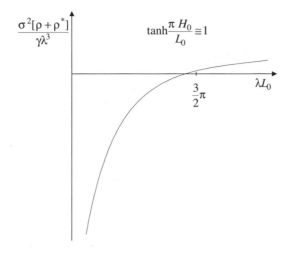

Figure 6.7. *A Graph of Equation (6.38)*

$\dfrac{A_0 - A_0^*}{\gamma\lambda^2}$, produces a formula for σ^2. It is

$$\frac{\sigma^2[\rho + \rho^*]}{\gamma\lambda^3} = \frac{1}{\lambda^3 L_0^3} \frac{\frac{1}{2}\,\pi\tanh\dfrac{\pi H_0}{L_0}[\lambda L_0 \cot \lambda L_0 - 1][\lambda^2 L_0^2 - \pi^2]}{\frac{1}{2}\,[\lambda L_0 \cot \lambda L_0 - 1] - \dfrac{\lambda^2 L_0^2}{\lambda^2 L_0^2 - \pi^2}} \tag{6.38}$$

and this formula tells us the following:

(i) $\sigma^2 = 0$ at the roots of $\lambda L_0 \cot \lambda L_0 - 1 = 0$

(ii) $\sigma^2 < 0$ at $\lambda L_0 = \pi$

(iii) σ^2 does not depend strongly on H_0 unless $\dfrac{\pi H_0}{L_0} < 1$

and

(iv) $\dfrac{\sigma^2[\rho + \rho^*]}{\gamma\lambda^3} \sim -\left[\dfrac{\pi}{\lambda L_0}\right]^3$ as λL_0 goes to zero

The formula for σ^2 is sketched in Figure 6.7.

6.7.3   The Fixed-Volume Requirement

Let two fluids lie one above the other as illustrated in Figure 6.2 in the essay. The surface dividing the two fluids is displaced from its reference position $z = 0$ to its current position $z = Z(x)$. Due to this, the volume[15]

---

[15]This is volume per unit depth into the page; potential energy will also be per unit depth into the page.

of the lower layer is increased by

$$\int_0^{L_0} \int_{-H_0^*}^{Z} dz \, dx \;-\; \int_0^{L_0} \int_{-H_0^*}^{0} dz \, dx \;=\; \int_0^{L_0} Z \, dx$$

and this must vanish in order that the displacement be carried out in such a way that the volumes of the two layers do not change.

In the case of Rayleigh's jet, everything turns on this requirement, but in the case of a plane surface, it seems to play a minor role. It is one aim of this discussion note to shed some light on this odd circumstance. A second aim is to indicate that a lot more than might be expected can be deduced from Rayleigh's work principle, especially if the displacement is not specified in advance.

The case of interest, then, is that in which the heavier fluid lies above the lighter. Then, a displacement of the dividing surface increases the potential energy associated with the surface area, while it decreases the potential energy associated with the gravitational field, thereby leaving the question of the stability of this arrangement of the fluids in doubt. To resolve this doubt, calculate the increase in the total potential energy caused by a displacement of the dividing surface from $z = 0$ to $z = Z(x)$.

First, the increase in the gravitational potential energy can be determined via

$$\int_0^{L_0} \int_{-H_0^*}^{Z} \rho^* g z \, dz \, dx \;+\; \int_0^{L_0} \int_{Z}^{H_0} \rho g z \, dz \, dx$$

$$-\; \int_0^{L_0} \int_{-H_0^*}^{0} \rho^* g z \, dz \, dx \;-\; \int_0^{L_0} \int_{0}^{H_0} \rho g z \, dz \, dx$$

which comes down to

$$-\int_0^{L_0} \frac{1}{2} g[\rho - \rho^*] Z^2 \, dx$$

and as this is less than zero whenever $\rho > \rho^*$, it is destabilizing.

Then, turning to the surface potential energy, notice that the area of the displaced surface can be obtained via

$$\int ds \;=\; \int_0^{L_0} [1 + [Z']^2]^{1/2} \, dx$$

The increase in the surface area then leads to a formula for the increase in the surface potential energy. It is

$$\gamma \int_0^{L_0} \left[[1 + [Z']^2]^{1/2} - 1 \right] \, dx$$

and as this is always positive, it is stabilizing.

The increase in the total potential energy is then

$$\gamma \int_0^{L_0} \left[[1 + [Z']^2]^{1/2} - 1 \right] \, dx \; - \; \frac{1}{2}[\rho - \rho^*]g \int_0^{L_0} Z^2 \, dx$$

which holds for arbitrary functions Z satisfying specified end conditions and the integral condition

$$\int_0^{L_0} Z \, dx \; = \; 0$$

To draw some conclusions from this expression for the increase in the potential energy, let the amplitude of the displacement be small enough that $[1 + [Z']^2]^{1/2}$ can be estimated to sufficient accuracy by $1 + \frac{1}{2}[Z']^2$, whereupon the increase in the potential energy can be obtained as

$$\frac{1}{2} \gamma \int_0^{L_0} \left[[Z']^2 - \lambda^2 Z^2 \right] \, dx$$

where $\lambda^2 = \dfrac{[\rho - \rho^*]g}{\gamma} > 0$.

Now, integrate the first term by parts and require the end conditions to be fixed or free so that the term ZZ' vanishes at both ends. By this, the potential energy increase on displacement turns out to be

$$-\frac{1}{2} \gamma \int_0^{L_0} Z[Z'' + \lambda^2 Z] \, dx$$

It is easy to see how to find displacements to which the potential energy is neutral (i.e., to which its change vanishes). To any displacement satisfying

$$\frac{d^2 Z}{dx^2} \; + \; \lambda^2 Z = 0$$

as well as the end conditions of interest, the potential energy cannot change and the corresponding value of λ^2 must be a possible critical value.

Notice, first, that had ρ been less than ρ^* and, hence, had λ^2 been negative, no solutions to this equation could have been found, other than $Z = 0$, and no critical values would have been found. This is not surprising.

Then, notice that not all solutions to this equation need satisfy

$$\int_0^{L_0} Z \, dx \; = \; 0$$

only those such that

$$Z'(x_0 = L_0) - Z'(x_0 = 0) \; = \; 0$$

can meet the fixed-volume requirement.

Now, to find displacements that are neutral and λ^2's that may be critical, write

$$Z = A \cos \lambda x + B \sin \lambda x$$

which satisfies

$$\frac{d^2 Z}{dx^2} + \lambda^2 Z = 0$$

and then introduce the end conditions. Turn first to the case of fixed-fixed end conditions. In this case, A must be zero and, by this, the interesting values of λL_0 are the positive roots of $\sin \lambda L_0 = 0$. The candidates for $\lambda^2_{critical}$ are then π^2/L_0^2, $4\pi^2/L_0^2$, $\cdots$. The least of these is π^2/L_0^2 and this comes as something of a surprise due to the fact that by solving the perturbation equations, it is already known that $\lambda^2_{critical}$ must be $4\pi^2/L_0^2$. But nowhere has the fixed-volume requirement been enforced, and to see what it tells us, notice that

$$\int_0^{L_0} \sin \frac{\pi x}{L_0} \, dx = \frac{2L_0}{\pi}$$

whereas

$$\int_0^{L_0} \sin \frac{2\pi x}{L_0} \, dx = 0$$

and by this, the critical value of λ^2 must then be $4\pi^2/L_0^2$ at constant volume, which is indeed correct. Surprisingly, the two layers of fluid might be thought to be unstable, at a lower value of λ^2, to a displacement which requires the addition of fluid to the lower layer.

It is here, and only here, that the fixed-volume requirement plays an important role in this problem. This is quite unlike its role in the jet problem.

Turning now to fixed-free end conditions, again A must be zero, but now the values of λL_0 must be the positive roots of $\cos \lambda L_0 = 0$, whereupon $\lambda L_0 = \frac{1}{2}\pi$, $\frac{3}{2}\pi$, $\cdots$. For none of these values of λ is the corresponding displacement volume neutral; in fact, what we find is

$$\int_0^{L_0} \sin \lambda x \, dx = -\frac{1}{\lambda}[\cos \lambda L_0 - 1] = \frac{1}{\lambda}$$

In every case it is potential-energy neutral. Rayleigh's work principle, then, turns up a large number of potential-energy neutral displacements; yet, in this case, none of them are volume neutral.

In working out the stability of a jet using the work principle, it was easy to include the fixed-volume requirement in the calculation due to the presence in the jet problem of a geometrical variable, the diameter of the jet, whose value can be determined only by bringing in an outside condition. To set the diameter, the fixed-volume requirement was used which means that the surface of the jet was displaced by imposing a kick rather than by adding fluid. There seems to be no comparably easy way to get a requirement such as fixed volume into the problem of this essay where two layers are divided by a planar interface.

6.7.4   The Effect of Viscosity

To take the viscosities of the two fluids into account, first look back at Figure 6.2 in the essay, drawn to indicate free-free end conditions where the interface dividing the two fluids comes into contact with the side walls. It presents a problem. The no-slip condition, which the fluids themselves now must satisfy at these walls, is not simply reconciled with the motion of the contact line, the line where the interface touches the walls. Yet, if the end conditions were fixed instead of free, no such technical difficulty would present itself and it is easy to run an experiment at fixed end conditions. The trick is to introduce a sharp edge. Then, it is an experimental fact that an interface can take up residence at this edge and not depart from it. This is illustrated by a sketch and a photograph in Figure 6.1 in the essay. In fact, this experiment is so nice that the reference interface dividing the fluids at rest in the base case can be made as flat as need be by controlling the pressure above the heavy fluid.

Now, the critical width of two fluid layers, the heavier layer lying above the lighter, will depend on the end conditions, just as does the critical length of a liquid bridge. In case the end conditions are fixed-fixed, the critical width,[16] scaled[17] by $\sqrt{\dfrac{\gamma}{g[\rho - \rho^*]}}$, is 2π, that is,

$$\frac{L_{0_{critical}}}{\sqrt{\dfrac{\gamma}{g[\rho - \rho^*]}}} = 2\pi$$

not π as in the case of free-free end conditions. It is this critical condition that can be proved, by elementary methods, to be independent of viscosity.

Figure 6.8 is presented to remind the readers what the problem is. It is an idealization of an experiment, but at neutral conditions this will not matter. The perturbation equations are defined on the reference configuration and only the surface dividing the two fluids can be displaced. All other surfaces are walls upon which the no-slip condition is required to hold, while the surface dividing the two fluids must now satisfy a fixed end condition where it meets the no-slip walls. These conditions on $\vec{v}_1$, $\vec{v}_1^*$ and Z_1 at the rigid walls will not be restated later, but they will be used in the calculation.

The nonlinear equations, taking into account the viscosities of the two fluids, have the simple solution

$$\vec{v}_0 - \vec{0}$$

$$p_0 = -\rho g z_0$$

---

[16]This result is obtained in the third discussion note.

[17]The length scale in the liquid bridge problem is R_0. Here, it is $\sqrt{\dfrac{\gamma}{g[\rho - \rho^*]}}$, not H_0 or H_0^*. Again, H_0 and H_0^* do not play the role here that R_0 plays there.

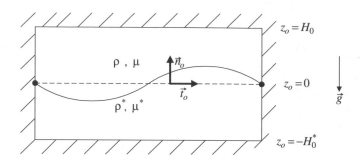

Figure 6.8. *An Idealization of an Experiment: Fixed End Conditions*

$$\vec{v}_0^* = \vec{0}$$

$$p_0^* = -\rho^* g z_0$$

and

$$Z_0 = 0$$

and it is this solution, which is as it was before, when viscosity was not taken into account, whose stability is to be determined. The reference surface then is the surface $z_0 = Z_0 = 0$, $0 < x_0 < L_0$. It is denoted S_0.

Due to the fact that the fluids are at rest in their base state, the perturbation equations are quite simple. The eigenvalue problem is equally simple, coming from the perturbation equations on introducing the time dependence $e^{\sigma t_0}$. On the reference domain, it is

$$\rho \sigma \vec{v}_1 = -\nabla_0 p_1 + \mu \nabla_0^2 \vec{v}_1 \tag{6.39}$$

and

$$\nabla_0 \cdot \vec{v}_1 = 0 \tag{6.40}$$

on V_0,

$$\rho^* \sigma \vec{v}_1^* = -\nabla_0 p_1^* + \mu^* \nabla_0^2 \vec{v}_1^* \tag{6.41}$$

and

$$\nabla_0 \cdot \vec{v}_1^* = 0 \tag{6.42}$$

on V_0^* and

$$\vec{n}_0 \cdot \vec{v}_1 = \sigma Z_1 \tag{6.43}$$

$$\vec{n}_0 \cdot \vec{v}_1^* = \sigma Z_1 \tag{6.44}$$

$$\vec{t}_0 \cdot \vec{v}_1 = \vec{t}_0 \cdot \vec{v}_1^* \tag{6.45}$$

$$p_1 - p_1^* + \left[\frac{\partial p_0}{\partial z_0} - \frac{\partial p_0^*}{\partial z_0} \right] Z_1 \; - 2\mu \vec{n}_0 \vec{n}_0 : \vec{\vec{D}}_1 + 2\mu^* \vec{n}_0 \vec{n}_0 : \vec{\vec{D}}_1^*$$

$$= \gamma \frac{d^2 Z_1}{dx_0^2} \tag{6.46}$$

and

$$-2\mu \vec{t}_0 \vec{n}_0 : \vec{\vec{D}}_1 + 2\mu^* \vec{t}_0 \vec{n}_0 : \vec{\vec{D}}_1^* = 0 \tag{6.47}$$

on S_0, where $\vec{n}_0 = \vec{k}$, $\vec{t}_0 = \vec{i}$ and where taking account of the viscosities of the two fluids introduces the third and fifth equations on S_0 and the additional terms in the fourth equation.

Again, the fact that $\vec{v}_0 = \vec{0} = \vec{v}_0^*$ and hence that $\vec{\vec{D}}_0 = \vec{0} = \vec{\vec{D}}_0^*$ leads to these simple equations. The base solution turns up only via $\left| \frac{\partial p_0}{\partial z_0} - \frac{\partial p_0^*}{\partial z_0} \right|$ and this is $-[\rho - \rho^*]g$ whether or not the viscosities of the fluids are included in the calculation.

To draw some conclusions about the solution to this eigenvalue problem, multiply equation (6.39) by $\vec{v}_1 \cdot$ and equation (6.41) by $\vec{v}_1^* \cdot$ and use equations (6.40) and (6.42). Then, integrate each product over its part of the reference domain and add the two results. In doing this, all the terms evaluated at the rigid walls vanish and there obtains

$$\sigma \left[\rho \int_{V_0} dV_0 \, |\vec{v}_1|^2 \; + \rho^* \int_{V_0^*} dV_0 \, |\vec{v}_1^*|^2 \right]$$

$$= \int_{S_0} dA_0 \, \vec{n}_0 \cdot \left[\vec{v}_1 p_1 - \vec{v}_1^* p_1^* \right] \; - 2\mu \int_{V_0} dV_0 \, \vec{\vec{D}}_1 : \vec{\vec{D}}_1$$

$$- 2\mu^* \int_{V_0^*} dV_0^* \, \vec{\vec{D}}_1^* : \vec{\vec{D}}_1^* - 2\mu \int_{S_0} dA_0 \, \vec{n}_0 \cdot \vec{\vec{D}}_1 \cdot \vec{v}_1$$

$$+ 2\mu^* \int_{S_0} dA_0 \, \vec{n}_0 \cdot \vec{\vec{D}}_1^* \cdot \vec{v}_1^* \tag{6.48}$$

The signs can be accounted for by the fact that $\vec{n}_0$ points out of V_0^* into V_0. Now, equations (6.43), (6.44), (6.45) and (6.47), on S_0, along with the formulas

$$\vec{v}_1 = \vec{n}_0 [\vec{n}_0 \cdot \vec{v}_1] + \vec{t}_0 [\vec{t}_0 \cdot \vec{v}_1]$$

and

$$\vec{v}_1^* = \vec{n}_0 [\vec{n}_0 \cdot \vec{v}_1^*] + \vec{t}_0 [\vec{t}_0 \cdot \vec{v}_1^*]$$

can be used to simplify the surface integrals on the right-hand side of equation (6.48). Doing this, there obtains

$$\sigma \left[\rho \int_{V_0} dV_0 \; |\vec{v}_1|^2 \; + \rho^* \int_{V_0^*} dV_0^* \; |\vec{v}_1^*|^2 \right]$$

$$= \bar{\sigma} \int_{S_0} dA_0 \; \bar{Z}_1 [p_1 - p_1^*] - 2\mu \int_{V_0} dV_0 \; \vec{\vec{D}}_1 : \vec{\vec{D}}_1 - 2\mu^* \int_{V_0^*} dV_0^* \; \vec{\vec{D}}_1^* : \vec{\vec{D}}_1^*$$

$$-2\mu \int_{S_0} dA_0 \; \vec{n}_0 \vec{n}_0 : \vec{\vec{D}}_1 \bar{\sigma} \bar{Z}_1 + 2\mu^* \int_{S_0} dA_0 \; \vec{n}_0 \vec{n}_0 : \vec{\vec{D}}_1^* \bar{\sigma} \bar{Z}_1 \qquad (6.49)$$

whence the integrals over S_0 can be added and equation (6.46) used to produce

$$\int_{S_0} dA_0 \; \bar{\sigma} \bar{Z}_1 \left[\gamma \frac{d^2 Z_1}{dx_0^2} + [\rho - \rho^*] g Z_1 \right]$$

which eliminates the viscosities of the fluids from the integrals over the dividing surface.

Then, introducing the shorthand notation

$$[KE] = \rho \int_{V_0} dV_0 \; |\vec{v}_1|^2 \; + \rho^* \int_{V_0^*} dV_0^* \; |\vec{v}_1^*|^2$$

and

$$[VISC] = 2\mu \int_{V_0} dV_0 \; \vec{\vec{D}}_1 : \vec{\vec{D}}_1 \; + 2\mu^* \int_{V_0^*} dV_0^* \; \vec{\vec{D}}_1^* : \vec{\vec{D}}_1^*$$

where both of these new variables are real and not negative, equation (6.49) can be written

$$\sigma[KE] = -[VISC] + \bar{\sigma} \int_{S_0} dA_0 \; \bar{Z}_1 \left[\gamma \frac{d^2 Z_1}{dx_0^2} + [\rho - \rho^*] g Z_1 \right] \qquad (6.50)$$

Now, S_0 is the interval $0 < x_0 < L_0$ and $dA_0 = dx_0$ when the problem is made two dimensional. This being the case, equation (6.50) can be simplified by integrating the first term by parts and introducing fixed end conditions. The result is

$$\sigma[KE] = -[VISC] + \bar{\sigma} \int_0^{L_0} dx_0 \left[-\gamma \left| \frac{dZ_1}{dx_0} \right|^2 \; + [\rho - \rho^*] g |Z_1|^2 \right] \qquad (6.51)$$

To see what this tells us, put

$$\sigma \; = \; a \; + \; ib$$

and resolve equation (6.51) into its real and imaginary parts. This leads to

$$a[KE] \; = \; - [VISC] \; - \; a[PE] \qquad (6.52)$$

and

$$b[KE] \; = \; b[PE] \qquad (6.53)$$

where

$$[PE] = -\int_0^{L_0} dx_0 \left[-\gamma \left| \frac{dZ_1}{dx_0} \right|^2 + [\rho - \rho^*] g |Z_1|^2 \right]$$

and $[PE]$ accounts for all the potential energy terms due to the surface deflection.

These are not equations determining the eigenvalue σ (i.e., a and b), because the integrals cannot be evaluated unless the corresponding eigenfunction is available, yet some conclusions can be drawn.

First, let $\mu = 0 = \mu^*$, which corresponds to two inviscid liquids. Then, equations (6.52) and (6.53) reduce to

$$a[KE] = -a[PE]$$

and

$$b[KE] = b[PE]$$

whence a and b cannot both be non zero and σ must be either real or purely imaginary. If $b = 0$ and $a \neq 0$ is to be a solution to the eigenvalue problem, then $[PE]$ must be non positive and this is impossible if $\rho < \rho^*$. If, on the other hand, $b \neq 0$ and $a = 0$ is to be a solution, then $[PE]$ must be non negative. This is certain if $\rho < \rho^*$, and it is also possible if $\rho > \rho^*$. To make it so, write

$$[PE] = \gamma \int_0^{L_0} dx_0 \left| \frac{dZ_1}{dx_0} \right|^2 - [\rho - \rho^*] g \int_0^{L_0} dx_0 \, |Z_1|^2$$

and use

$$\int_0^{L_0} dx_0 \left| \frac{dZ_1}{dx_0} \right|^2 \geq \lambda_1^2 \int_0^{L_0} dx_0 \, |Z_1|^2$$

to conclude that

$$[PE] \geq [\lambda_1^2 \gamma - [\rho - \rho^*] g] \int_0^{L_0} dx_0 \, |Z_1|^2$$

where λ_1^2 is the least eigenvalue of $-d^2/dx_0^2$ corresponding to fixed end conditions. Now, λ_1^2 is a multiple of $1/L_0^2$, and by making L_0 small enough, it becomes certain that $[PE]$ is not negative. By so limiting L_0 to small enough values, only solutions such that $b \neq 0$ and $a = 0$ are possible and then all displacements of the dividing surface are stable (i.e., an adverse density gradient can be stabilized by surface tension if long wavelength displacements are ruled out). Again, we learn that the value of γ is important to the critical width.

Now, turn to the case where μ and μ^* are not both zero and let $[VISC]$, and hence $[KE]$, be greater than zero. Then, a and b must satisfy

$$a[KE] = -[VISC] - a[PE]$$

and

$$b[KE] = b[PE]$$

First, let b be other than zero. Then, $[PE]$ must be non negative and

$$a = \frac{-[VISC]}{[KE] + [PE]}$$

whence a must be non positive. Again, as above, this is always so if $\rho < \rho^*$ and it can be made so if $\rho > \rho^*$ by requiring L_0 to be small enough. This establishes a stable range of widths in the face of an adverse density gradient. Now, by this, if b is not zero, then a cannot be zero whence at a critical point, where a must be zero, so too σ must be zero. If $\sigma = 0$ (i.e., if $a = 0 = b$), then $[VISC]$ must be zero and, in the light of the requirements at the rigid wall, $\vec{v}_1$ and $\vec{v}_1^*$ must be zero. The critical point is then a rest point; hence, the solutions to the eigenvalue problem at the critical point must be the same whether or not the viscosity of either or both fluids is zero. By this, the critical condition does not depend on the viscosities of the two fluids and, hence, it can be determined via inviscid calculations.

6.7.5   The Three-Dimensional Case

In this discussion note, Rayleigh's work principle is used to find out what the two-dimensional calculations presented in the essay tell us about three-dimensional problems.

Let two fluids lie one above the other in a gravitational field. The two layers are bounded above and below by the planes $z = H_0$ and $z = -H_0^*$. Their densities are denoted ρ and ρ^* and γ denotes the tension acting in the surface separating the one from the other. The equilibrium interface is planar and can be taken to be the domain D of the $x, y-$ plane. Figure 6.9 illustrates this.

A displacement of this interface is then denoted by

$$z = f(x, \ y)$$

where the point $(x, \ y)$ lies in D and we must determine the increase in both the surface and the gravitational potential energy corresponding to this displacement.

Turn first to the area of the displaced surface. Let x and y be surface coordinates. Then, in the notation of Appendix B, we have

$$\vec{r}_1 = \vec{i} + f_x \vec{k}$$

and

$$\vec{r}_2 = \vec{j} + f_y \vec{k}$$

whence

$$g_{11} = 1 + f_x^2$$

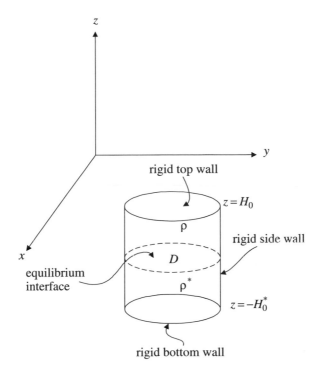

Figure 6.9. *Rayleigh–Taylor Instability: The Three-Dimensional Case*

$$g_{22} = 1 + f_y^2$$

and

$$g_{12} = f_x f_y$$

Hence, the area of the displaced surface is

$$\iint\limits_{D} [g_{11}g_{22} - g_{12}^2]^{1/2} \, dx \, dy = \iint\limits_{D} [1 + f_x^2 + f_y^2]^{1/2} \, dx \, dy$$

and, for small displacements, the increase in the surface area on displacement is, to sufficient accuracy,

$$\frac{1}{2} \iint\limits_{D} [f_x^2 + f_y^2] \, dx \, dy$$

The gravitational potential energy in the displaced configuration is

$$\iint_D dx\,dy\left[\int_{-H_0^*}^{f(x,\,y)} g\rho^* z\,dz + \int_{f(x,\,y)}^{H_0} g\rho z\,dz\right]$$

$$= \iint_D dx\,dy\left[\frac{1}{2} g\rho^*[f^2 - H_0^{*2}] + \frac{1}{2} g\rho[H_0^2 - f^2]\right]$$

whence its increase on displacement is

$$\iint_D \frac{1}{2} g[\rho^* - \rho]f^2\,dx\,dy$$

This requires no approximation and is independent of H_0 and H_0^*.

The work required to carry out the displacement is then the increase in the potential energy and this is

$$\frac{1}{2}\iint_D dx\,dy\left[\gamma[f_x^2 + f_y^2] + g[\rho^* - \rho]f^2\right]$$

The first term is always positive; the second is positive or negative as $\rho^* - \rho$ is positive or negative.

To illustrate the use of this formula in a simple way, let D be a rectangle of sides L_0 and W_0. Then, taking f to be

$$\epsilon \cos k_x x \, \cos k_y y$$

where

$$k_x = \frac{\pi}{L_0} m$$

and

$$k_y = \frac{\pi}{W_0} n$$

where $m, n = 0, 1, 2, \cdots$ and where either m or n may be zero, but not both, the work required is

$$\frac{1}{2}\epsilon^2 \int_0^{L_0}\int_0^{W_0} dx\,dy\left[\gamma k_x^2 \sin^2 k_x x \cos^2 k_y y + \gamma k_y^2 \cos^2 k_x x \sin^2 k_y y\right.$$

$$\left. + g[\rho^* - \rho]\cos^2 k_x x \cos^2 k_y y\right]$$

On carrying out the integration, this is

$$\frac{1}{2}\epsilon^2 \frac{L_0}{2}\frac{W_0}{2}\left[\gamma[k_x^2 + k_y^2] + g[\rho^* - \rho]\right]$$

where $k_x^2 = m^2\dfrac{\pi^2}{L_0^2}$ and $k_y^2 = n^2\dfrac{\pi^2}{W_0^2}$.

The two layers are stable to all displacements so long as $\rho^* - \rho > 0$. But when $\rho^* - \rho < 0$, they may be unstable to certain displacements if L_0 and W_0 are large.

Let $\rho^* - \rho$ be less than zero (i.e., heavy over light). Then, the layers will be stable to displacements corresponding to large values of m and n. If $L_0 > W_0$, the most dangerous displacement corresponds to $m = 1$, $n = 0$. If the layers are stable to this displacement, they are stable to all displacements, including $m = 0$, $n = 1$. The critical value of L_0 is then determined by

$$\gamma \frac{\pi^2}{L_0^2} + g[\rho^* - \rho] = 0$$

or

$$L_0 = \pi^2 \frac{\gamma}{g[\rho - \rho^*]}$$

This, of course, is the result obtained in the essay. The interface is stable, when $L_0 > W_0$, so long as $L_0 < \pi^2 \frac{\gamma}{g[\rho - \rho^*]}$.

6.7.6 What Rayleigh's Work Principle Tells Us About Three Fluid Layers

In this discussion note, a calculation is presented that is reminiscent of an earlier calculation made to decide the stability of a two-fluid jet. It, too, is inconclusive.

In this case, three fluid layers of densities ρ, ρ^* and ρ^{**} lie one over another in a gravitational field as illustrated in Figure 6.10. The small displacements of the surfaces indicated in the figure must maintain the volumes of the layers at their undisplaced values.

Again, this is a two-dimensional calculation, the displacement of each surface depending only on x, not on y. The surface potential energy as well as the gravitational potential energy are then determined per unit depth into the page. The width of the layers is denoted by L_0.

So, per unit depth, the area of a displaced surface is

$$\int_0^{L_0} ds = \int_0^{L_0} \left[1 + \left[\frac{dz}{dx} \right]^2 \right]^{1/2} dx$$

which to sufficient accuracy, taking the displacement to be small, is

$$\int_0^{L_0} \left[1 + \frac{1}{2} \left[\frac{dz}{dx} \right]^2 \right] dx$$

whence the area increase is

$$\frac{1}{2} \int_0^{L_0} \left[\frac{dz}{dx} \right]^2 dx$$

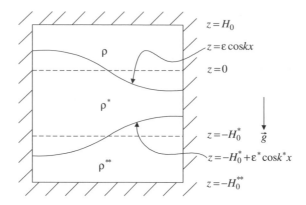

Figure 6.10. *Rayleigh–Taylor Instability: Three-Fluid Layers*

which is positive, hence stabilizing.

This leads to an expression for the increase in the surface potential energy when the small displacements indicated in Figure 6.10 are imposed on the surfaces $z = -H_0^*$ and $z = 0$. It is

$$\frac{1}{4} L_0 \left[\epsilon^{*2} \gamma^* k^{*2} + \epsilon^2 \gamma k^2 \right]$$

and the amount of the increase depends upon the values of ϵ, k, ϵ^* and k^*.

To this must be added the increase in the gravitational potential energy. Per unit depth, the gravitational potential energy in the displaced configuration is

$$g \int_0^{L_0} dx \left[\int_{-H_0^{**}}^{-H_0^*+\epsilon^* \cos k^* x} \rho^{**} z \, dz + \int_{-H_0^*+\epsilon^* \cos k^* x}^{\epsilon \cos kx} \rho^* z \, dz \right.$$

$$\left. + \int_{\epsilon \cos kx}^{H_0} \rho z \, dz \right]$$

whence its increase is

$$\frac{1}{4} L_0 \left[g[\rho^{**} - \rho^*] \epsilon^{*2} + g[\rho^* - \rho] \epsilon^2 \right]$$

This requires no approximation and is independent of k^2 and k^{*2}.

The work required is then

$$\frac{1}{4} L_0 \left[\left[g[\rho^{**} - \rho^*] + \gamma^* k^{*2} \right] \epsilon^{*2} + \left[g[\rho^* - \rho] + \gamma k^2 \right] \epsilon^2 \right]$$

and this is the sum of what would obtain if each surface were to be displaced by itself, the other remaining planar.

In this calculation, ϵ and ϵ^*, as well as k^2 and k^{*2}, appear to be variables whose values can be freely assigned, except that k^2 and k^{*2} must fit in with the value of L_0 in the usual way.

The first thing to observe is that the depths of the layers play no role in this problem. The variables H_0, H_0^* and H_0^{**} are not important. They are not like the variables R_0 and R_0^* in the jet problem.

Some conclusions can be drawn: If both terms are positive, the system is stable to the proposed displacements; if both are negative, it is unstable. Both will be positive for all k^2 and k^{*2} if and only if $\rho^* > \rho$ and $\rho^{**} > \rho^*$. If $\rho^* < \rho$, the first term will be negative for low values of k and positive for high values of k. If $\rho^{**} < \rho^*$, the second term will be negative for low values of k^* and positive for high values of k^*.

Another conclusion would seem to be this: An upper surface, unstable by itself to displacements corresponding to some range of wave number, could be stabilized by introducing a second surface below it, stable to all displacements, and then exercising a small amount of control. Any reader who believes this needs to move on and carry out a perturbation calculation. Every reader ought to speculate whether the two surfaces remain coupled in such a calculation or whether the perturbation problem splits into two single-surface calculations.[18]

If the readers carry out the suggested perturbation calculation, they will learn at least one important fact. They would have learned this already had they done the two-fluid jet calculation suggested earlier and they will learn it once again in the ninth essay. It is this: In any two-surface problem, whenever σ^2 is zero, one surface will be displaced but the other will not and whenever σ^2 is not zero, both surfaces will be displaced.

This said, it might be of some interest to speculate on the long-time arrangement of three fluids of densities ρ, ρ^* and ρ^{**}, initially lying as in Figure 6.10, where

$$\rho > \rho^*, \quad \rho > \rho^{**}, \quad \rho^{**} > \rho^*$$

and where the width of the layers is set so that the most dense fluid lying over the less dense fluid would be a stable arrangement but the most dense fluid lying over the least dense fluid would not.

6.7.7 The Loss of Stability When an Oscillation Is Imposed on Two Fluids: A Light Fluid Lying Above a Heavy Fluid

Let two liquids lie in a stable configuration, light over heavy. Then, in view of Taylor's observation presented in the essay, it might be thought that this stability could be overturned by imposing an oscillatory acceleration on the two fluids in the direction of gravity. That it is possible to do this

---

[18]By doing this, the reader will find that the problem does not split, that k and k^* must be equal, and that ϵ and ϵ^* are not independent.

in the case of a pendulum at rest in its stable position adds encouragement to this speculation.

Introducing a simple harmonic motion of the entire system parallel to the direction of gravity and then describing the motion of the fluids in terms of an observer to which this motion is invisible does not add much to the equations used in the essay.

To make the changes required, d/dt must be written in place of σ and the contribution of the harmonic acceleration to the base pressures, p_0 and p_0^*, must be taken into account. Doing this, the equations determining A, A^* and $\hat{Z}_1$, where $\hat{Z}_1$ is now written Z_1 and where A, A^* and $\hat{Z}_1$ retain their earlier meaning, turn out to be

$$-Ak \sinh kH_0 = -\rho \frac{d^2 Z_1}{dt^2}$$

$$A^* k \sinh kH_0^* = -\rho^* \frac{d^2 Z_1}{dt^2}$$

and

$$A \cosh kH_0 - A^* \cosh kH_0^* = \left[-a(t)[\rho - \rho^*] + g[\rho - \rho^*] - \gamma k^2 \right] Z_1$$

where $a(t)$ denotes the acceleration imposed on the two fluids and where ρ is less than ρ^*.

This leads to

$$\frac{d^2 Z_1}{dt^2} + \frac{\gamma k^2 - [\rho - \rho^*]g + a(t)[\rho - \rho^*]}{\rho \coth kH_0 + \rho^* \coth kH_0^*} k Z_1 = 0$$

and hence to

$$\frac{d^2 Z_1}{dt^2} + \omega_0^2 Z_1 + \delta \cos(\omega t) Z_1 = 0$$

on introducing ω_0^2 and $\delta \cos \omega t$ via

$$\omega_0^2 = \frac{\gamma k^2 - [\rho - \rho^*]g}{\rho \coth kH_0 + \rho^* \coth kH_0^*} k$$

and

$$\delta \cos \omega t = \frac{a(t)k}{\rho \coth kH_0 + \rho^* \coth kH_0^*} [\rho - \rho^*]$$

Then, on writing

$$Z_1 = Z_{10} + \delta Z_{11} + \frac{1}{2} \delta^2 Z_{12} + \cdots$$

there obtains

$$\frac{d^2 Z_{10}}{dt^2} + \omega_0^2 Z_{10} = 0$$

$$\frac{d^2 Z_{11}}{dt^2} + \omega_0^2 Z_{11} + \cos \omega t \; Z_{10} = 0$$

etc.

and using

$$\cos \omega t \; \cos \omega_0 t = \frac{1}{2} \Big[\cos[\omega + \omega_0]t + \cos[\omega - \omega_0]t \Big]$$

resonance can be found at first order if

$$\omega = 2\omega_0$$

This is an indication that an otherwise stable configuration can be destabilized by introducing small oscillations. It might be noticed that resonant frequencies can continue to be found at every order beyond the first.

7
Solidification

This essay retains the flavor of the earlier essays insofar as displaced surfaces remain important. But now, the surfaces of interest divide two phases of a pure material: a low-temperature phase and a high-temperature phase. In the case of interest, an advancing solidification front undergoes a displacement to which it may or may not be stable. Our job in this essay is to determine whether or not a solidification front, if it is planar, can remain planar in the face of a small disturbance.

The simplest case, and the one of interest to us, occurs when solidification is controlled entirely by diffusion.

In some ways, the solidification problem is like the Saffman–Taylor problem introduced at the end of the sixth essay. In that problem where one fluid displaces another, the fluids retain their identity, the one invading the other's space. This problem is a little different; now one phase turns into the other, and as liquid turns into solid, the speed of the solidification front at any point depends on how fast the latent heat released there can be conducted through one phase or the other before being rejected to the surroundings. But again, the speed and direction of the front determine its stability.

Figure 7.1 illustrates two ways of rejecting this heat. In both cases, a liquid lies in a vessel whose walls are held at a temperature below the freezing point of the liquid. Heat is removed at these walls. In the left-hand sketch, the liquid is above its freezing point; solid forms at the walls and grows into the liquid. The latent heat is conducted through the solid and rejected to the walls. This is a stable configuration; the solidification front moves smoothly toward the center of the container. In the right-

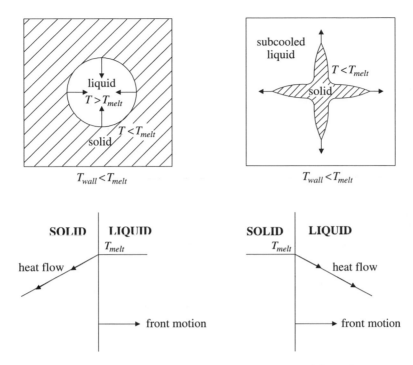

Figure 7.1. *Two Ways of Rejecting the Latent Heat of Solidification*

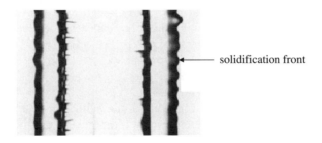

a freezing cylinder of ice whose surface is shown at two times

Figure 7.2. *A Photograph by Hardy and Coriell Illustrating the Instability*
Reprinted from Journal of Crystal Growth, Vol. 3, No. 4, S. C. Hardy and S. R.
Coriell, "Morphological Stability and the Ice-Water Interfacial Free Energy", p.
569, (1968), with permission from Elsevier Science

hand sketch the liquid is subcooled[1] to a temperature below its freezing point. Solidification then begins upon the introduction of a seed crystal at the center of the container. The latent heat is now conducted through the liquid before being rejected to the walls. This is an unstable configuration. Instead of moving smoothly outward, the solidification front takes on a snowflake-like shape with dendrites growing rapidly in several directions from the seed particle. Figure 7.2 reproduces a photograph by Hardy and Coriell illustrating this instability.

7.1   The Physics of the Instability

It is possible to form some idea why rejecting the latent heat in one way should be stable, but rejecting it in the other may not. Let the solid lie to the left, the liquid to the right, as illustrated in Figure 7.3. Let the two phases be divided by a planar interface. The solid advances to the right as liquid turns into solid, releasing its latent heat. In the upper sketches, this heat is conducted through the solid; in the lower sketches, it is conducted through the liquid. Both of the sketches in the left half of Figure 7.3 indicate a displacement of the planar front to the right, a local region where the speed of the front runs above the speed of the planar front. To sustain this increased speed, the rate at which latent heat is rejected must also increase. In the upper left sketch, the displacement weakens the temperature gradient available to do this, in the lower left sketch, it strengthens it. A local increase in the speed of the front then acts to oppose itself in the upper sketch; it acts to reinforce itself in the lower sketch. This is also the case if the displacement were to the left (viz., a local region where the speed runs below the speed of the planar front). In this case, the displacement acts to strengthen the temperature gradient in the upper right sketch, reversing the decrease in speed, while it weakens the temperature gradient in the lower right sketch, reinforcing the slowing down.

The two upper sketches predict that a planar front must surely be stable in case the latent heat is rejected to the solid. Then, a displacement of the front to the right produces a new temperature gradient insufficient to support even its original speed, while a displacement to the left produces a new temperature gradient more than sufficient to support its original speed. The two lower sketches predict only that a front rejecting heat to the liquid may be unstable; the reinforcement may or may not be sufficient to sustain the displacement.

The sketches in Figure 7.3 do not include the effect of surface tension on the stability of the front. It is stabilizing. It acts in this problem to adjust

---

[1] The term *subcooled liquid* denotes liquid cooled below its freezing point. The term *superheated solid* will denote solid heated above its melting point.

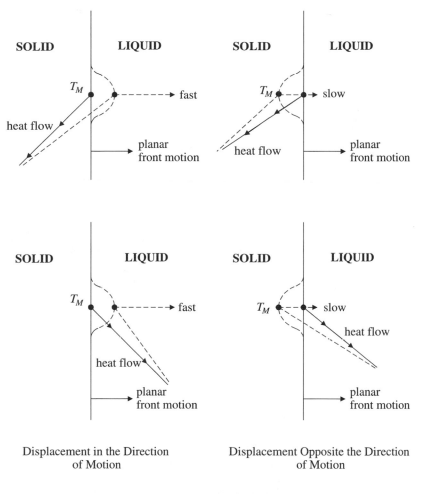

Figure 7.3. *Displacements of a Solidification Front*

the melting point of the solid. Although the interface is certainly taken to be at equilibrium, the equilibrium temperature depends on the shape of the interface.

Let T_M denote the melting point of a planar solid surface; then, the melting temperature is reduced at a crest where the solid projects into the liquid, whereas it is increased at a trough where the liquid projects into the solid. The formula for this, worked out in the first endnote, is

$$T = T_M + \frac{\gamma T_M}{\mathcal{L}} 2H$$

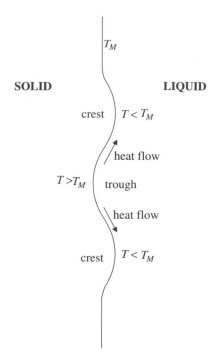

Figure 7.4. *One Stabilizing Effect of Surface Tension*

where γ denotes the surface tension, $\mathcal{L}$ denotes the latent heat released on solidification and $2H$ denotes twice the mean curvature of the surface, taken to be negative if the solid projects into the liquid (i.e., at a crest the curvature is negative, at a trough, it is positive).

The surface tension exercises its influence on the stability of a solidification front in two ways. The first is illustrated in Figure 7.4, where a wavy displacement of the interface is indicated. At a crest, the temperature falls below T_M; at an adjacent trough it rises above T_M. The faster-moving crest needs to reject heat faster than the slower-moving trough, yet the transverse temperature gradient induced by the displacement carries heat out of the trough into the crest, speeding up the trough while slowing down the crest. This is stabilizing, but how important it is depends on the wavelength of the displacement. For short wavelengths, the effect is strong; for long wavelengths, it is weak. This then is much like the effect of surface tension seen in all the earlier essays. It acts most strongly to stabilize the surface to its shortest wavelength displacements.

The adjustment of the melting point required by surface tension also plays a stabilizing role via its modification of the conclusions drawn from Figure 7.3. Redrawing that figure, now taking into account the effect of surface tension on the melting point, produces Figure 7.5, which indicates

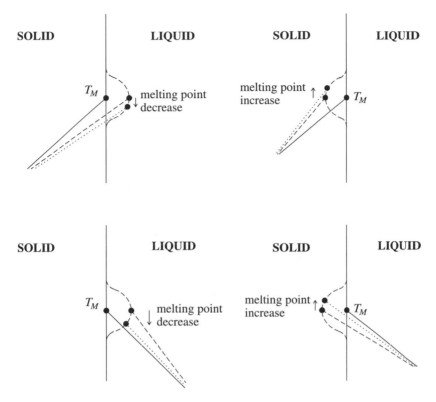

Figure 7.5. *Figure 7.3 Redrawn to Take Account of Surface Tension*

that a displacement to the right, a crest, weakens the temperature gradient more than before if it is in the solid, whereas it does not strengthen it as much as before, and may even weaken it, if it is in the liquid. Again, the strength of the stabilizing effect depends on the wavelength of the disturbance. By the same sort of reasoning, surface tension is also stabilizing if the displacement is to the left, a trough, for then the speed of the front is reduced, but the surface temperature is increased.

What we ought to expect, then, on taking surface tension into account, is just what we have found before: A displacement, otherwise unstable, will be stabilized by surface tension if it is of short wavelength, but not if it is of long wavelength.

There is no need to work through the case where the solid is melting and the front is moving into the solid. In this case, latent heat must be supplied and this can be done either by a temperature gradient in the liquid or by a temperature gradient in the solid. The two cases are illustrated in Figure 7.6, where the right-hand sketch corresponds to superheating the solid. The same picture reasoning as above would lead to the conclusion

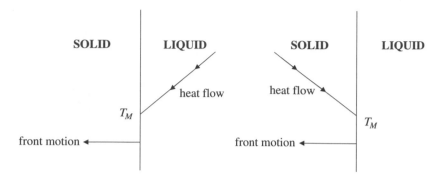

Figure 7.6. *Two Ways of Supplying the Latent Heat of Melting*

that at the left is a picture of a stable melting front, while at the right is a picture of a possibly unstable melting front.

The pictures in the two cases, solidification and melting, indicate that instability corresponds either to heat rejection to a subcooled liquid or to heat supplied by a superheated solid.

7.2   Solidification: Rejecting the Latent Heat to the Solid

7.2.1   The Nonlinear Equations

To make this problem simple enough to work out by hand, the densities of the two phases will, in the end, be taken to be equal and hence no flow will be induced by the front motion in either the base or the perturbation problems. Likewise, no account will be taken of the effect of surface curvature on anything beyond the melting point of the solid. The latent heat, for instance, will be that corresponding to a planar interface. The temperature dependence of the thermophysical properties will be ignored. Taking into account a changing latent heat turns out to be interesting, but doing this will be put off to the end of the essay, where it will serve as a bridge to the problem of precipitation taken up in the next essay.

Two cases are of interest. In the first case, the latent heat is rejected to the solid. To make this definite, take the solid to be at rest in the laboratory frame and let the liquid lying a great distance from the front be at its ordinary melting temperature, T_M. Then, introduce a new frame in uniform translation at the velocity $\vec{U} = U\vec{k}$, where U is to be set so that the front lies near the plane $z = 0$. All of the equations we need are form invariant under this change of observer, so every variable we introduce henceforth will be a moving-frame variable.

The latent heat will be removed by supposing that a plane at $z = -L_0$, $L_0 > 0$, can be maintained at a fixed temperature T_C, $T_C < T_M$, by some imaginative cooling device.

Let the shape of the front be specified by

$$z = Z(x, t)$$

where all our work will be taken to be two dimensional. To determine the function Z, which tells us the position of the interface dividing the solid and the melt, we need to write the equations describing the conduction of heat in the two phases and the equations which tie this to the local motion of the front. These are the nonlinear equations.

They are as follows: First, in the solid phase,

$$\frac{\partial T}{\partial t} = \alpha \nabla^2 T - \vec{v} \cdot \nabla T \tag{7.1}$$

must hold on $-L_0 < z < Z$, while in the liquid phase,

$$\frac{\partial T^*}{\partial t} = \alpha^* \nabla^2 T^* - \vec{v}^* \cdot \nabla T^* \tag{7.2}$$

must hold on $Z < z < \infty$. Then, at the points along the phase boundary (i.e., at $z = Z$), two demands must be met. The first is that phase equilibrium be maintained whence

$$T = T_M + \frac{\gamma T_M}{\mathcal{L}} \, 2H = T^* \tag{7.3}$$

must hold. The second is that the net rate of heat conduction across the surface account for the rate at which latent heat is being released due to the motion of the front, namely

$$\lambda \vec{n} \cdot \nabla T - \lambda^* \vec{n} \cdot \nabla T^* = \mathcal{L} \vec{n} \cdot [-\vec{v} + \vec{u}] \tag{7.4}$$

must hold. This equation is derived in the second endnote, where $\mathcal{L}$ is defined. It denotes the difference in the specific enthalpy across the surface, liquid less solid, multiplied by the solid-phase density. The superscript * denotes a liquid-phase variable, λ denotes the thermal conductivity and α denotes the thermal diffusivity. The surface normal, speed and mean curvature can be obtained in terms of the function Z via the formulas

$$\vec{n} = \frac{-Z_x \vec{i} + \vec{k}}{\sqrt{Z_x^2 + 1}} \tag{7.5}$$

$$u = \frac{Z_t}{\sqrt{Z_x^2 + 1}} \tag{7.6}$$

and

$$2H = \frac{Z_{xx}}{[Z_x^2 + 1]^{3/2}} \tag{7.7}$$

To these equations must be added the far-field equations. Away from the interface, then,

$$T(z = -L_0) = T_C < T_M \tag{7.8}$$

and

$$T^*(z \to \infty) = T_M \tag{7.9}$$

must be satisfied. It is by setting the value of T_C that control over the speed of the solidification front can be achieved.

We introduce no side wall conditions, taking the solidification front to range from $-\infty$ to $+\infty$ in the transverse variable x. Due to this, all non zero transverse wave numbers must be admitted.

Turning to the velocities in the solid and the liquid phases, we can go a long way by declaring only that $\vec{v}$ and $\vec{v}^*$ must satisfy

$$\nabla \cdot \vec{v} = 0 = \nabla \cdot \vec{v}^*$$

in the solid and in the melt and

$$\vec{n} \cdot [\vec{v} - \vec{u}]\rho = \vec{n} \cdot [\vec{v}^* - \vec{u}]\rho^*$$

along the interface dividing the two phases. The solid-phase velocity is set once and for all by requiring the solid to be at rest in the laboratory frame, namely

$$\vec{v} = -\vec{U} = -U\vec{k}$$

Then, at the front, there obtains

$$\vec{n} \cdot \vec{v}^* = -\vec{n} \cdot \vec{k} U \frac{\rho}{\rho^*} - u \frac{[\rho - \rho^*]}{\rho^*} \tag{7.10}$$

where $u = \vec{n} \cdot \vec{u}$, and this determines the liquid-phase velocity normal to the front in terms of U and u, u being the local speed of the front in the moving frame.

7.2.2   The Base Solution

The nonlinear equations have a steady one-dimensional solution in which a planar solidification front remains at rest in the moving frame. This solution is called the base solution.

In the laboratory frame, let a planar front be moving normal to itself at a constant speed U in the $\vec{k}$ direction, where U now denotes the base state front speed in the laboratory frame. Then, in the moving frame and at the front, u must be zero and v_z^* must be $-\dfrac{\rho}{\rho^*} U$. Hence, $\vec{v}^*$ must be

$$-\frac{\rho}{\rho^*} U \vec{k}$$

throughout the subcooled liquid, as $\nabla \cdot \vec{v}^*$ must be zero there.

To find a steady one-dimensional solution, where ρ and ρ^* are taken to be equal, whence $\vec{v}$ and $\vec{v}^*$ both must be $-\vec{U}$, equations (7.1) and (7.2) must be solved and they reduce in this case to

$$0 = \alpha \frac{d^2 T}{dz^2} + U \frac{dT}{dz} \qquad (7.11)$$

and

$$0 = \alpha^* \frac{d^2 T^*}{dz^2} + U \frac{dT^*}{dz} \qquad (7.12)$$

To fix the coordinate system, let the plane dividing the solid and the liquid in the base state be the $x, y-$ plane. Denoting the base-state variables by the subscript zero, the interface is then the plane

$$z_0 = Z_0 = 0$$

and, at the interface, the normal, the normal speed and the mean curvature are given by

$$\vec{n}_0 = \vec{k}, \; u_0 = 0 \text{ and } 2H_0 = 0$$

The solutions to equations (7.11) and (7.12), which also satisfy equations (7.3), (7.8) and (7.9), are

$$T_0 = A_0 + B_0 e^{-\frac{U}{\alpha} z_0}$$

and

$$T_0^* = T_M$$

where

$$A_0 = T_M - B_0 \text{ and } B_0 = \frac{T_C - T_M}{e^{\frac{U L_0}{\alpha}} - 1} < 0$$

These formulas satisfy all the nonlinear equations, save equation (7.4), the heat balance equation across the surface. This equation can be used to determine the surface speed, U, which then turns out to be given by

$$\frac{U L_0}{\alpha} = \ln \left[1 - \frac{\lambda}{\alpha \mathcal{L}} [T_C - T_M] \right] > 0 \qquad (7.13)$$

Two derivatives of T_0, on the solid side of the reference surface, will turn out to be important in determining the stability of a planar solidification front when heat is rejected to the solid. They are

$$\frac{dT_0}{dz_0}(z_0 = 0) = -\frac{U}{\alpha} B_0 > 0 \qquad (7.14)$$

and

$$\frac{d^2 T_0}{dz_0^2}(z_0 = 0) = \frac{U^2}{\alpha^2} B_0 < 0 \qquad (7.15)$$

7.2.3   The Perturbation Equations

The base solution determines the reference domain and the solution to the perturbation equations, on the reference domain, determines the stability of the base solution. To obtain the perturbation equations, a mapping of the reference domain into the present domain must be introduced. By this, the function Z determining the position of the solidification front in the present configuration acquires its expansion. It is

$$Z(x, \ t, \ \epsilon) = Z_0(x_0, \ t_0) + \epsilon Z_1(x_0, \ t_0) + \ \cdots$$

where ϵ may be a measure of the size of the disturbance imposed on the base solution, where Z_0, which defines the reference domain, is zero and where Z_1 must be determined as a part of the solution to the perturbation problem. The function Z_1 establishes the position of the displaced solidification front to first order in ϵ.

The only other surface of interest lies some distance from the interface. It is the plane upon which the temperature is held at the fixed value T_C. This surface is denoted by

$$z_0 = -L_0$$

in the reference configuration and by

$$z = -L_0$$

in the present configuration. It is not displaced.

To obtain the perturbation equations, the rules explained in the second essay can be used. To do this, first turn to the domain. On the solid side of the interface, the term $\vec{v} \cdot \nabla T$ in equation (7.1) simply leads to $-U \dfrac{\partial T_1}{\partial z_0}$. But on the liquid side, the term $\vec{v}^* \cdot \nabla T^*$ in equation (7.2) leads to the sum of two terms, namely, to

$$\vec{v}_1^* \cdot \nabla T_0^* + \vec{v}_0^* \cdot \nabla T_1^*$$

which, in view of the base solution, is just

$$v_{z_1}^* \frac{dT_0^*}{dz_0} - U \frac{\partial T_1^*}{\partial z_0}$$

To omit the term $v_{z_1}^* \dfrac{dT_0^*}{dz_0}$, notice that equation (7.10), namely

$$\vec{n} \cdot \vec{v}^* = -\vec{n} \cdot \vec{k} U \frac{\rho}{\rho^*} - u \frac{[\rho - \rho^*]}{\rho^*}$$

must hold at the interface, and substitute into this the expansions

$$\vec{n} = \vec{k} - \epsilon Z_{1x_0} \vec{i}$$

$$\vec{v}^* = \vec{v}_0^* + \epsilon \left[\vec{v}_1^* + Z_1 \frac{\partial \vec{v}_0^*}{\partial z_0} \right]$$

$$u = \epsilon u_1 = \epsilon Z_{1t_0}$$

and

$$\vec{v}_0^* = -U\vec{k}$$

to obtain

$$v_{z_1}^* = -Z_{1t_0} \frac{\rho - \rho^*}{\rho^*}$$

This leads to the important conclusion that $v_{z_1}^*$ must be zero at the interface in the case of interest where ρ and ρ^* are taken to be equal.

Now, the perturbation velocity throughout the subcooled liquid is driven by the motion of the front, as $v_{z_1}^*$ at the front acquires its value $-Z_{1t_0} \frac{\rho - \rho^*}{\rho^*}$. By taking $\rho = \rho^*$, this source of flow is eliminated, and although this, by itself, does not eliminate convection, as $\vec{v}^*$ can still be perturbed, such a perturbation dies out on its own, receiving no input from the rest of the problem. Our work, then, begins once this contribution is small enough to be omitted. As $\vec{v}^*$ is not then perturbed, the perturbation equations on the domain come down to

$$\frac{\partial T_1}{\partial t_0} = \alpha \nabla_0^2 T_1 + U \frac{\partial T_1}{\partial z_0}, \quad -L_0 < z_0 < 0 \tag{7.16}$$

and

$$\frac{\partial T_1^*}{\partial t_0} = \alpha^* \nabla_0^2 T_1^* + U \frac{\partial T_1^*}{\partial z_0}, \quad 0 < z_0 < \infty \tag{7.17}$$

At the control surface, $z_0 = -L_0$,

$$T_1 = 0 \tag{7.18}$$

must hold, while as $z_0 \to \infty$,

$$T_1^* \to 0 \tag{7.19}$$

must be satisfied.

Now, turn to the interface dividing the two fluids. Equations (7.3) and (7.4) must be satisfied there (i.e., the two phases must be in equilibrium across the front and the latent heat released there must be conducted away). Satisfying these requirements, to first order, requires

$$T_1 + Z_1 \frac{dT_0}{dz_0} = \frac{\gamma T_M}{\mathcal{L}} 2H_1 = T_1^* + Z_1 \frac{dT_0^*}{dz_0} \tag{7.20}$$

and

$$\lambda \left[\frac{dT_1}{dz_0} + Z_1 \frac{d^2 T_0}{dz_0^2} \right] - \lambda^* \left[\frac{dT_1^*}{dz_0} + Z_1 \frac{d^2 T_0^*}{dz_0^2} \right] = \mathcal{L} u_1 = \mathcal{L} \frac{\partial Z_1}{\partial t_0} \tag{7.21}$$

to hold where $\frac{dT_0^*}{dz_0} = 0 = \frac{d^2 T_0^*}{dz_0^2}$, but $\frac{dT_0}{dz_0}$ is positive while $\frac{d^2 T_0}{dz_0^2}$ is negative.

The simple form of the two heat conduction terms is due, first, to the fact that $Z_0 = 0$ and then to the fact that T_0 depends only on z_0, whereupon

$$\vec{n} \cdot \nabla T = [\, \vec{k} - \epsilon Z_{1_{x_0}} \vec{i}\,] \cdot \left[\vec{k} \frac{dT_0}{dz_0} + \epsilon \left[\nabla T_1 + Z_1 \vec{k} \frac{d^2 T_0}{dz_0^2} \right] \right]$$

$$= \frac{dT_0}{dz_0} + \epsilon \left[\frac{dT_1}{dz_0} + Z_1 \frac{d^2 T_0}{dz_0^2} \right]$$

and

$$\vec{n} \cdot \vec{U} = -\epsilon Z_{1_{x_0}} \vec{i} \cdot \vec{U} = 0$$

Equations (7.20) and (7.21), holding at the reference interface, introduce the base solution into the perturbation problem, and hence they lie at the heart of the stability question.

Equations (7.16) through (7.21) present the perturbation problem as an initial-value problem for T_1, T_1^* and Z_1. To turn it into an eigenvalue problem, substitute

$$T_1 = \hat{T}_1(z_0) e^{ikx_0} e^{\sigma t_0}$$

$$T_1^* = \hat{T}_1^*(z_0) e^{ikx_0} e^{\sigma t_0}$$

and

$$Z_1 = \hat{Z}_1 e^{ikx_0} e^{\sigma t_0}$$

into it to obtain the equations that $\hat{T}_1$, $\hat{T}_1^*$ and $\hat{Z}_1$ must satisfy. They are

$$\sigma \hat{T}_1 = \alpha \left[\frac{d^2}{dz_0^2} - k^2 \right] \hat{T}_1 + U \frac{d\hat{T}_1}{dz_0}, \quad -L_0 < z_0 < 0 \tag{7.22}$$

$$\sigma \hat{T}_1^* = \alpha^* \left[\frac{d^2}{dz_0^2} - k^2 \right] \hat{T}_1^* + U \frac{d\hat{T}_1^*}{dz_0}, \quad 0 < z_0 < \infty \tag{7.23}$$

$$\hat{T}_1 = 0, \quad z_0 = -L_0 \tag{7.24}$$

$$\hat{T}_1^* = 0, \quad z_0 \to \infty \tag{7.25}$$

$$\hat{T}_1 + \hat{Z}_1 \frac{dT_0}{dz_0} = -\frac{\gamma T_M}{\mathcal{L}} k^2 \hat{Z}_1 = \hat{T}_1^* + \hat{Z}_1 \frac{dT_0^*}{dz_0}, \quad z_0 = 0 \tag{7.26}$$

and

$$\lambda \left[\frac{d\hat{T}_1}{dz_0} + \hat{Z}_1 \frac{d^2 T_0}{dz_0^2} \right] - \lambda^* \left[\frac{d\hat{T}_1^*}{dz_0} + \hat{Z}_1 \frac{d^2 T_0^*}{dz_0^2} \right]$$

$$= \mathcal{L} \hat{u}_1 = \mathcal{L} \sigma \hat{Z}_1, \quad z_0 = 0 \tag{7.27}$$

The two equations denoted by (7.26), are the link in this essay to a common thread of equations running through these essays. In all these equations, which hold at an interface, most of them being phase-equilibrium equations, gradients of the base solution appear, as does the factor γk^2. These are the ingredients out of which pictures explaining the instability can be constructed.

This eigenvalue problem has solutions other than $\hat{T}_1$, $\hat{T}_1^*$ and $\hat{Z}_1$ all zero only for certain values of σ. These values of σ, the eigenvalues, determine whether a disturbance of wave number k grows or dies out.

The simplest thing to do is to determine whether or not there are any values of k^2 for which $\sigma = 0$ is a solution. These values of k^2 are then critical, dividing the disturbances into two classes: those to which the system is stable and those to which it is not.

Putting $\sigma = 0$, the eigenvalue problem in σ turns into an eigenvalue problem in k^2. It is

$$\alpha \left[\frac{d^2}{dz_0^2} - k^2 \right] \hat{T}_1 + U \frac{d\hat{T}_1}{dz_0} = 0, \quad -L_0 < z_0 < 0 \tag{7.28}$$

$$\alpha^* \left[\frac{d^2}{dz_0^2} - k^2 \right] \hat{T}_1^* + U \frac{d\hat{T}_1^*}{dz_0} = 0, \quad 0 < z_0 < \infty \tag{7.29}$$

$$\hat{T}_1 = 0, \quad z_0 = -L_0 \tag{7.30}$$

$$\hat{T}_1^* = 0, \quad z_0 = \infty \tag{7.31}$$

$$\hat{T}_1 + \hat{Z}_1 \frac{dT_0}{dz_0} = -\frac{\gamma T_M}{\mathcal{L}} k^2 \hat{Z}_1, \quad z_0 = 0 \tag{7.32}$$

$$\hat{T}_1^* = -\frac{\gamma T_M}{\mathcal{L}} k^2 \hat{Z}_1, \quad z_0 = 0 \tag{7.33}$$

and

$$\lambda \left[\frac{d\hat{T}_1}{dz_0} + \hat{Z}_1 \frac{d^2 T_0}{dz_0^2} \right] - \lambda^* \frac{d\hat{T}_1^*}{dz_0} = 0, \quad z_0 = 0 \tag{7.34}$$

These equations are sufficient to determine $\hat{T}_1$, $\hat{T}_1^*$ and $\hat{Z}_1$. The special values of k^2, where solutions other than $\hat{T}_1$, $\hat{T}_1^*$ and $\hat{Z}_1$ all zero can be found, admit the solution $\sigma = 0$ to the perturbation eigenvalue problem.

Equations (7.28), (7.29) and (7.31) can be satisfied by writing

$$\hat{T}_1 = A_1 e^{m_+ z_0} + B_1 e^{m_- z_0}$$

and

$$\hat{T}_1^* = B_1^* e^{m_-^* z_0}$$

where

$$2m_\pm = -\frac{U}{\alpha} \pm \left[\frac{U^2}{\alpha^2} + 4k^2 \right]^{1/2}$$

where

$$2m_{\pm}^* = -\frac{U}{\alpha^*} \pm \left[\frac{U^2}{\alpha^{*2}} + 4k^2\right]^{1/2}$$

and where m_-^* must be negative, while M, a combination of m_+ and m_- given by $\dfrac{-m_+ e^{-[m_- - m_+]L_0} + m_-}{1 - e^{-[m_- - m_+]L_0}}$, must be positive. Both m_-^* and M will appear in equation (7.35), where knowing their signs is important.

The remaining four equations, equations (7.30), (7.32), (7.33) and (7.34), form a set of linear, homogeneous, algebraic equations for the determination of A_1, B_1, B_1^* and $\hat{Z}_1$. The first three of these can be used to eliminate A_1, B_1^* and B_1 in favor of $\hat{Z}_1$, namely

$$A_1 = -B_1 e^{-[m_- - m_+]L_0}$$

$$B_1^* = -\frac{\gamma T_M}{\mathcal{L}} k^2 \hat{Z}_1$$

and

$$B_1 = \frac{\left[-\dfrac{\gamma T_M}{\mathcal{L}} k^2 - \dfrac{dT_0}{dz_0}\right] \hat{Z}_1}{1 - e^{-[m_- - m_+]L_0}}$$

and these lead, upon substitution into the fourth equation, to

$$\left[\lambda M\left[-\frac{\gamma T_M}{\mathcal{L}} k^2 - \frac{dT_0}{dz_0}\right] + \lambda \frac{d^2 T_0}{dz_0^2} + \lambda^* m_-^* \frac{\gamma T_M}{\mathcal{L}} k^2\right] \hat{Z}_1 = 0 \qquad (7.35)$$

The left-hand side of equation (7.35) is the product of two factors; the right-hand side is zero. The factor multiplying $\hat{Z}_1$ cannot be zero. In fact, by equations (7.14) and (7.15), each of its four terms must be negative for all positive values of k^2. Hence, equation (7.35) has only the solution $\hat{Z}_1 = 0$ and this means that A_1, B_1 and B_1^* must also vanish. Due to this, $\sigma = 0$ cannot satisfy the perturbation eigenvalue problem for any value of k^2, whence the solidification front is stable. That a solidification front is stable, in case the latent heat is rejected to the cold solid behind the front, is a prediction of the pictures drawn at the outset.

Things are a little different in case the latent heat is rejected to subcooled liquid ahead of the front and it is to this that we now turn.

7.3 Solidification: Rejecting the Latent Heat to the Liquid

The nonlinear equations remain as they were before, save the far-field conditions. Now, the liquid a distance L_0 ahead of the front has its temperature fixed at T_C, where $T_C < T_M$. Then, away from the interface, T and T^* must satisfy

$$T(z \to -\infty) = T_M$$

and

$$T^*(z = L_0) = T_C$$

instead of equations (7.8) and (7.9)

Again, there is a simple solution to the nonlinear equations. It is

$$T_0 = T_M$$

and

$$T_0^* = A_0^* + B_0^* e^{-\frac{U}{\alpha^*} z_0}$$

where

$$A_0^* = T_M - B_0^* \quad \text{and} \quad B_0^* = \frac{T_C - T_M}{e^{-\frac{U}{\alpha^*} L_0} - 1} > 0$$

and, again, the solid–liquid interface lies in the plane $z_0 = Z_0 = 0$. It is at rest in the moving frame.

Two derivatives of the base solution carry over to the perturbation problem. They are

$$\frac{dT_0^*}{dz_0}(z_0 = 0) = -\frac{U}{\alpha^*} B_0^* < 0 \tag{7.36}$$

and

$$\frac{d^2 T_0^*}{dz_0^2}(z_0 = 0) = \frac{U^2}{\alpha^{*2}} B_0^* > 0 \tag{7.37}$$

The heat balance across the front can be used to determine U in terms of T_C, and in this case, there obtains

$$\frac{U L_0}{\alpha^*} = -\ln\left[1 + \frac{T_C - T_M}{\mathcal{L}\alpha^*/\lambda^*}\right]$$

whereupon U increases from zero to infinity as T_C decreases from T_M to $T_M - \mathcal{L}\alpha^*/\lambda^*$. By using this result, a simple formula for B_0^* can be obtained. It is

$$B_0^* = \frac{\mathcal{L}\alpha^*}{\lambda^*}$$

To determine the stability of the base solution, the perturbation equations must be solved. These equations are just as before, save that the far-field equations must now be

$$T_1(z_0 \to -\infty) = 0$$

and

$$T_1^*(z_0 = L_0) = 0$$

while at the interface $\dfrac{dT_0}{dz_0}$ and $\dfrac{d^2T_0}{dz_0^2}$ now vanish, but $\dfrac{dT_0^*}{dz_0}$ and $\dfrac{d^2T_0^*}{dz_0^2}$ do not.

Again putting

$$T_1 = \hat{T}_1(z_0)e^{ikx_0}e^{\sigma t_0}$$

$$T_1^* = \hat{T}_1^*(z_0)e^{ikx_0}e^{\sigma t_0}$$

and

$$Z_1 = \hat{Z}_1 e^{ikx_0}e^{\sigma t_0}$$

and then setting $\sigma = 0$ to determine whether a critical value of k^2 can be found, we must solve

$$\alpha \left[\frac{d^2}{dz_0^2} - k^2\right]\hat{T}_1 + U\frac{d\hat{T}_1}{dz_0} = 0, \ -\infty < z_0 < 0$$

$$\alpha^* \left[\frac{d^2}{dz_0^2} - k^2\right]\hat{T}_1^* + U\frac{d\hat{T}_1^*}{dz_0} = 0, \ 0 < z_0 < L_0$$

$$\hat{T}_1 = 0, \ z_0 = -\infty$$

$$\hat{T}_1^* = 0, \ z_0 = L_0$$

$$\hat{T}_1 = -\frac{\gamma T_M}{\mathcal{L}} k^2 \hat{Z}_1, \ z_0 = 0$$

$$\hat{T}_1^* + \hat{Z}_1\frac{dT_0^*}{dz_0} = -\frac{\gamma T_M}{\mathcal{L}} k^2 \hat{Z}_1, \ z_0 = 0$$

and

$$\lambda\frac{d\hat{T}_1}{dz_0} - \lambda^*\left[\frac{d\hat{T}_1^*}{dz_0} + \hat{Z}_1\frac{d^2T_0^*}{dz_0^2}\right] = 0, \ z_0 = 0$$

This is an eigenvalue problem. Where it has a solution, other than $\hat{T}_1 = 0$, $\hat{T}_1^* = 0$ and $\hat{Z}_1 = 0$, determines a value of k^2 such that $\sigma = 0$ is a solution.

By writing

$$\hat{T}_1 = A_1 e^{m_+ z_0}$$

and

$$\hat{T}_1^* = A_1^* e^{m_+^* z_0} + B_1^* e^{m_-^* z_0}$$

where, again,

$$2m_\pm = -\frac{U}{\alpha} \pm \left[\frac{U^2}{\alpha^2} + 4k^2\right]^{1/2}$$

and

$$2m_{\pm}^* = -\frac{U}{\alpha^*} \pm \left[\frac{U^2}{\alpha^{*2}} + 4k^2\right]^{1/2}$$

the first three equations can be satisfied and the remaining four equations reduce to a set of linear, homogeneous, algebraic equations in A_1, A_1^*, B_1^* and $\hat{Z}_1$. The first three of these can be solved for A_1, A_1^* and B_1^* in terms of $\hat{Z}_1$, namely

$$A_1^* = -B_1^* e^{[m_-^* - m_+^*]L_0}$$

$$A_1 = -\frac{\gamma T_M}{\mathcal{L}} k^2 \hat{Z}_1$$

and

$$B_1^* = \frac{-\dfrac{\gamma T_M}{\mathcal{L}} k^2 - \dfrac{dT_0^*}{dz_0}}{1 - e^{[m_-^* - m_+^*]L_0}} \hat{Z}_1$$

and the results can be substituted into the fourth equation to obtain[2]

$$\left[-\lambda m_+ \frac{\gamma T_M}{\mathcal{L}} k^2 + \lambda^* M^* \left[\frac{-\gamma T_M}{\mathcal{L}} k^2 - \frac{dT_0^*}{dz_0}\right] - \lambda^* \frac{d^2 T_0^*}{dz_0^2}\right] \hat{Z}_1 = 0 \quad (7.38)$$

where m_+ is positive and where, for all values of k^2, the factor M^*, where M^* is given by $-\dfrac{\left[-m_+^* e^{[m^* - m_1^*]L_0} + m_-^*\right]}{1 - e^{[m_-^* - m_+^*]L_0}}$, is of one sign and it is positive.

A solution to equation (7.38), other than $\hat{Z}_1 = 0$, can be found if and only if the factor multiplying $\hat{Z}_1$ vanishes. This factor is the sum of four terms, and, in view of equations (7.36) and (7.37), the first, second and fourth are negative, whereas the third is positive. This, then, presents the possibility that a value of k^2 may be found such that the four terms add to zero, whereupon the factor multiplying $\hat{Z}_1$ vanishes and, by this, $\hat{Z}_1$ itself need not vanish. Denote this factor by $F(k^2)$.

To see that a positive value of k^2 exists at which $F(k^2)$ vanishes, let U, and all the other input variables, be fixed, with $U > 0$. Then, if k^2 is zero, $F(k^2)$ is determined by its last two terms. Using $m_+^* = 0$, $m_-^* = -\dfrac{U}{\alpha^*}$ and

---

[2]In the discussion note, the case $U = 0$ is worked out. Then, the case $U > 0$, $k^2 = 0$ is taken up. This will be important in the ninth essay. The reader can decide that taking the surface tension to depend on temperature would add nothing to this problem.

$$M^* = \frac{U/\alpha^*}{1 - e^{-\frac{U}{\alpha^*}L_0}}, \text{ at } k^2 = 0, F(k^2 = 0) \text{ reduces to}$$

$$\frac{\frac{\lambda^* U}{\alpha^*}}{1 - e^{-\frac{U}{\alpha^*}L_0}} \left[-\frac{dT_0^*}{dz_0} \right] - \lambda^* \frac{d^2 T_0^*}{dz_0^2}$$

and this is positive due to the fact that

$$\lambda^* \frac{U^2}{\alpha^{*2}} B_0^* \left[\frac{1}{1 - e^{-\frac{U}{\alpha^*}L_0}} - 1 \right]$$

is positive. So, when $k^2 = 0$, F must be positive.

But as $k^2 \to \infty$, we find $m_+^* \to \sqrt{k^2}$, $m_-^* \to -\sqrt{k^2}$ and $M^* \to \sqrt{k^2}$, whence $F(k^2 \to \infty)$ is determined by its first two terms and now it is negative.

As $F(k^2)$ is positive at $k^2 = 0$ and then turns negative as $k^2 \to \infty$, there must be a positive value of k^2 at which it vanishes.[3] The critical wave number, then, is the square root of this value of k^2. It is the wave number at which $\sigma = 0$ is a solution to the perturbation eigenvalue problem and it divides the wave numbers of the disturbances into their unstable and their stable ranges. Such a division obtains for any assignment of values to the input variables.

The critical value of k^2 is the root of the equation

$$-\lambda m_+ \frac{\gamma T_M}{\mathcal{L}} k^2 + \lambda^* M^* \left[-\frac{\gamma T_M}{\mathcal{L}} k^2 - \frac{dT_0^*}{dz_0} \right] - \lambda^* \frac{d^2 T_0^*}{dz_0^2} = 0 \qquad (7.39)$$

and, although it is not possible to get an explicit formula for k_{critical}^2 out of this, it is possible to determine how the critical wave number depends on U in case U is small.

To do this, put equations (7.36) and (7.37) into equation (7.39) and notice that it begins to appear as though k^2 may be proportional to U as U tends to zero. Let this be so. Then, as U tends to zero, both m_+^* and m_-^* go to zero, M^* can be replaced by $-1/L_0$ and $F(k^2)$ takes the limiting form

$$-\lambda m_+ \frac{\gamma T_M}{\mathcal{L}} k^2 - \frac{\lambda^*}{L_0} \left[-\frac{\gamma T_M}{\mathcal{L}} k^2 + \frac{U}{\alpha^*} B_0^* \right] - \lambda^* \frac{U^2}{\alpha^{*2}} B_0^*$$

where

$$B_0^* = \frac{\mathcal{L}\alpha^*}{\lambda^*}$$

In this, as U goes to zero, the first and the fourth terms can be omitted, but the second and the third terms must be retained. The critical value of

---

[3]There cannot be more than one value of k^2 at which $F(k^2)$ vanishes. The proof is left to the reader.

k^2 in this limit is then

$$\frac{\mathcal{L}^2}{\lambda^* \gamma T_M} U$$

Hence, if U is small, there is a range of unstable wave numbers running from zero to the critical value of k^2 and it is interesting to notice that the length of this range depends inversely on the magnitude of the surface tension. This is not only the fluid-displacement result found earlier, but it will be found again.

To get some idea what the graph of k^2_{critical} versus U might look like, drop the stabilizing influence of lateral diffusion from the model. This requires replacing $4k^2$ in the formulas for m^*_+ and m^*_- by zero, whereupon

$$m^*_+ = 0$$

and

$$m^*_- = -\frac{U}{\alpha^*}$$

By doing this, the term accounting for diffusion in the liquid is lost and M^* reduces to

$$\frac{\frac{U}{\alpha^*}}{1 - e^{-\frac{U}{\alpha^*} L_0}}$$

Then, equation (7.39) is simply

$$\frac{\lambda^* \frac{U}{\alpha^*}}{1 - e^{-U L_0/\alpha^*}} \left[-\frac{\gamma T_M}{\mathcal{L}} k^2 + \frac{U}{\alpha^*} B_0^* \right] - \lambda^* \frac{U^2}{\alpha^{*2}} B_0^* = 0$$

and its solution is

$$k^2_{\text{critical}} = \frac{\mathcal{L}^2}{\gamma T_M \lambda^*} U e^{-\frac{U L_0}{\alpha^*}}$$

This reduces, at small values of U, to the low U result found before lateral diffusion was omitted. As k^2_{critical} is small when U is small, this is not surprising, as lateral diffusion cannot be important at small values of k^2. Lateral diffusion can only correct this formula in the mid-range of U, as it can only lower the largest values of k^2_{critical}. What comes into play and calls for the increased stability that this formula predicts as U grows large remains a mystery, especially as all the stabilizing influences of surface tension would appear to be lost as $U \to \infty$ and $k^2 \to 0$.

Now, whether the latent heat is rejected to the solid or, instead, to the liquid, the base solution comes into the calculation of $k^2_{critical}$ via both the first and the second derivatives of the temperature at the interface. In both cases, the front is moving to the right in the laboratory frame, whereupon the convection is to the left in the moving frame. This makes the second derivative negative when the heat is rejected to the solid, positive when it is rejected to the liquid. Convection aids conduction in the removal of the latent heat from the interface in one case and opposes it in the other. But in neither case is the second derivative a source of the instability. In terms of its algebraic sign, it joins the terms multiplied by the surface tension in the equation determining $k^2_{critical}$. It is the first derivative of the temperature at the interface that determines whether or not an advancing solidification front can be unstable. As the source of this derivative can be traced back to the phase-equilibrium requirement across the interface, this not only confirms, but also justifies, the predictions drawn from the pictures at the outset.

This brings the solidification problem to a close. The reader can work out the case of melting. We turn our attention to the case where the latent heat depends on the shape of the interface. Both solidification and melting are now of interest, as they differ in an odd way.

7.4   Solidification and Melting: Is It Important to Account for the Change in the Latent Heat as a Solidification Front Is Displaced?

While attention was directed earlier to changes in the equilibrium temperature as the way in which surface tension acts to stabilize an advancing solidification front, no account was taken of changes in the latent heat.

These changes are no longer ignored and a surprise is in store for the reader: The action of surface tension via the latent heat may be stabilizing, but it may also be destabilizing. To see this, both solidification and melting must be worked out, and the two cases can be taken up together by introducing planes at both $z = -L_0$ and $z = L_0^*$ whereupon temperatures can be specified and held fixed. There are then four cases: solidification, with heat rejected to either the solid or the liquid, and melting, with heat supplied by either the solid or the liquid.

Before presenting the calculation, we remind the reader that at a crest in the solid, the equilibrium temperature must be lower than it would have been had the interface remained flat. So, too, the latent heat must be lower, whereas at a trough, it must be higher. This is worked out in the first endnote.

The formula accounting for this is

$$\mathcal{L} = \mathcal{L}_M + \gamma\, 2H$$

where $\mathcal{L}_M$ denotes the latent heat at a planar surface. Then, in case the base state interface is a plane, whereupon $2H_0$ vanishes, the latent heat at a displaced front takes the form

$$\mathcal{L} = \mathcal{L}_0 + \epsilon \mathcal{L}_1$$

where $\mathcal{L}_0 = \mathcal{L}_M$ and $\mathcal{L}_1 = \gamma \, 2H_1$.

Before turning to the calculation, let us indicate where $\mathcal{L}_1$ will play its role. To do this, return to the two equations that must be satisfied across the front, equations (7.3) and (7.4), namely

$$T = T_M + \frac{T_M \gamma}{\mathcal{L}} \, 2H = T^*$$

and

$$\lambda \vec{n} \cdot \nabla T - \lambda^* \vec{n} \cdot \nabla T^* = \mathcal{L}[\vec{n} \cdot \vec{u} + \vec{n} \cdot \vec{U}]$$

and write their expansions along a mapping of the reference configuration into the present configuration using $\mathcal{L} = \mathcal{L}_0 + \epsilon \mathcal{L}_1$, where $\vec{U}$ is given by $U\vec{k}$ and where, due to the fact that Z_0 is zero, the expansions of $\vec{n}$, u and $2H$ are given by

$$\vec{n} = \vec{k} - \epsilon Z_{x_0} \vec{i}$$

$$\vec{n} \cdot \vec{u} = u = \epsilon u_1$$

and

$$2H = \epsilon 2H_1$$

Doing this, turns equations (7.3) and (7.4) into[4]

$$T_0 + \epsilon \left[T_1 + Z_1 \frac{dT_0}{dz_0} \right] = T_M + \frac{\epsilon T_M \gamma 2H_1}{\mathcal{L}_0 + \epsilon \mathcal{L}_1} = T_0^* + \epsilon \left[T_1^* + Z_1 \frac{dT_0^*}{dz_0} \right]$$

and

$$\lambda \vec{k} \cdot \left[\vec{k} \frac{dT_0}{dz_0} + \epsilon \left[\nabla_0 T_1 + Z_1 \vec{k} \frac{d^2 T_0}{dz_0^2} \right] \right] - \lambda^* \vec{k} \cdot \left[\vec{k} \frac{dT_0^*}{dz_0} + \epsilon \left[\nabla_0 T_1^* + Z_1 \vec{k} \frac{d^2 T_0^*}{dz_0^2} \right] \right]$$

$$= [\mathcal{L}_0 + \epsilon \mathcal{L}_1][\epsilon u_1 + U]$$

whereupon

$$T_1 + Z_1 \frac{dT_0}{dz_0} = \frac{T_M \gamma}{\mathcal{L}_0} \, 2H_1 = T_1^* + Z_1 \frac{dT_0^*}{dz_0}$$

and

$$\lambda \left[\frac{dT_1}{dz_0} + Z_1 \frac{d^2 T_0}{dz_0^2} \right] - \lambda^* \left[\frac{dT_1^*}{dz_0} + Z_1 \frac{d^2 T_0^*}{dz_0^2} \right] = \mathcal{L}_0 u_1 + \mathcal{L}_1 U$$

must hold at first order at the reference interface (viz., at $z_0 = Z_0 = 0$).

The first of these two equations was used earlier. It is equation (7.20). Indeed, the change in the latent heat plays no role in the perturbed phase-equilibrium equation due to the fact that the interface in the base state is flat.

---

[4]The base solution is one dimensional.

It is the second equation, the perturbed heat balance across the interface, that leads to new results. It replaces equation (7.21). In this equation, the second term on the right-hand side is new. It is by this term that the perturbed latent heat comes into the problem.

7.4.1   The Nonlinear Equations

To learn the effect of perturbing the latent heat, begin by writing the nonlinear equations. Taking $\vec{v} = -U\vec{k} = \vec{v}^*$ from the start, these equations are

$$\frac{\partial T}{\partial t} - U\frac{\partial T}{\partial z} = \alpha\nabla^2 T \tag{7.40}$$

in the solid and

$$\frac{\partial T^*}{\partial t} - U\frac{\partial T^*}{\partial z} = \alpha^*\nabla^2 T^* \tag{7.41}$$

in the liquid, while at their interface,

$$T = T_M + \frac{\gamma T_M}{\mathcal{L}}2H = T^* \tag{7.42}$$

and

$$\lambda\vec{n}\cdot\nabla T - \lambda^*\vec{n}\cdot\nabla T^* = \mathcal{L}\vec{n}\cdot[\vec{u} + \vec{U}] \tag{7.43}$$

must hold.

These are moving-frame equations and the assumptions upon which they rest have already been explained.

Our control over the motion of the front is exercised by setting, and holding fixed, the temperatures at two planes, at fixed distances L_0 and L_0^* on either side of the front. By this, the equations

$$T = T_{L_0}, \quad z_0 = -L_0 \tag{7.44}$$

and

$$T^* = T_{L_0^*}, \quad z_0 = L_0^* \tag{7.45}$$

complete the nonlinear problem.

7.4.2   The Base Solution

This problem has a steady one-dimensional solution wherein the interface is flat and lies at $z_0 = Z_0 = 0$. Denoting it by the subscript zero, observe first that T_0 and T_0^* must satisfy equations (7.40) and (7.41), namely

$$-\frac{U}{\alpha}\frac{dT_0}{dz_0} = \frac{d^2 T_0}{dz_0^2}, \quad -L_0 < z_0 < 0$$

and

$$-\frac{U}{\alpha^*}\frac{dT_0^*}{dz_0} = \frac{d^2T_0^*}{dz_0^2}, \quad 0 < z_0 < L_0^*$$

whereupon

$$T_0 = A_0 + B_0 e^{-\frac{U}{\alpha}z_0}$$

and

$$T_0^* = A_0^* + B_0^* e^{-\frac{U}{\alpha^*}z_0}$$

obtain. Then, to satisfy the equilibrium requirement at $z_0 = 0$ [viz., equation (7.42) with $2H_0 = 0$], and the far-field requirements at $z_0 = -L_0$ and at $z_0 = L_0^*$ [viz., equations (7.44) and (7.45)], the constants A_0, B_0, A_0^* and B_0^* must be given by

$$A_0 = T_M - B_0$$

$$B_0 = \frac{T_{L_0} - T_M}{e^{\frac{UL_0}{\alpha}} - 1}$$

$$A_0^* = T_M - B_0^*$$

and

$$B_0^* = \frac{T_{L_0}^* - T_M}{e^{-\frac{UL_0^*}{\alpha^*}} - 1}$$

By this, the four derivatives at the interface by which the base solution comes into the perturbation problem can be evaluated. They are

$$\frac{dT_0}{dz_0} = -\frac{U}{\alpha}B_0 \tag{7.46}$$

$$\frac{d^2T_0}{dz_0^2} - \frac{U^2}{\alpha^2}B_0 \tag{7.47}$$

$$\frac{dT_0^*}{dz_0} = -\frac{U}{\alpha^*}B_0^* \tag{7.48}$$

and

$$\frac{d^2T_0^*}{dz_0^2} = \frac{U^2}{\alpha^{*2}}B_0^* \tag{7.49}$$

Equation (7.43), with $\vec{n}_0 = \vec{k}$ and $u_0 = 0$, then remains to determine the speed of the front and its direction.

7.4.3   The Perturbation Problem and Its Solution

To determine the stability of the base solution, the perturbation equations must be solved. By writing the perturbation equations in the usual way, introducing $\hat{T}_1$, $\hat{T}_1^*$ and $\hat{Z}_1$ via

$$T_1 = \hat{T}_1(z_0)e^{ikx_0}e^{\sigma t_0}$$

$$T_1^* = \hat{T}_1^*(z_0)e^{ikx_0}e^{\sigma t_0}$$

and

$$Z_1 = \hat{Z}_1 e^{ikx_0}e^{\sigma t_0}$$

and then setting σ to zero, the eigenvalue problem at neutral conditions is obtained. Taking into account the base solution, it can be written

$$-\frac{U}{\alpha}\frac{d\hat{T}_1}{dz_0} = \left[\frac{d^2}{dz_0^2} - k^2\right]\hat{T}_1, \quad -L_0 < z_0 < 0$$

$$-\frac{U}{\alpha^*}\frac{d\hat{T}_1^*}{dz_0} = \left[\frac{d^2}{dz_0^2} - k^2\right]\hat{T}_1^*, \quad 0 < z_0 < L_0^*$$

$$\hat{T}_1 = 0, \quad z_0 = -L_0$$

$$\hat{T}_1 + \hat{Z}_1\frac{dT_0}{dz_0} = -k^2\frac{\gamma T_M}{\mathcal{L}_0}\hat{Z}_1, \quad z_0 = 0$$

$$\hat{T}_1^* + \hat{Z}_1\frac{dT_0^*}{dz_0} = -k^2\frac{\gamma T_M}{\mathcal{L}_0}\hat{Z}_1, \quad z_0 = 0$$

$$\hat{T}_1^* = 0, \quad z_0 = L_0^*$$

and

$$\lambda\left[\frac{d\hat{T}_1}{dz_0} + \hat{Z}_1\frac{d^2T_0}{dz_0^2}\right] - \lambda^*\left[\frac{d\hat{T}_1^*}{dz_0} + \hat{Z}_1\frac{d^2T_0^*}{dz_0^2}\right] = \hat{\mathcal{L}}_1 U, \quad z_0 = 0 \qquad (7.50)$$

where equation (7.50) is the new version of equation (7.34). In writing these equations, $2H_1 = Z_{1x_0x_0}$ has been used and, as σ has been set to zero, they are equations determining whether or not a neutral condition can be found and, if it can, the value of k^2 at which it appears.

To solve these seven equations, notice that

$$\hat{T}_1 = A_1 e^{m_+ z_0} + B_1 e^{m_- z_0}$$

and

$$\hat{T}_1^* = A_1^* e^{m_+^* z_0} + B_1^* e^{m_-^* z_0}$$

satisfy the first two, whereupon

$$A_1 = -B_1 e^{-[m_- - m_+]L_0}$$

and

$$B_1 = \frac{-\dfrac{k^2 \gamma T_M}{\mathcal{L}_0} - \dfrac{dT_0}{dz_0}}{1 - e^{-[m_- - m_+]L_0}} \hat{Z}_1$$

satisfy the next two and

$$A_1^* = -B_1^* e^{[m_-^* - m_+^*]L_0^*}$$

and

$$B_1^* = -\frac{-\dfrac{k^2 \gamma T_M}{\mathcal{L}_0} - \dfrac{dT_0^*}{dz_0}}{1 - e^{[m_-^* - m_+^*]L_0^*}} \hat{Z}_1$$

satisfy the fifth and the sixth. In this, $m_\pm$ and $m_\pm^*$ take the values

$$2m_\pm = -\frac{U}{\alpha} \pm \left[\frac{U^2}{\alpha^2} + 4k^2\right]^{1/2}$$

and

$$2m_\pm^* = -\frac{U}{\alpha^*} \pm \left[\frac{U^2}{\alpha^{*2}} + 4k^2\right]^{1/2}$$

and A_1, B_1, A_1^* and B_1^* are all multiples of $\hat{Z}_1$.

Before turning to the last equation, equation (7.50), observe that $\hat{T}_1$ and $\hat{T}_1^*$ are multiples of $\hat{Z}_1$ and so too $\hat{\mathcal{L}}_1$, due to

$$\mathcal{L}_1 = \gamma 2 H_1 = \gamma Z_{1 x_0 x_0}$$

Then, the derivatives of $\hat{T}_1$ and $\hat{T}_1^*$ that will be needed in equation (7.50) are

$$\frac{d\hat{T}_1}{dz_0}(z_0 = 0) = m_+ A_1 + m_- B_1 = M\left[-\frac{k^2 \gamma T_M}{\mathcal{L}_0} - \frac{dT_0}{dz_0}\right]\hat{Z}_1$$

and

$$\frac{d\hat{T}_1^*}{dz_0}(z_0 = 0) = m_+^* A_1^* + m_-^* B_1^* = -M^*\left[-\frac{k^2 \gamma T_M}{\mathcal{L}_0} - \frac{dT_0^*}{dz_0}\right]\hat{Z}_1$$

where

$$M = \frac{-m_+ e^{-[m_- - m_+]L_0} + m_-}{1 - e^{-[m_- - m_+]L_0}} > 0$$

and

$$M^* = -\frac{-m_+^* e^{[m_-^* - m_+^*]L_0^*} + m_-^*}{1 - e^{[m_-^* - m_+^*]L_0^*}} > 0$$

Now, turning to the last equation, it can be arranged as a product of two factors and this product must be zero. One of these factors is $\hat{Z}_1$, and in order that $\hat{Z}_1$ not be zero, the other factor must be zero. This factor is

$$\lambda \left[M \left[-\frac{k^2 T_M \gamma}{\mathcal{L}_0} - \frac{dT_0}{dz_0} \right] + \frac{d^2 T_0}{dz_0^2} \right]$$

$$-\lambda^* \left[M^* \left[\frac{k^2 T_M \gamma}{\mathcal{L}_0} + \frac{dT_0^*}{dz_0} \right] + \frac{d^2 T_0^*}{dz_0^2} \right] + U \gamma k^2$$

and either it is never zero, whereupon $\hat{Z}_1$ must be zero, and the eigenvalue problem has only the solution $\hat{T}_1 = 0$, $\hat{T}_1^* = 0$ and $\hat{Z}_1 = 0$, or it vanishes for certain values of k^2. It is for these values of k^2, where $\hat{Z}_1$ need not be zero, that the perturbation eigenvalue problem has the solution $\sigma = 0$.

To look at this second factor more closely and to see how it depends upon k^2, write it as

$$-\frac{\gamma T_M}{\mathcal{L}_0} k^2 [\lambda M + \lambda^* M^*] + \left[-\lambda M \frac{dT_0}{dz_0} - \lambda^* M^* \frac{dT_0^*}{dz_0} \right]$$

I   II

$$+ \left[\lambda \frac{d^2 T_0}{dz_0^2} - \lambda^* \frac{d^2 T_0^*}{dz_0^2} \right] + \gamma U k^2$$

III   IV

and number the terms in the order they are written (i.e., as indicated above).

Now, this prepares us to look into any number of cases, yet, it is only the four simple cases illustrated in Figure 7.7 that draw our present attention. Two of these are the solidification cases of the essay; the other two are their melting counterparts. In each case, one of the phases is held at T_M, a great distance from the front. The signs of the four terms are indicated in the figure for each case.

The terms marked I, II and III are terms discovered earlier, but the term marked IV is new. It expresses the new effect of the change in the latent heat on the stability of the front.

Refer to the cases by the quadrants in which they lie in the figure. Then look only at the first three terms and notice that the terms marked I and III are always negative, while the term marked II is sometimes negative, sometimes positive. It is this term that introduces the possibility of an instability. It comes from the equilibrium condition at the interface and it introduces the first derivative of the temperature there. It is negative in the first and third quadrants; it is positive in the second and fourth quadrants. Due to this, a critical value of k^2 can be found in the second

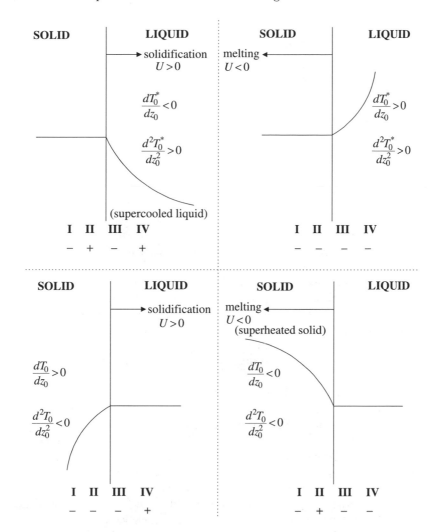

Figure 7.7. *The Base Solution in the Simplest Solidification and Melting Problems and the Signs of the Terms Adding to the Factor Multiplying* $\hat{Z}_1$

and fourth quadrants, but not in the first and third quadrants. Our job is to see whether or not the term marked IV alters these conclusions.

Now, look at the fourth term and notice that its sign is positive in solidification, while it is negative in melting. By this, it is stabilizing in melting, but destabilizing in solidification. Although its sign depends on the direction of the front motion, its strength depends on the value of γ, being stronger the larger the value of γ. Surface tension, then, plays a role not seen before: For the first time, it can be destabilizing.

To see what is going on, take the two solidification cases (i.e., the second and third quadrants), where the signs of the terms are

	I	II	III	IV
second quadrant :	−	+	−	+
third quadrant :	−	−	−	+

In the second quadrant, where heat is rejected to subcooled liquid, the signs of the first three terms make it possible to establish that a value of k^2 can be found at which these terms add to zero. Now, taking account of the new fourth term, the critical value of k^2 shifts to the right, making displacements, earlier thought to be stable, unstable!

In the third quadrant, where the latent heat is rejected to the solid, the signs of the first three terms, being all the same, make it impossible to determine a value of k^2 at which these terms add to zero. Now, the fourth term comes in, in the third quadrant, in just the same way as it did in the second quadrant. It adds a positive term to the sum of the first three terms and, again, this produces something new, now the possibility of instability in an otherwise stable case.

These results can be explained by carrying our picture reasoning one step beyond what was presented earlier. Take the case of solidification, where the latent heat is rejected to subcooled liquid (viz., the two sketches in Figure 7.3 at the bottom and the diagram in Figure 7.7, the second quadrant). Then, the front is moving to the right, as illustrated in Figure 7.3, whereupon two displacements are imposed: one to the right, a crest, moving a little faster, the other to the left, a trough, moving a little slower. The new temperature gradients are illustrated in the figure, strengthened at the crest, supporting the new increased speed, weakened at the trough, but again supporting the new speed, now decreased.

It remains to add to this picture the fact that the latent heat must be a little lower at a crest than it is at a planar front, whereas at a trough, it must be a little higher. This helps the new gradient support the increased speed at a crest, as the product of the latent heat and the new speed is a little smaller than it would have been had the latent heat not been decreased. So, too, it helps the new gradient support the decreased speed at a trough where the product of the latent heat and the new speed is now a little higher. By this, the latent heat adjustment is destabilizing.

The readers can work out this kind of explanation in the other three cases.

Our conclusion is this: Latent heat changes due to surface deflections destabilize solidification fronts, while they stabilize melting fronts.

7.5 ENDNOTES

7.5.1 A Few Equilibrium Formulas

In this essay, and the next, formulas which hold at equilibrium across a planar phase boundary must be adjusted to take into account a deformation of the interface into a non planar shape. All these changes stem from a change in the pressure across a phase boundary due to surface tension and curvature. The pressure difference is given by

$$p^{liquid} - p^{solid} = \gamma \, 2H$$

Three formulas must be derived, one in the endnote to the eighth essay for the increase in the solubility of a solid at a crest, and two in this endnote: one for the decrease in the melting temperature at a crest and one for the decrease in the latent heat at a crest.[5]

• The Change in the Latent Heat as a Solid–Liquid Interface is Displaced

Figure 7.8 indicates a solid and a liquid in phase equilibrium. At a flat surface, the equilibrium temperature is denoted T_M; at a curved surface, it is denoted $T_M + \Delta T$. The pressure in the liquid is always p, whereas in the solid it is p at a flat surface and $p + \Delta p$ at a curved surface.

The latent heat difference across the curved surface is then

$$\Delta \hat{H}_{curved} = \hat{H}^{liquid}_{curved} - \hat{H}^{solid}_{curved}$$

where the right-hand side can be written

$$\hat{H}^{liquid}_{curved} - \hat{H}^{liquid}_{flat} - \hat{H}^{solid}_{curved} + \hat{H}^{solid}_{flat} + \Delta \hat{H}_{flat}$$

To estimate the two differences omit the contribution due to the coefficient of thermal expansion, use

$$d\hat{H} = \hat{c}_p \, dT + \hat{V} \, dp$$

and do not take into account the variation in $\hat{c}_p$ and $\hat{V}$ due to small changes in T and p. Then, there obtains

$$\Delta \hat{H}_{curved} = \hat{c}^{liquid}_p \Delta T - \hat{c}^{solid}_p \Delta T - \hat{V}^{solid} \Delta p^{solid} + \Delta \hat{H}_{flat}$$

---

[5]The reader may notice that in certain of these formulas the correction to an equilibrium variable, whether it be a temperature or a solute concentration, turns out to be a nonlinear function of the surface curvature. Then, the reader may notice that only the linear terms in these formulas are carried back to the essay. Nothing is lost by doing this due to the fact that perturbations of a plane surface are of interest. In this case, $2H$ can be written

$$2H = \epsilon 2H_1$$

due to the fact that $2H_0$ is zero. Hence, the curvature is already of order ϵ. Neither the base equations nor the perturbation equations can sense the nonlinear dependence on curvature; it would first come in at order two.

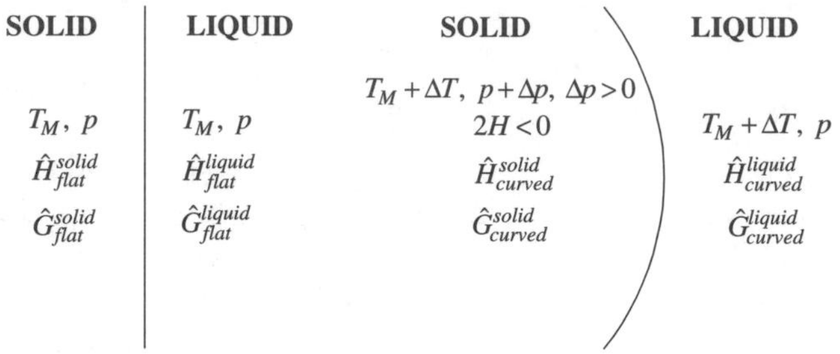

SOLID	LIQUID	SOLID	LIQUID
		$T_M + \Delta T,\ p + \Delta p,\ \Delta p > 0$	
$T_M,\ p$	$T_M,\ p$	$2H < 0$	$T_M + \Delta T,\ p$
$\hat{H}^{solid}_{flat}$	$\hat{H}^{liquid}_{flat}$	$\hat{H}^{solid}_{curved}$	$\hat{H}^{liquid}_{curved}$
$\hat{G}^{solid}_{flat}$	$\hat{G}^{liquid}_{flat}$	$\hat{G}^{solid}_{curved}$	$\hat{G}^{liquid}_{curved}$

Figure 7.8. *A Solid and a Liquid in Equilibrium: Left, a Flat Surface; Right, a Curved Surface*

and, by this, using

$$\Delta p^{solid} = -\gamma\,2H$$

and[6]

$$\Delta T = \frac{T_M}{\mathcal{L}_{flat}}\,\gamma\,2H$$

the formula for the change in $\mathcal{L}$ due to curvature appears, where $\mathcal{L} = \rho^{solid}\Delta H$. It is

$$\mathcal{L}_{curved} = \rho^{solid}\left[\hat{c}_p^{liquid} - \hat{c}_p^{solid}\right]\frac{T_M}{\mathcal{L}_{flat}}\,\gamma\,2H + \gamma\,2H + \mathcal{L}_{flat}$$

where $\rho^{solid} = \dfrac{1}{\hat{V}^{solid}}$.

In the case of ordinary materials, the first term on the right-hand side is not important and the formula reduces to

$$\mathcal{L}_{curved} = \mathcal{L}_{flat} + \gamma\,2H$$

This is used in the essay.

• The Change in the Melting Point as a Solid–Liquid Interface is Displaced

This calculation is very much like that just carried out and, in fact, the result obtained here is used there. However, as this calculation is a little less straightforward, it is placed second. Again, Figure 7.8 indicates the important variables. But now, use is made of

$$\hat{G}^{liquid}_{flat} - \hat{G}^{solid}_{flat} = 0 = \hat{G}^{liquid}_{curved} - \hat{G}^{solid}_{curved}$$

---

[6]The result $\Delta T = \dfrac{T_M}{\mathcal{L}_{flat}}\gamma\,2H$ is obtained in the next subsection.

where changes in $\hat{G}$ due to changes in T and p are evaluated via

$$d\hat{G} = -\hat{S}\ dT + \hat{V}\ dp$$

Again, the pressure does not change on the liquid side as the surface is displaced, this change being entirely confined to the solid, and, again, $\hat{V}^{solid}$, where $\hat{V}^{solid} = \dfrac{1}{\rho^{solid}}$, is taken to be insensitive to small changes in temperature and pressure.

By this, there obtains

$$-\hat{S}_{flat}^{*solid}\Delta T + \hat{V}^{solid}\Delta p = -\hat{S}_{flat}^{*liquid}\Delta T$$

where $\hat{S}_{flat}^{*}$ must be evaluated at some temperature intermediate between T and $T + \Delta T$, but still at the pressure p. Then, using

$$\Delta p = -\gamma\ 2H$$

there obtains

$$\Delta T = \frac{1}{\rho^{solid}}\ \gamma\ 2H\ \frac{1}{\left[\hat{S}_{flat}^{*liquid} - \hat{S}_{flat}^{*solid}\right]}$$

Now, the last factor on the right-hand side can be written

$$\frac{1}{\left[\hat{S}_{flat}^{liquid} - \hat{S}_{flat}^{solid}\right]}$$

upon omitting terms of order ΔT, and this is

$$\frac{T_M}{\left[\hat{H}_{flat}^{liquid} - \hat{H}_{flat}^{solid}\right]}$$

whereupon

$$\Delta T = \frac{T_M}{\mathcal{L}_{flat}}\ \gamma\ 2H$$

This is the formula used in the essay.

7.5.2 Equations Holding Across Phase Boundaries

The basic laws of physics lead to difference equations across surfaces where variables undergo discontinuities just as they lead to differential equations throughout regions where variables are smooth.

The difference equations of interest to us in thermal problems are three in number and they are

$$\vec{n} \cdot \left[[\vec{v} - \vec{u}]\rho\right]^S = \vec{n} \cdot \left[[\vec{v} - \vec{u}]\rho\right]^L \qquad (7.51)$$

$$\vec{n} \cdot \left[[\vec{v} - \vec{u}]\rho\vec{v} - \vec{\vec{T}} \right]^{S} + \gamma \, 2H\vec{n} = \vec{n} \cdot \left[[\vec{v} - \vec{u}]\rho\vec{v} - \vec{\vec{T}} \right]^{L} \tag{7.52}$$

and

$$\vec{n} \cdot \left[[\vec{v} - \vec{u}]\rho\left[\hat{U} + \frac{1}{2} \, \vec{v}^{\,2}\right] + \vec{q} - \vec{\vec{T}} \cdot \vec{v} \right]^{S} + \gamma \, 2H\vec{n} \cdot \vec{u}$$

$$= \vec{n} \cdot \left[[\vec{v} - \vec{u}]\rho\left[\hat{U} + \frac{1}{2} \, \vec{v}^{\,2}\right] + \vec{q} - \vec{\vec{T}} \cdot \vec{v} \right]^{L} \tag{7.53}$$

Notice that $\vec{n}$ points out of phase S into phase L and $2H$ is taken to be negative at a crest, looking along $\vec{n}$.

The first of these is form invariant under a change of observer; the second and third are not. The second and third equations are taken to hold in a laboratory frame and our aim in this endnote is to use the second equation to rewrite the mechanical terms in the third equation, thereby producing a form invariant difference equation that accounts for the latent heat released as a solidification front moves into a freezing liquid.

To do this, work only on the left-hand side, multiply equation (7.52) by $\vec{u}\cdot$, subtract the result from equation (7.53) to obtain

$$\vec{n} \cdot \left[[\vec{v} - \vec{u}]\rho\hat{U} + [\vec{v} - \vec{u}]\rho\left[\frac{1}{2} \, \vec{v}^{\,2} - \vec{v} \cdot \vec{u}\right] + \vec{q} - \vec{\vec{T}} \cdot [\vec{v} - \vec{u}] \right]^{S}$$

and use $\vec{\vec{T}} = -p\vec{\vec{I}} + \vec{\vec{S}}$ and $\hat{H} = \hat{U} + \dfrac{p}{\rho}$ to get

$$\vec{n} \cdot \left[[\vec{v} - \vec{u}]\rho\hat{H} + [\vec{v} - \vec{u}]\rho\left[\frac{1}{2} \, [\vec{v} - \vec{u}]^2 - \frac{1}{2} \, \vec{u}^2\right] + \vec{q} - \vec{\vec{S}} \cdot [\vec{v} - \vec{u}] \right]^{S}$$

The resulting difference equation is then

$$\vec{n} \cdot \left[[\vec{v} - \vec{u}]\rho\left[\hat{H} + \frac{1}{2}[\vec{v} - \vec{u}]^2\right] + \vec{q} - \vec{\vec{S}} \cdot [\vec{v} - \vec{u}] \right]^{S}$$

$$= \vec{n} \cdot \left[[\vec{v} - \vec{u}]\rho\left[\hat{H} + \frac{1}{2}[\vec{v} - \vec{u}]^2\right] + \vec{q} - \vec{\vec{S}} \cdot [\vec{v} - \vec{u}] \right]^{L} \tag{7.54}$$

and it is form invariant.

In the ordinary case, where the kinetic energy and the viscous terms do not play an important role, this equation reduces to

$$\vec{n} \cdot \left[[\vec{v} - \vec{u}]\rho\hat{H} + \vec{q} \right]^{S} = \vec{n} \cdot \left[[\vec{v} - \vec{u}]\rho\hat{H} + \vec{q} \right]^{L}$$

which can be written, using equation (7.51), as

$$\vec{n} \cdot [\vec{v}^{S} - \vec{u}]\rho^{S}[\hat{H}^{S} - \hat{H}^{L}] = \vec{n} \cdot \vec{q}^{\,L} - \vec{n} \cdot \vec{q}^{\,S}$$

Then, a laboratory observer seeing a solid at rest while its surface is advancing into a freezing liquid would write

$$u\rho^S[\hat{H}^L - \hat{H}^S] = \vec{n} \cdot \vec{q}^{\,L} - \vec{n} \cdot \vec{q}^{\,S}$$

where $\vec{n} \cdot \vec{q}^{\,S}$ would be zero if the solid were always at its melting point. This is the equation needed by this observer to determine the speed of the solidification front.

In the essay, it is $\rho^S[\hat{H}^L - \hat{H}^S]$ that is denoted by the symbol $\mathcal{L}$.

7.6   DISCUSSION NOTE

7.6.1   Solidification, Heat Rejected to the Liquid, Two Special Cases: Equilibrium and $k^2 = 0$

The reader's attention is directed in this discussion note to the case where the base state is an equilibrium state (i.e., where $T_C = T_M$) and then to the case, equilibrium or not, where k^2 is zero.

The base solution at equilibrium is simple. It is given by

$$T_0 = T_M$$

$$T_0^* = T_M$$

and

$$U = 0$$

whereupon $\dfrac{dT_0^*}{dz_0}(z_0 = 0)$ and $\dfrac{d^2 T_0}{dz_0^2}(z_0 = 0)$ both vanish. In this case, the perturbation problem simplifies to

$$\alpha \left[\frac{d^2}{dz_0^2} - k^2 \right] \hat{T}_1 = \sigma \hat{T}_1, \quad -\infty < z_0 < 0 \tag{7.55}$$

$$\alpha^* \left[\frac{d^2}{dz_0^2} - k^2 \right] \hat{T}_1^* = \sigma \hat{T}_1^*, \quad 0 < z_0 < L_0 \tag{7.56}$$

$$\hat{T}_1 = 0, \quad z_0 = -\infty \tag{7.57}$$

$$\hat{T}_1^* = 0, \quad z_0 = L_0 \tag{7.58}$$

$$\hat{T}_1 = -\frac{\gamma T_M}{\mathcal{L}} k^2 \hat{Z}_1, \quad z_0 = 0 \tag{7.59}$$

$$\hat{T}_1^* = -\frac{\gamma T_M}{\mathcal{L}} k^2 \hat{Z}_1, \quad z_0 = 0 \tag{7.60}$$

and

$$\lambda \frac{d\hat{T}_1}{dz_0} - \lambda^* \frac{d\hat{T}_1^*}{dz_0} = \sigma \mathcal{L} \hat{Z}_1, \quad z_0 = 0 \tag{7.61}$$

First, let k^2 be zero. Then, it is easy to see that $\sigma = 0$ is a solution corresponding to $\hat{T}_1 = 0 = \hat{T}_1^*$ and $\hat{Z}_1 \neq 0$. It is also easy to see that $\hat{T}_1$ must always be zero, whence the perturbation problem is just

$$\alpha^* \frac{d^2 \hat{T}_1^*}{dz_0^2} = \sigma \hat{T}_1^*, \quad 0 < z_0 < L_0$$

$$\hat{T}_1^* = 0, \quad z_0 = L_0$$

$$\hat{T}_1^* = 0, \quad z_0 = 0$$

and

$$-\lambda^* \frac{d\hat{T}_1^*}{dz_0} = \sigma \mathcal{L} \hat{Z}_1, \quad z_0 = 0$$

where the first three equations determine $\hat{T}_1^*$ and σ, while the last equation determines $\hat{Z}_1$. In view of this, all non zero σ's must be negative, and, hence, at equilibrium and at $k^2 = 0$, only non positive values of σ turn up.

Now, let k^2 be greater than zero. Then, equation (7.55) can be satisfied by writing

$$\hat{T}_1 = Ae^{m_+ z_0} + Be^{m_- z_0}$$

where m_+ and m_- are given by

$$m_\pm = \pm\sqrt{k^2 + \frac{\sigma}{\alpha}}$$

and where $k^2 + \frac{\sigma}{\alpha}$ must be positive[7] in order to make $\hat{T}_1$ vanish at $z_0 = -\infty$. The constants A and B are determined by equations (7.57) and (7.59) to be

$$A = -\frac{\gamma T_M}{\mathcal{L}} k^2 \hat{Z}_1$$

and

$$B = 0$$

Likewise, equations (7.56), (7.58) and (7.60) are satisfied by writing

$$\hat{T}_1^* = A^* e^{m_+^* z_0} + B^* e^{m_-^* z_0}$$

where $m_\pm^*$, A^* and B^* are given by

$$m_\pm^* = \pm\sqrt{k^2 + \frac{\sigma}{\alpha^*}}$$

$$A^* = \frac{\frac{\gamma T_M}{\mathcal{L}} k^2 \hat{Z}_1 e^{[m_-^* - m_+^*]L_0}}{1 - e^{[m_-^* - m_+^*]L_0}}$$

and

$$B^* = \frac{-\frac{\gamma T_M}{\mathcal{L}} k^2 \hat{Z}_1}{1 - e^{[m_-^* - m_+^*]L_0}}$$

---

[7]The readers can easily prove that σ must be real in case $\alpha = \alpha^*$ and $\lambda = \lambda^*$. Still, σ can be taken to be real on the grounds that an equilibrium state is being perturbed.

These results give $\hat{T}_1$ and $\hat{T}_1^*$ in terms of $\hat{Z}_1$ and they can be substituted into equation (7.61) to determine the values of σ for which $\hat{Z}_1$ is not zero. On doing this, there obtains

$$\lambda\left[-\frac{\gamma T_M}{\mathcal{L}}k^2 m_+\right] - \lambda^*\left[\frac{\frac{\gamma T_M}{\mathcal{L}}k^2 e^{[m_-^* - m_+^*]L_0}m_+^*}{1 - e^{[m_-^* - m_+^*]L_0}} - \frac{\frac{\gamma T_M}{\mathcal{L}}k^2 m_-^*}{1 - e^{[m_-^* - m_+^*]L_0}}\right]$$

$$= \sigma\mathcal{L}$$

Then, taking the signs of m_+, m_+^* and m_-^* to be those indicated by their subscripts, the left-hand side of this equation must be negative and, hence, only negative values of σ can be obtained.[8]

Consequently, at equilibrium, the facts are these: At $k^2 = 0$, $\sigma = 0$ is certainly a solution; all other values of σ must be negative. At $k^2 > 0$, all values of σ must be negative. The equilibrium state is stable, as the reader may have guessed, but what is going on at $k^2 = 0$ as the system departs from equilibrium is of interest. It has to do with one of our main themes: the introduction of a control variable, which, it turns out, cannot be done unless zero can be eliminated as a possible value taken by k^2.

To see why $k^2 = 0$ presents a problem, let U be greater than or equal to zero and set k^2 to zero in the perturbation problem. Turn first to the perturbation equations for $\hat{T}_1$. They are

$$\alpha\frac{d^2\hat{T}_1}{dz_0^2} + U\frac{d\hat{T}_1}{dz_0} = \sigma\hat{T}_1, \quad -\infty < z_0 < 0$$

$$\hat{T}_1 = 0, \quad z_0 = -\infty$$

and

$$\hat{T}_1 = 0, \quad z_0 = 0$$

and their solution is $\hat{T}_1 = 0$. The perturbation problem then simplifies to

$$\alpha^*\frac{d^2\hat{T}_1^*}{dz_0^2} + U\frac{d\hat{T}_1^*}{dz_0} = \sigma\hat{T}_1^*, \quad 0 < z_0 < L_0 \tag{7.62}$$

$$\hat{T}_1^* = 0, \quad z_0 = L_0 \tag{7.63}$$

$$\hat{T}_1^* = -\hat{Z}_1\frac{dT_0^*}{dz_0}, \quad z_0 = 0 \tag{7.64}$$

and

$$\frac{d\hat{T}_1^*}{dz_0} = -\hat{Z}_1\frac{d^2T_0^*}{dz_0^2} - \sigma\frac{\mathcal{L}}{\lambda^*}\hat{Z}_1, \quad z_0 = 0 \tag{7.65}$$

---

[8]Take σ to be real. Then by looking at the imaginary part of this equation, the reader can decide that $\dfrac{\sigma}{\alpha^*}$ must be greater than $-k^2$.

where we have

$$\frac{d\hat{T}_0^*}{dz_0}(z_0 = 0) = -\frac{U}{\alpha^*}B_0^*$$

and

$$\frac{d^2\hat{T}_0^*}{dz_0^2}(z_0 = 0) = \frac{U^2}{\alpha^{*2}}B_0^*$$

where $B_0^* = \dfrac{\mathcal{L}\alpha^*}{\lambda^*}$. Our aim is to show that there must be solutions where σ is greater than zero. This should not be surprising due to the fact that the stabilizing role of surface tension vanishes at $k^2 = 0$.

The first thing to notice is that $\sigma = 0$ cannot be a solution so long as U is not zero. If it were, equation (7.62) would be satisfied by

$$T_1^* = A^* + B^* e^{-\frac{U}{\alpha^*}z_0}$$

where A^* and B^* would be given by

$$A^* = \frac{\hat{Z}_1 \dfrac{dT_0^*}{dz_0} e^{-\frac{U}{\alpha^*}L_0}}{1 - e^{-\frac{U}{\alpha^*}L_0}}$$

and

$$B^* = \frac{-\hat{Z}_1 \dfrac{dT_0^*}{dz_0}}{1 - e^{-\frac{U}{\alpha^*}L_0}}$$

in order to satisfy equations (7.63) and (7.64). But then, equation (7.65) can only be satisfied by taking $\hat{Z}_1$, and hence $\hat{T}_1^*$, to be zero, due to the fact that at $z_0 = 0$ we have $\dfrac{d^2T_0^*}{dz_0^2} \Big/ \dfrac{dT_0^*}{dz_0} = -\dfrac{U}{\alpha^*}$.

Consequently, at $k^2 = 0$, $\sigma = 0$ cannot be a solution if U is positive, yet we know that $\sigma = 0$ is a solution if U is zero. The question then is this: Does the solution $\sigma = 0$ at $U = 0$ turn positive as U increases? The answer is that it does.

To see that this is so, notice that the eigenvalue $\sigma = 0$ at $U = 0$ corresponds to the eigenfunction $\hat{T}_1^* = 0$, $\hat{Z}_1 \neq 0$. Then, differentiate equations (7.62), (7.63), (7.64) and (7.65) with respect to U, denoting these derivatives by a superscript prime, to obtain

$$\alpha^* \frac{d^2\hat{T}_1^{*'}}{dz_0^2} + \frac{d\hat{T}_1^*}{dz_0} + U\frac{d\hat{T}_1^{*'}}{dz_0} = \sigma\hat{T}_1^{*'} + \sigma'\hat{T}_1^*, \quad 0 < z_0 < L_0$$

$$\hat{T}_1^{*'} = 0, \quad z_0 = L_0$$

$$\hat{T}_1^{*'} = -\hat{Z}_1\left[-\frac{1}{\alpha^*}B_0^*\right] - \hat{Z}_1'\left[-\frac{U}{\alpha^*}B_0^*\right], \quad z_0 = 0$$

and

$$\frac{d\hat{T}_1^{*'}}{dz_0} = -\hat{Z}_1 \left[\frac{2U}{\alpha^{*2}} B_0^*\right] - \hat{Z}_1' \left[\frac{U^2}{\alpha^{*2}} B_0^*\right] - \sigma \frac{\mathcal{L}}{\lambda^*} \hat{Z}_1' - \sigma' \frac{\mathcal{L}}{\lambda^*} \hat{Z}_1, \quad z_0 = 0$$

which simplifies, on setting U, σ and $\hat{T}_1^*$ to zero, to

$$\alpha^* \frac{d^2 \hat{T}_1^{*'}}{dz_0^2} = 0, \quad 0 < z_0 < L_0$$

$$\hat{T}_1^{*'} = 0, \quad z_0 = L_0$$

$$\hat{T}_1^{*'} = \hat{Z}_1 \frac{B_0^*}{\alpha^*}, \quad z_0 = 0$$

and

$$\frac{d\hat{T}_1^{*'}}{dz_0} = -\sigma' \frac{\mathcal{L}}{\lambda^*} \hat{Z}_1, \quad z_0 = 0$$

The first three equations determine $\hat{T}_1^{*'}$ in terms of $\hat{Z}_1$ and then the last equation determines σ'. The result is that σ', where $\sigma' = \frac{d\sigma}{dU}(U = 0)$, is given by

$$\sigma' = \frac{1}{L_0}$$

In view of this result, we conclude that positive values of σ can be found if U is positive and k^2 is zero.

8
Precipitation

In this essay, we work out a problem that is very much like the problem worked out in the last essay. It differs in one small way and this will make itself known as our work progresses.

Let a solid be slightly soluble in a solvent, but do not let the solvent dissolve in the solid. Let c_S denote the density of the solid and let c denote the solid concentration as a solute in the solution. Denote by c_{EQ} the value of c when the solution and the solid are in equilibrium. Then, if c is less than c_{EQ}, the solid dissolves in an unsaturated solution, whereas if c is greater than c_{EQ}, solute precipitates out of a supersaturated solution. The two cases are illustrated in Figure 8.1 where at the left, the solution–solid phase boundary is moving to the right as the solid dissolves in the solution; at the right it moves to the left as solute precipitates out of the solution. The solid is placed to the right in this and subsequent figures, it being the rich phase.

Now, the saturation concentration of a solution in contact with a solid depends on the curvature of the surface dividing the two phases. It is higher at a crest, where the solid projects into the solution; it is lower at a trough, where the solution projects into the solid. The formula predicting this is obtained in the endnote and it is

$$c_{EQ} = c_{SAT} - A\gamma \, 2H, \quad A > 0$$

In this formula, c_{SAT} denotes the saturation concentration at a plane interface and $2H$ denotes the mean curvature of the surface, taken to be negative at a crest, positive at a trough.

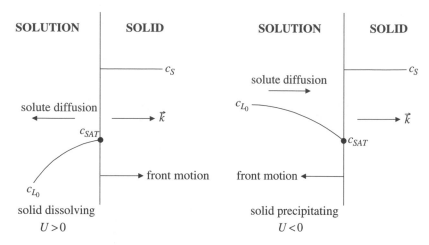

Figure 8.1. *A Solid Dissolving in an Unsaturated Solution (Left) and a Solute Precipitating out of a Supersaturated Solution (Right)*

Our job is to determine whether a planar interface, dividing a solid and the solution out of which it is growing or into which it is dissolving, is or is not stable to small displacements. Both cases are worked out, control being exercised by specifying the solute concentration some distance from the front on the solution side.

8.1   The Physics of the Instability

All of the essays start out either with a Rayleigh work calculation or with a picture illustrating the instability. Both lead to predictions. In this essay, as in the sixth, seventh and ninth essays, a picture is presented and it tells us this: In the case of precipitation from a supersaturated solution, the front might be unstable; in the case of a solid dissolving into an unsaturated solution, it ought to be stable.

Figure 8.2 illustrates a planar precipitation front, moving to the left at a certain speed, upon which two displacements are imposed: a crest where the speed of the front is a little higher and a trough where it is a little lower. A displacement producing a crest strengthens the solute gradient on the solution side, whereas a displacement producing a trough weakens the solute gradient. Both the strengthened gradient at the crest and the weakened gradient at the trough support the new speeds created by the displacements. This tells us that a planar precipitation front might be unstable.

The stabilizing influence of surface tension is also indicated in the figure. At a crest, the saturation concentration increases; at a trough, it

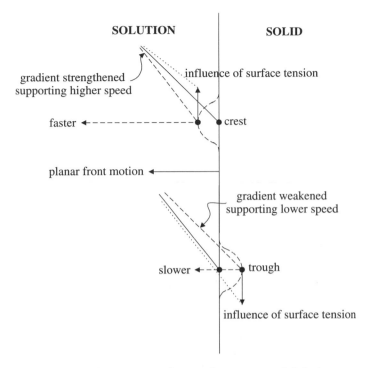

SOLUTION   **SOLID**

gradient strengthened   influence of surface tension
supporting higher speed

faster ◄ ─ ─ ─ ─ ─ ─ ─ ─ ─ ● ● crest

planar front motion ◄ ─────────

gradient weakened
supporting lower speed

slower ◄ ─ ─ ● ─ ─► trough

influence of surface tension

Figure 8.2. *Precipitation from a Supersaturated Solution*

decreases. Due to this, surface tension acts to reverse the changes in the
solute gradient, both at a crest and at a trough, and, as the strength of
this effect depends upon curvature, precipitation fronts must be stable to
displacements of short enough wavelength.

Figure 8.3 illustrates the case where a planar front is moving to the right
as a solid dissolves in an unsaturated solution. Again, two displacements
are indicated: a crest where the speed of the front is now a little lower and
a trough where it is now a little higher. Again, the displacement producing
a crest strengthens the solute gradient in the solution, while the displacement producing a trough weakens the solute gradient. But now, both the
strengthened gradient at the crest and the weakened gradient at the trough
oppose the new speeds caused by these displacements. This is a recipe for
stability, whence a solid dissolving into an unsaturated solution ought to
do so across a stable interface.

The effect of surface tension is also indicated in the figure. It adjusts the
solute gradients in the solution so as to reinforce the stability prediction
already obtained.

To establish these conclusions, let the solid remain at rest in the laboratory frame and denote the velocity of the front, when it is planar and

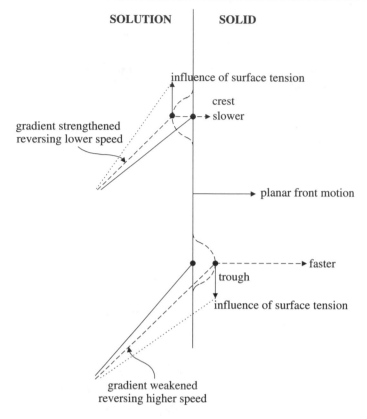

SOLUTION | SOLID

influence of surface tension

crest
slower

gradient strengthened
reversing lower speed

planar front motion

faster
trough

influence of surface tension

gradient weakened
reversing higher speed

Figure 8.3. *A Solid Dissolving in an Unsaturated Solution*

moving steadily, by $\vec{U}$, where $\vec{U} = U\vec{k}$. Then, introduce a moving observer.[1] To this observer, the surface, when it is planar and moving steadily, appears to be at rest. By this, the solid acquires the velocity $\vec{v}^{solid}$, where

$$\vec{v}^{solid} = -U\vec{k}$$

But the solution velocity, denoted by $\vec{v}$, need not be so simple. Yet, setting $\vec{v}$ also to $-U\vec{k}$ is what we do. The reasons for this are explained in the seventh essay. A necessary condition to make this so is that the densities of the two phases be the same.

---

[1] Again, the work is carried out in a moving frame, again the equations needed are form invariant under uniform translation, again all variables introduced are moving-frame variables and again it is by doing this that a simple steady solution can be found.

8.2   The Nonlinear Equations

The solid phase is very simple. It excludes the solvent and, due to this, the solute concentration there is c_S, everywhere and for all time. Turning then to the solution, the nonlinear equations determining the solute concentration there are, first,

$$\frac{\partial c}{\partial t} = D\nabla^2 c + U\frac{\partial c}{\partial z} \tag{8.1}$$

where $z < Z(x, t)$, $-\infty < x < \infty$ and $t > 0$. The function Z indicates the position of the solution–solid interface and its determination is part of solving the problem. In terms of Z, the normal to the surface, the normal speed and the mean curvature can be determined by the formulas

$$\vec{n} = \frac{Z_x\vec{i} - \vec{k}}{\sqrt{Z_x^2 + 1}} \tag{8.2}$$

$$\vec{n} \cdot \vec{u} = u = \frac{-Z_t}{\sqrt{Z_x^2 + 1}} \tag{8.3}$$

and

$$2H = \frac{-Z_{xx}}{[Z_x^2 + 1]^{3/2}} \tag{8.4}$$

where, as these formulas indicate, we work out a two-dimensional version of a three-dimensional problem.

Then, to complete the specification of our problem, the solute concentration must satisfy certain conditions at the solution–solid interface and at some distance from this interface. Turn first to the interface [viz., to $z = Z(x, t)$]. The solute concentration there must meet two requirements: the requirement of phase equilibrium and the requirement that the solute surplus or deficit induced by the motion of the interface be made good by solute diffusion into or out of the solution. These are

$$c = c_{EQ} = c_{SAT} - A\gamma\,2H \tag{8.5}$$

and

$$\vec{n} \cdot [c\vec{v} - c\vec{u}]^{solid} = \vec{n} \cdot [-D\nabla c + c\vec{v} - c\vec{u}]^{solution} \tag{8.6}$$

where diffusion in the solid does not take place and where $u = \vec{n} \cdot \vec{u}$ denotes the surface speed. Using $\vec{v}^{solid} = -\vec{U} = \vec{v}^{solution}$, equation (8.6) reduces to

$$[c_S - c][U\vec{n} \cdot \vec{k} + u] = D\vec{n} \cdot \nabla c \tag{8.7}$$

and, in case the solid is dissolving in the solution, the difference $c_S - c$ indicates that solute must be rejected to the solution. This difference plays the role of $\mathcal{L}$ in the last essay, but $c_S - c$ must be perturbed and this introduces a new term into the perturbation equations, a term not seen in the last essay until the very end where displacements of a phase boundary lead also to perturbations of $\mathcal{L}$.

It remains only to define the conditions on the solute concentration away from the phase boundary. To do this, we introduce a control surface [viz., the plane $z = -L_0$], lying a distance L_0 from the front on the solution side. There, we imagine the solute concentration to be held fixed at the value c_{L_0}, whereupon, at $z = -L_0$, c must satisfy

$$c = c_{L_0} \tag{8.8}$$

It is the value of c_{L_0} that determines the direction and the speed at which the front moves. If c_{L_0} exceeds c_{SAT}, the solution would be supersaturated and the front when it is viewed by a laboratory observer, would move to the left (i.e., U would be negative). On the other hand, if c_{SAT} exceeds c_{L_0}, the solution would be unsaturated and the front would move to the right as the solid dissolves in the solution.

8.3   The Base Solution

The nonlinear equations have a steady one-dimensional solution in which the phase boundary is a plane. This is the base solution and all the base variables will be denoted by the subscript zero. Taking the front to lie in the x_0, y_0- plane and setting

$$Z_0 = 0, \quad u_0 = 0, \quad \vec{n}_0 = -\vec{k} \text{ and } 2H_0 = 0$$

c_0, as it depends on z_0, must satisfy

$$0 = D\frac{d^2 c_0}{dz_0^2} + U\frac{dc_0}{dz_0}, \quad -L_0 < z_0 < 0 \tag{8.9}$$

$$c_0 = c_{L_0}, \quad z_0 = -L_0 \tag{8.10}$$

$$c_0 = c_{SAT}, \quad z_0 = 0 \tag{8.11}$$

and

$$[c_S - c_0]U = D\frac{dc_0}{dz_0}, \quad z_0 = 0 \tag{8.12}$$

The solution to equations (8.9), (8.10) and (8.11) is simply

$$c_0 = A_0 + B_0 e^{-Uz_0/D}$$

where

$$A_0 = c_{SAT} - B_0$$

and

$$B_0 = \frac{c_{L_0} - c_{SAT}}{e^{UL_0/D} - 1}$$

Equation (8.12) then determines U as the solution to

$$[c_S - c_{SAT}]U = D\left[-\frac{U}{D} B_0\right]$$

whereupon

$$B_0 = c_{SAT} - c_S < 0$$

and

$$\frac{UL_0}{D} = \ln\left[1 - \frac{c_{L_0} - c_{SAT}}{c_S - c_{SAT}}\right]$$

There are two cases: If the solid is dissolving in the solution, c_{L_0} must be less than c_{SAT}, U must be positive, and the front must be moving to the right, viewed from the laboratory frame. Then, the value of UL_0/D increases from zero, when $c_{L_0} = c_{SAT}$, to

$$\ln\left[1 + \frac{c_{SAT}}{c_S - c_{SAT}}\right] \simeq \frac{c_{SAT}}{c_S - c_{SAT}}$$

when $c_{L_0} = 0$. If, on the other hand, the solute is precipitating out of the solution, c_{L_0} must be greater than c_{SAT}, U must be negative, and the front must be moving to the left. Then, UL_0/D decreases, without bound, from zero as c_{L_0} increases above c_{SAT}.

Two results will be needed later. They are the first and second derivatives of c_0 at $z_0 = 0$. These derivatives turn out to be

$$\frac{dc_0}{dz_0} = -\frac{U}{D} B_0 \tag{8.13}$$

and

$$\frac{d^2 c_0}{dz_0^2} = \frac{U^2}{D^2} B_0 \tag{8.14}$$

The second derivative is always negative; the sign of the first derivative is the sign of U.

8.4   The Perturbation Problem and Its Solution

Let the displaced surface in the present configuration be denoted by

$$z = Z(x, t, \epsilon)$$

and let the undisplaced surface in the reference configuration be denoted by $z_0 = Z_0(x_0, t_0) = 0$. Then, to write the perturbation equations, introduce a mapping of the reference domain, expand Z in powers of ϵ via

$$Z = Z_0(x_0, t_0) + \epsilon Z_1(x_0, t_0) + \cdots$$

and notice that, as $Z_0 = 0$, the normal to the displaced surface, its normal speed and its mean curvature are given, to first order, by

$$\vec{n} = \vec{n}_0 + \epsilon \vec{n}_1 = -\vec{k} + \epsilon Z_{1x_0} \vec{i}$$

$$\vec{n} \cdot \vec{u} = u = \epsilon u_1 = -\epsilon Z_{1t_0}$$

and

$$2H = \epsilon 2H_1 = -\epsilon Z_{1x_0 x_0}$$

where $u_0 = 0 = 2H_0$ and where $\vec{n}_0$ is perpendicular to $\vec{n}_1$.

The perturbation equations obtain in the usual way. They are the equations to be satisfied by c_1 on the reference domain. On $-L_0 < z_0 < Z_0 = 0$, c_1 must satisfy

$$\frac{\partial c_1}{\partial t_0} = D \nabla_0^2 c_1 + U \frac{\partial c_1}{\partial z_0} \tag{8.15}$$

while at the left-hand boundary (i.e., at $z_0 = -L_0$), it must satisfy

$$c_1 = 0 \tag{8.16}$$

as the solute concentration must remain fixed there. Turning to the right-hand boundary (i.e., to the solution–solid interface), phase equilibrium requires,[2] at first order,

$$c_1 + Z_1 \frac{dc_0}{dz_0} = A\gamma Z_{1x_0 x_0} \tag{8.17}$$

and it remains only to work out the first-order term in the expansion of the solute balance across this surface. To do this, substitute the expansions of c, $\vec{n}$, etc. into equation (8.7), that is, into

$$[c_S - c][U\vec{n} \cdot \vec{k} + u] = D\vec{n} \cdot \nabla c$$

and retain only terms up to first order to get

$$\left[c_S - \left[c_0 + \epsilon \left[c_1 + Z_1 \frac{dc_0}{dz_0} \right] \right] \right] [U + \epsilon Z_{1t_0}]$$

$$= D\vec{k} \cdot \left[\frac{dc_0}{dz_0} \vec{k} + \epsilon \left[\nabla_0 c_1 + Z_1 \frac{d^2 c_0}{dz_0^2} \vec{k} \right] \right]$$

where $\vec{n} \cdot \vec{k} = -1$ to zeroth and first orders in ϵ. Then, at first order, c_1 must satisfy

$$-\left[c_1 + Z_1 \frac{dc_0}{dz_0} \right] U + [c_S - c_0] Z_{1t_0} = D \left[\frac{\partial c_1}{\partial z_0} + Z_1 \frac{d^2 c_0}{dz_0^2} \right]$$

---

[2]The perturbation of A is not introduced. It would come in multiplied by the base value of $2H$ and $2H_0 = 0$.

which reduces to

$$-A\gamma U Z_{1x_0x_0} + [c_S - c_{SAT}]Z_{1t_0} = D\left[\frac{\partial c_1}{\partial z_0} + Z_1 \frac{d^2 c_0}{dz_0^2}\right] \qquad (8.18)$$

upon replacing c_0 and $c_1 + Z_1 \dfrac{dc_0}{dz_0}$ at $z_0 = 0$ by c_{SAT} and $A\gamma Z_{1x_0x_0}$. Equations (8.17) and (8.18) hold at $z_0 = 0$ and complete the statement of the perturbation problem, save for specifying an initial condition.

To turn this problem into an eigenvalue problem, substitute

$$c_1 = \hat{c}_1(z_0)e^{ikx_0}e^{\sigma t_0}$$

and

$$Z_1 = \hat{Z}_1 e^{ikx_0}e^{\sigma t_0}$$

into equations (8.15), (8.16), (8.17) and (8.18), whereupon $\hat{c}_1$ and $\hat{Z}_1$ must satisfy

$$\sigma \hat{c}_1 = D\left[\frac{d^2\hat{c}_1}{dz_0^2} - k^2\hat{c}_1\right] + U\frac{d\hat{c}_1}{dz_0}, \quad -L_0 < z_0 < 0$$

$$\hat{c}_1 = 0, \quad z_0 = -L_0$$

$$\hat{c}_1 + \hat{Z}_1\frac{dc_0}{dz_0} = -A\gamma k^2 \hat{Z}_1, \quad z_0 = 0$$

and

$$A\gamma U k^2 \hat{Z}_1 + \sigma[c_S - c_{SAT}]\hat{Z}_1 = D\left[\frac{d\hat{c}_1}{dz_0} + \hat{Z}_1\frac{d^2 c_0}{dz_0^2}\right], \quad z_0 = 0$$

This eigenvalue problem determines solutions other than $\hat{c}_1 = 0 = \hat{Z}_1$ only for special values of σ. These are the eigenvalues and they determine whether or not the base solution is stable to a displacement of wave number k.

Again, just as in the seventh essay, the simplest thing to do is to try to find out if $\sigma = 0$ can be a solution to this problem. To do this, set $\sigma = 0$ and determine whether, for any value of k^2, a solution, other than $\hat{c}_1 = 0 = \hat{Z}_1$, can be found to

$$0 = \left[\frac{d^2}{dz_0^2} - k^2\right]\hat{c}_1 + \frac{U}{D}\frac{d\hat{c}_1}{dz_0}, \quad -L_0 < z_0 < 0 \qquad (8.19)$$

$$\hat{c}_1 = 0, \quad z_0 = -L_0 \qquad (8.20)$$

$$\hat{c}_1 + \frac{dc_0}{dz_0}\hat{Z}_1 = -A\gamma k^2 \hat{Z}_1, \quad z_0 = 0 \qquad (8.21)$$

and

$$A\gamma U k^2 \hat{Z}_1 = D\left[\frac{d\hat{c}_1}{dz_0} + \frac{d^2 c_0}{dz_0^2}\hat{Z}_1\right], \quad z_0 = 0 \qquad (8.22)$$

Equation (8.19) can be satisfied by

$$\hat{c}_1 = A_1 e^{m_+ z_0} + B_1 e^{m_- z_0}$$

where

$$2m_\pm = -\frac{U}{D} \pm \sqrt{\frac{U^2}{D^2} + 4k^2}$$

and then equations (8.20) and (8.21) can be used to determine both A_1 and B_1 in terms of $\hat{Z}_1$. Doing this produces

$$A_1 = -B_1 e^{-[m_- - m_+]L_0}$$

and

$$B_1 = \frac{-\left[A\gamma k^2 + \dfrac{dc_0}{dz_0} \right]}{1 - e^{-[m_- - m_+]L_0}} \hat{Z}_1$$

whereupon, at $z_0 = 0$, the derivative of $\hat{c}_1$ is given by

$$\frac{d\hat{c}_1}{dz_0} = m_+ A_1 + m_- B_1$$

$$= \frac{-m_+ e^{-[m_- - m_+]L_0} + m_-}{1 - e^{-[m_- - m_+]L_0}} \left[-A\gamma k^2 - \frac{dc_0}{dz_0} \right] \hat{Z}_1$$

Now, turn to the remaining equation, equation (8.22), and substitute for $\dfrac{d\hat{c}_1}{dz_0}$ to get

$$\left[D\left[M\left[-A\gamma k^2 - \frac{dc_0}{dz_0} \right] + \frac{d^2 c_0}{dz_0^2} \right] - A\gamma U k^2 \right] \hat{Z}_1 = 0 \qquad (8.23)$$

where

$$M = \frac{-m_+ e^{-[m_- - m_+]L_0} + m_-}{1 - e^{-[m_- - m_+]L_0}}$$

and where M is positive for all values of k^2 and for all values of U/D. Equation (8.23), then, tells us that the product of two factors, the second being $\hat{Z}_1$, must be zero. By this, unless the first factor can be made to vanish, the only solution to the eigenvalue problem, when σ is set to zero, is $\hat{c}_1 = 0 = \hat{Z}_1$. Save this, zero cannot be a solution to the eigenvalue problem.

Turn then to the first factor in equation (8.23), denote it by F, and examine the signs of its four terms. There are two cases. In the first, solute precipitates out of a supersaturated solution, whereupon

$$U < 0, \quad \frac{dc_0}{dz_0} < 0 \quad \text{and} \quad \frac{d^2 c_0}{dz_0^2} < 0$$

In the second, the solid dissolves in an unsaturated solution, whereupon

$$U > 0, \quad \frac{dc_0}{dz_0} > 0 \quad \text{and} \quad \frac{d^2 c_0}{dz_0^2} < 0$$

The first factor, and the signs of its terms in the two cases, is presented below:

$$-A\gamma k^2 DM + DM \left[-\frac{dc_0}{dz_0} \right] + D\frac{d^2 c_0}{dz_0^2} - A\gamma U k^2$$

| case 1 | $-$ | $+$ | $-$ | $+$ |
| case 2 | $-$ | $-$ | $-$ | $-$ |

The two cases are illustrated in Figure 8.4, drawn to remind the reader of Figure 7.7 in the seventh essay. In this problem, the first and second quadrants do not appear due to the fact that a solid can only dissolve in one way, but it can melt in two ways. The third and fourth quadrants here correspond to the third and fourth quadrants in the earlier figure.

The first three terms in F would predict stability in case 2, the third quadrant; they would predict the possibility of an instability in case 1, the fourth quadrant. This is in accord with the predictions of the pictures drawn at the beginning of this essay.

The fourth term is interesting. It comes from $[c_S - c][\vec{n} \cdot U\vec{k} + u]$, a term in the solute balance across the front, and it must be perturbed, as c must be perturbed. A corresponding term is obtained in the solidification–melting problem if the latent heat is perturbed. The fourth terms in the two problems have to do, in the previous essay, with the latent heat that must be supplied or rejected by conduction, but, in this essay, with the solute that must be supplied or rejected by diffusion. The factors $\mathcal{L}$ and $[c_S - c]$, then, play the same role.

The fourth terms do not come into the two problems with the same sign. In this essay, the fourth term stabilizes case 2, but destabilizes case 1. In the earlier problem, it was the reverse. In fact, the perturbation of $\mathcal{L}$ in the seventh essay leads to the term $-\gamma k^2 U \hat{Z}_1$; the perturbation of $c_S - c$ in this essay leads to the term $+A\gamma k^2 U \hat{Z}_1$. These terms, which enter the calculation in the same way, have opposite signs.

To see that the fourth term appears in F correctly, look at Figure 8.5, where the crest in Figure 8.3 has been redrawn.

In the new figure, surface tension not only strengthens the new gradient beyond what can be accounted for by the displacement itself, but it also reduces the rate at which solute needs to be rejected to sustain the front speed. Both of these factors tend to reverse the lower speed corresponding to the introduction of a crest upon the surface of a solid dissolving into an unsaturated solution and both are stabilizing. This accounts for the signs indicated in the third quadrant of Figure 8.4.

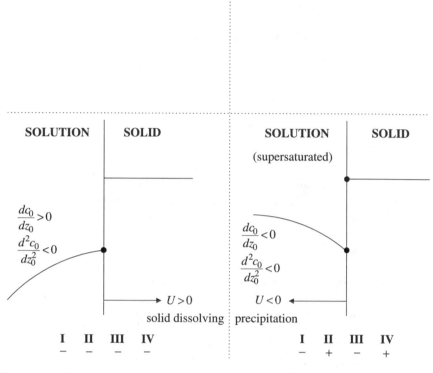

Figure 8.4. *The Base Solution in the Two Cases, a Dissolving Solid to the Left and a Precipitating Solute to the Right, and the Signs of the Terms Adding to the Factor F*

The reader can try his hand at this kind of explanation by deducing the signs of the four terms in F indicated in the fourth quadrant.

The equation $F(k^2) = 0$ determines, by its roots, the critical values of k^2. In case 1, critical values can be found; in case 2, they cannot. As in the seventh essay, when there is a root, there is only one, but an explicit formula for it cannot be written.

To estimate the critical value of k^2 in case 1, set F to zero and substitute into it

$$\frac{dc_0}{dz_0} = -\frac{U}{D}\,B_0$$

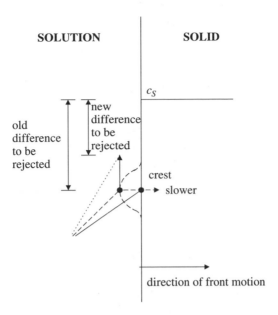

Figure 8.5. *The Crest of Figure 8.3 Redrawn to Illustrate the Two Effects of Surface Tension: It Strengthens the Solute Gradient in the Liquid While Also Reducing the Amount of Solute That Needs to Be Rejected*

and

$$\frac{d^2 c_0}{dz_0^2} = \frac{U^2}{D^2} B_0$$

as given by equations (8.13) and (8.14), where B_0, which is $c_{SAT} - c_S$, and U must both be negative. By doing this, there obtains

$$\left[- A\gamma Dk^2 - |U|[c_{SAT} - c_S] \right] M + \frac{U^2}{D} [c_{SAT} - c_S] + A\gamma |U| k^2 = 0 \quad (8.24)$$

It is possible to derive an explicit formula for the critical value of k^2 when $|U|$ is small by observing first that if $|U|$ is small, then, so too, k^2 must be small whence m_+ and m_- must both be close to zero. The factor M can then be estimated via

$$M = -\frac{-m_+ e^{-[m_- - m_+]L_0} + m_-}{1 - e^{-[m_- - m_+]L_0}} \simeq \frac{1}{L_0}$$

and, when this is used, equation (8.24) for the critical value of k^2 comes down to

$$\frac{-A\gamma Dk^2 - |U|[c_{SAT} - c_S]}{L_0} + \frac{U^2}{D} [c_{SAT} - c_S] + A\gamma |U| k^2 = 0$$

which leads to the result

$$k^2_{critical} = \frac{[c_S - c_{SAT}]}{A\gamma D}|U|$$

This low-speed estimate of $k^2_{critical}$ is not surprising. It is one of many such low-speed results obtained in these essays, and all these results look alike.

To get a simple estimate of the critical value of k^2 that does not require $|U|$ to be small, omit the stabilizing influence of lateral diffusion. This requires only that $4k^2$ be omitted in the formulas for m_+ and m_-, whereupon they reduce to

$$m_+ = 0$$

and

$$m_- = -\frac{U}{D}$$

and, by this, M reduces to

$$\frac{-U/D}{1 - e^{UL_0/D}}$$

Again, M is positive no matter the sign of U. As U is negative in the case at hand, write

$$M = \frac{|U|/D}{1 - e^{-|U|L_0/D}}$$

and put this into equation (8.24) to obtain

$$\left[-A\gamma Dk^2 - |U|[c_{SAT} - c_S]\right] \frac{|U|/D}{1 - e^{-|U|L_0/D}}$$

$$+\frac{U^2}{D}[c_{SAT} - c_S] + A\gamma|U|k^2 = 0$$

whereupon there results

$$\left[A\gamma|U|k^2 + \frac{U^2}{D}[c_{SAT} - c_S]\right] \frac{-e^{-|U|L_0/D}}{1 - e^{-|U|L_0/D}} = 0$$

Then, for all but infinitely large values of $|U|$ where $\dfrac{-e^{-|U|L_0/D}}{1 - e^{-|U|L_0/D}}$ vanishes, the estimate of $k^2_{critical}$, obtained by omitting transverse diffusion, turns out to be

$$\frac{c_S - c_{SAT}}{A\gamma D}|U|$$

This does not require $|U|$ to be small.

Now, it is not surprising that this estimate of $k^2_{critical}$ agrees with the small $|U|$ result obtained earlier, for, then, k^2 is small and transverse diffusion, although stabilizing, is weak. What is a little surprising is that, on omitting transverse diffusion, the small $|U|$ formula can be extrapolated to all values of $|U|$, predicting increasing instability as $|U|$ increases (i.e., increasing values of $k^2_{critical}$).

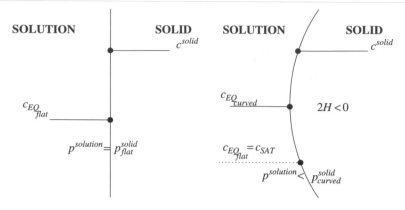

Figure 8.6. *A Solid and a Solution in Phase Equilibrium*

8.5   ENDNOTE

8.5.1   *The Solubility of a Solid*

The derivation in this endnote is like the two derivations presented in the first endnote to the seventh essay. Again, the change, now in solubility, stems from a change in the pressure as a surface is displaced from its reference planar shape.

The derivation is just like an osmotic pressure calculation. To carry it out, let the solution remain at a fixed pressure and look at two states at the same temperature, differing only in whether the phase boundary is flat or curved. This is illustrated in Figure 8.6. In this figure, a solid phase is in equilibrium with a solution containing dissolved solid as a solute. Let c_{EQ} denote the equilibrium concentration of the solid in the solution. This concentration is taken to be small. No solvent invades the solid phase and c^{solid}, the solid phase density, is taken to be fixed.

Then, in order that the chemical potential of the solid species remain the same in both phases as the surface is displaced, the pressure increase on the solid side must be offset by a concentration increase on the solution side. By this, there obtains

$$\frac{1}{c^{solid}} \, dp^{solid} = RT \, d\ln c_{EQ}$$

which leads, on integration, to

$$\frac{1}{c^{solid}} \left[p^{solid}_{curved} - p^{solid}_{flat} \right] = RT \, \ln\left(\frac{c_{EQ_{curved}}}{c_{EQ_{flat}}} \right)$$

Into this, substitute $-\gamma \, 2H$ for the pressure difference across the phase boundary to obtain

$$c_{EQ_{curved}} = c_{EQ_{flat}} e^{-\frac{\gamma \, 2H}{RT c^{solid}}}$$

which, to sufficient accuracy, is

$$c_{EQ_{curved}} = c_{EQ_{flat}} \left[1 - \frac{\gamma\, 2H}{RT\, c^{solid}} \right]$$

This last formula requires $\left| \dfrac{\gamma\, 2H}{RT\, c^{solid}} \right|$ to be small.

Now, replacing $c_{EQ_{flat}}$ by c_{SAT} and $c_{EQ_{curved}}$ by c_{EQ}, there obtains

$$c_{EQ} = c_{SAT} - A\gamma\, 2H$$

where

$$A = \frac{c_{SAT}}{RT\, c^{solid}} > 0$$

This formula is used in the essay.

9

Electrodeposition

The reader's attention is drawn in this essay to the following question: During the operation of an electrolytic cell, do the shapes of the electrodes remain stable? Our main aim is to establish the fact that an equilibrium electrolytic cell is stable to all small disturbances of finite wavelength and then to predict the advance of the critical wave number upward from zero as current is drawn and the cell departs from equilibrium.

Figure 9.1 reproduces a photograph of López-Salvans et al. illustrating the instability.

To begin, we present a short outline of what the reader will find in this essay. Here, as in the seventh essay, the stability of a surface dividing a solid and an adjacent liquid is in question, but in this essay, the dependence of the phase-equilibrium formulas on the shape of the surface is not the issue; instead, reactions are now taking place at the surface and they determine its stability. Our first job, then, is to explain the factors which are important in determining the rates of these reactions, the oxidation and reduction reactions by which electrical current passes between an ionic solution and a metal electrode. Based on this comes a prediction that the origins of the instability observed in the operation of an electrolytic cell lie in self-reinforcing displacements at the cathode where reduction runs ahead of oxidation and where the appearance of a small crest displaces the electrical potential and the metal ion concentration in the direction that favors reduction over oxidation. To turn the physical picture of the instability into a formula for the critical wave number, the nonlinear equations are introduced and then simplified by requiring the ionic solution to be

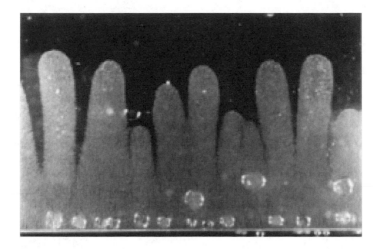

Figure 9.1. *A Photograph Illustrating the Instability*
M.-Q. López-Salvans, P.P. Trigueros, S. Vallmitjana, J. Claret, and F. Sagués,
Physical Review Letters, Vol. 76, No. 21, [May 1996], p. 4062, by the American
Physical Society

everywhere electrically neutral. A solution to the nonlinear equations is obtained and attention is directed to determining its stability.

To set the problem, let two metal electrodes be separated by a dilute solution of a completely ionized salt in water. To make this definite, the reader can take the electrodes to be copper and the electrolytic solution to be copper sulphate dissolved in water. Yet, we will not be so specific until graphs need to be drawn.

Some notation is indicated in Figure 9.2, which presents a picture of a cell in which metal is removed from the left-hand electrode and is then deposited at the right-hand electrode, after it traverses the intervening solution in the form of metal ions.

The way the figure is drawn, the electrical potential V^A is higher than the electrical potential V^C and a potential gradient then appears in the solution. The potential at a point in the solution is denoted ϕ and its gradient acts to move the metal ions in the solution from left to right.

A Cartesian coordinate system will be used where the axis Oy points across the cell from left to right in the direction of current flow; the axis Ox is perpendicular to this. This is illustrated in the figure and, as it suggests, we will take up no more than a two-dimensional problem.

Now, the current passing through an electrolytic solution is carried by moving ions while it is carried by moving electrons in the electrodes, the connection wires, etc. At the interface dividing the metal and the solution, an electron transfer reaction must take place to account for the change in

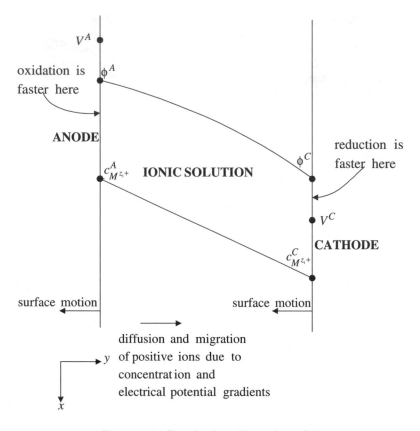

Figure 9.2. *Sketch of an Electrolytic Cell*

the current carriers. This reaction need not be instantaneous and its rate must be taken into account.

To see what this rate depends upon, let us start with a single reversible reaction in which z_1 electrons are transferred from the metal to the solution or vice versa. Denote the metal species by M and its ion by M^{z_1+} and write the reaction by which the metal goes into solution, the oxidation reaction, as

$$M \longrightarrow M^{z_1+} + z_1 e^-$$

This is called the backward reaction. Its reverse, the forward or reduction reaction, the reaction by which metal ions deposit on the electrode, is then

$$M^{z_1+} + z_1 e^- \longrightarrow M$$

Now, the rates at which these two steps proceed determine, by their difference, the rate of electron transfer and, by this, the rate at which current flows in the cell. At a point of the surface, these two rates depend most strongly on three factors: the species concentrations, the potential difference across the surface and the curvature of the surface.

First, the rate of the oxidation step is proportional to the density of the electrode, c_M; the rate of the reduction step is proportional to the density of metal ions, $c_{M^{z_1+}}$. Of these, c_M can be taken to be fixed.

Second, the electrical potential is ordinarily different on the two sides of a surface dividing a metal and the adjacent ionic solution. This difference, denoted $\Delta\phi$, is always taken to be $\phi^{metal} - \phi^{solution}$. It favors the rate of the electron transfer reaction in one direction over the other. If it is positive, it favors the oxidation step over the reduction step; if it is negative, it favors reduction over oxidation. The result is that a positive $\Delta\phi$ favors the metal dissolving in the solution; a negative $\Delta\phi$ favors the reverse.

Third, the curvature of the surface plays a role. At a crest, where the electrode projects into the solution, the oxidation step is favored over the reduction step, whereas at a trough, it is the reverse. Thus a crest favors the metal dissolving in the solution; a trough favors the metal ions plating out. By this, crests and troughs sow the seeds of their own destruction. It is in this way that the surface tension comes into play, being the factor that determines the strength of this stabilizing effect.

9.1   The Physics of the Instability

To see what the rate of electron transfer, and in particular these three factors, has to do with the possibility of running a cell which maintains parallel electrodes always parallel, begin by looking at Figure 9.2. It is important to notice that on the left, at the anode, the speed of the oxidation reaction is above that of the reduction reaction; at the right, at the cathode, the reverse is true. The surface of the anode then moves to the left as metal goes into solution and so, too, the surface of the cathode as metal ions come out of solution. The rates of the reactions and the speeds at which the surfaces move in this planar configuration then establish a reference for the figures to come.

In Figure 9.3, this same state of affairs is illustrated again, except that a small crest and a small trough are added to each surface.

To see what happens to these displacements, once they are introduced, let the electrical potentials of the metals remain as they were in Figure 9.2. Then, to estimate the electrical potential and the metal ion concentration on the solution side of an interface, at a trough or at a crest, extrapolate or interpolate the curves drawn in Figure 9.2. On the left, at the anode, $\phi^{solution}$ and $c_{M^{z_1+}}$ rise at a trough, fall at a crest. Both extrapolations favor reduction at a trough, both interpolations favor oxidation at a crest. Both then decrease the increased speed of a trough, building it up, while

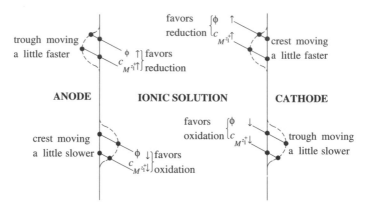

Figure 9.3. *A Small Crest and a Small Trough Added to the Surface of the Electrodes*

increasing the decreased speed of a crest, tearing it down. Due to this, the anode would appear to be stable to surface displacements.

On the right, at the cathode, the opposite takes place. Here, $\phi^{solution}$ and $c_{M^{z_1+}}$ fall at a trough, rise at a crest. Both favor oxidation at a trough, both favor reduction at a crest. Again, both decrease the speed at a trough, while increasing the speed at a crest. But now, a small trough deepens while a small crest builds up. By this, the cathode would appear to be unstable to surface displacements.

This leaves only surface curvature to be taken into account. Now troughs favor reduction; crests favor oxidation. So superimposed on Figure 9.3 is the surface curvature influence on the electron transfer reaction indicated in Figure 9.4. This tells us that surface curvature is stabilizing, reinforcing the anticipated stability of the anode, but opposing the anticipated instability of the cathode.

To carry this a little further, let a displacement be imposed at the cathode. Then, the greater its curvature, the stronger the stabilizing effect of surface tension on its growth. Offsetting this, the more current the cell draws, the steeper the base state gradients and the stronger the destabilizing effect of the perturbed metal ion concentration and the perturbed electrical potential on its growth. These two effects, in opposition, would lead to the prediction that increasing the current should shift the critical wave number to the right.

The reader might redraw these figures and conclude that a cell, at equilibrium, must be stable to displacements of all finite wavelengths. But once current is drawn, the cell ought to be unstable to displacements of long enough wavelength (i.e., of weak enough curvature), due to what is going on at its cathode. Our job, then, is to turn these pictures into formulas spelling out the conditions under which a cell is either stable or unsta-

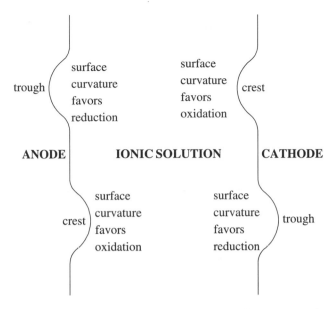

ANODE   **IONIC SOLUTION**   **CATHODE**

Figure 9.4. *The Effect of Surface Curvature on the Electron Transfer Reaction*

ble to a surface displacement. In particular, it is to determine the critical wavelength dividing the ranges of stable and unstable displacements.

9.2   The Electron Transfer Reaction

To turn a metal atom into a metal ion[1] at an electrode surface requires the transfer of electrons across this surface. However, electrons do not move freely across surfaces and the rate of this process depends upon the conditions under which it takes place. This rate is described by the Butler–Volmer equation, and to write it in the form that we need it, requires some preliminary information on electrical current and on reference concentrations.

Let two surfaces be moving as indicated in Figure 9.5. The important variables are indicated in the figure. The normals will always be taken to point out of the metal into the solution. A surface will be denoted by $y = f(x, t)$; formulas for its normal, normal speed and mean curvature are indicated in the figure. The mean curvature is negative at a crest, positive at a trough.

---

[1] Of course, the metal is a lattice of metal ions in a sea of electrons, but nothing is lost by talking about the electrode reaction as if it takes place between metal atoms and metal ions in solution.

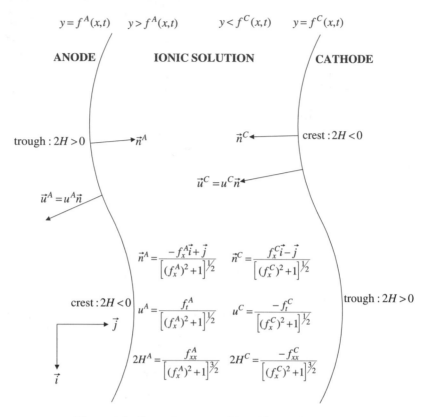

Figure 9.5. *Some Important Variables and Formulas*

The current in the circuit is measured by an ammeter. It is denoted by I. The electrical current density in the solution is denoted by $\vec{i}$, whereupon

$$\vec{i} = \sum F z_i \vec{N}_i$$

and

$$I = \int_S dA \; \vec{n} \cdot \left[\vec{i} - F \sum z_i c_i \vec{u} \right]$$

where S denotes the surface dividing an electrode and the solution, where c_i, $i = 1, 2, 3$, denote the molar densities of the metal ion, the non metallic ion and the solvent, where $|z_i|$ denotes the number of equivalents per mole and where $\vec{N}_i$ denotes the molar current density of species i in the solution.

Now, the current I read on the ammeter is the sum of the local currents produced by the electron transfer reactions at each point along the surface of an electrode and the net rate of these reactions is what the Butler–Volmer equation predicts. To be certain that the signs are correct, observe that $\vec{n} \cdot \vec{i}$ denotes the current crossing a surface at rest, due to electron transfer. This is positive if $\vec{i}$ is in the direction of $\vec{n}$, whence the current is leaving the electrode, entering the solution, otherwise it is negative. The oxidation reaction then produces positive current while the reduction reaction produces negative current. Due to this, the Butler–Volmer equation, taking into account the deflection of an electrode surface and hence its curvature, is written

$$\frac{\vec{n} \cdot [\vec{i} - F \sum z_i c_i \vec{u}]}{z_1 F}$$

$$= \overleftarrow{k} c_M \, e^{[1-\beta] \frac{z_1 F}{RT} \left[\Delta\phi - \frac{\gamma}{z_1 F c_M} 2H\right]} - \overrightarrow{k} c_1 \, e^{-\beta \frac{z_1 F}{RT} \left[\Delta\phi - \frac{\gamma}{z_1 F c_M} 2H\right]}$$

where c_M denotes the molar density of the metal, where γ denotes the surface tension and where β, $0 < \beta < 1$, denotes a factor having to do with the activation step. The proportionality factors $\overleftarrow{k}$ and $\overrightarrow{k}$ depend at most on the temperature at which the electron transfer reactions are taking place.

The first term on the right-hand side accounts for the oxidation step, where positive values of $\Delta\phi$ and negative values of $2H$ produce higher rates than do negative values of $\Delta\phi$ and positive values of $2H$. The second term accounts for the reduction step, where the reverse is true.

By introducing a reference metal ion concentration, the constants $\overleftarrow{k} c_M$ and $\overrightarrow{k}$ can be eliminated from the Butler–Volmer equation, which can then be written in terms of the exchange current density, denoted by i_0, and the equilibrium potential difference, denoted by $\Delta\phi_{EQ}$. As the average metal ion concentration[2] is a convenient scale for the ion concentrations appearing in our work, it is taken to be the reference concentration and it might be worthwhile to state here that as long as the density of the electrolytic solution does not depend on the ion concentration, and this will be one of our assumptions, the average concentration of the metal ions in the solution cannot change during an electrodeposition experiment.[3]

The elimination of $\overleftarrow{k} c_M$ and $\overrightarrow{k}$ in favor of i_0 and $\Delta\phi_{EQ}$ is presented in the first endnote. Both i_0 and $\Delta\phi_{EQ}$ refer to a reference surface, a surface which is flat and across which the electrode is in contact with a solution at the reference metal ion concentration. In terms of these variables, then, the Butler–Volmer equation can be written

$$\vec{n} \cdot \left[\vec{i} - F \sum z_i c_i \vec{u}\right]$$

$$= i_0 \left[e^{[1-\beta] \frac{z_1 F}{RT} \left[\Delta\phi - \Delta\phi_{EQ} - \frac{\gamma}{z_1 F c_M} 2H\right]} - \frac{c_1}{\bar{c}_1} e^{-\beta \frac{z_1 F}{RT} \left[\Delta\phi - \Delta\phi_{EQ} - \frac{\gamma}{z_1 F c_M} 2H\right]} \right]$$

---

[2] In this essay, an overbar is used to denote average values.

[3] This is explained in the second discussion note.

The values of $\Delta\phi_{EQ}$ and i_0 at a standard concentration must be measured. The value of $\Delta\phi_{EQ}$ at any reference concentration can then be obtained from the Nernst equation via

$$\Delta\phi_{EQ}(c_{1\ ref}) - \Delta\phi_{EQ}(c_{1\ std}) = \frac{RT}{z_1 F} \ln \frac{c_{1\ ref}}{c_{1\ std}}$$

whence i_0, given in terms of $\Delta\phi_{EQ}$ by

$$i_0 = F \overleftarrow{k} c_M e^{[1-\beta]\frac{z_1 F}{RT}\Delta\phi_{EQ}}$$

is

$$\frac{i_0(c_{1ref})}{i_0(c_{1std})} = \left[\frac{c_{1ref}}{c_{1std}}\right]^{[1-\beta]}$$

The values of $\Delta\phi_{EQ}$ and i_0 at a standard concentration, then, become parameters of the problem replacing $\overleftarrow{k} c_M$ and $\overrightarrow{k}$. The values of $\Delta\phi_{EQ}$ and i_0 at a specified reference concentration are then parameters of a specific experiment.

9.3   The Nonlinear Equations

A steady state can not be achieved when an electrodeposition process is viewed by a laboratory observer. But, in case the two electrode surfaces are moving at the same constant speed and in the same direction, they can be translated to rest and the process may then appear to be steady to an observer in uniform motion. Anticipating this to be so, our work will be carried out in a moving frame in uniform translation, at a velocity denoted $\vec{U}$, vis-à-vis a laboratory observer. As all of our equations will be form invariant under a change to a uniformly moving observer, this will hardly be noticed. To write the nonlinear equations, the surfaces will be taken to be in motion, even in the moving frame. To do otherwise would make it impossible to write these equations correctly.

Let the electrodes be perfect conductors lying at rest in the laboratory frame. They can be specified entirely by their electrical potentials, denoted by V^A and V^C. Hence, with the electrodes simply accounted for, the nonlinear equations need only account for what is going on in the electrolytic solution and at its interfaces with the two electrodes.[4] Then

$$\frac{\partial c_1}{\partial t} + \nabla \cdot \vec{N}_1 = 0 \tag{9.1}$$

---

[4]If there are boundaries in the transverse direction, homogeneous Neumann conditions will be assumed to hold there and these boundaries will not be given much attention.

$$\frac{\partial c_2}{\partial t} + \nabla \cdot \vec{N}_2 = 0 \tag{9.2}$$

and

$$\nabla \cdot \vec{v} = 0$$

must hold on the domain (i.e., throughout the ionic solution), where the salt dissociates completely into ions and no reaction takes place in the solution.

The molar current densities, $\vec{N}_1$ and $\vec{N}_2$, appearing in these equations can be obtained by adding the contributions due to diffusion, migration and convection. In a sufficiently dilute solution, this sum can be expressed via

$$\vec{N}_i = -D_i \nabla c_i - \frac{D_i}{RT} F z_i c_i \nabla \phi + c_i \vec{v}, \quad i = 1,\, 2 \tag{9.3}$$

where the correct velocity to use in the convection term ordinarily would be the solvent velocity, but, in a sufficiently dilute solution, the solvent velocity and the mass-average velocity cannot be distinguished and so the mass-average velocity is used in the foregoing. Henceforth, D_1 and D_2 are taken to be independent of ion concentration.

At the two electrode surfaces,

$$\left[\vec{n} \cdot [\vec{N}_1 - c_1 \vec{u}]\right]^{metal} - \left[\vec{n} \cdot [\vec{N}_1 - c_1 \vec{u}]\right]^{solution} = 0$$

$$\left[\vec{n} \cdot [\vec{N}_2 - c_2 \vec{u}]\right]^{metal} - \left[\vec{n} \cdot [\vec{N}_2 - c_2 \vec{u}]\right]^{solution} = 0$$

and

$$\left[\vec{n} \cdot \rho[\vec{v} - \vec{u}]\right]^{metal} - \left[\vec{n} \cdot \rho[\vec{v} - \vec{u}]\right]^{solution} = 0$$

must hold,[5] where on the metal side of either surface

$$\vec{v} = -\vec{U}, \quad \vec{N}_1 = -c_1 \vec{U} \quad \text{and} \quad \vec{N}_2 = -c_2 \vec{U}$$

must hold and, so too, $c_1 = c_M$ and $c_2 = 0$. These equations express the fact that the two electrodes, but not their surfaces, remain at rest in the laboratory frame and the fact that each is composed entirely of species 1. This is where $\vec{U}$ comes into the moving-frame equations.

---

[5]Nothing is lost by failing to distinguish metal atoms and metal ions. The only place where such a distinction might even be thought to be necessary would be in doing the mass accounting, in mole units, across the electrode-solution interfaces. If the distinction is made, then metal atoms do not appear in the solution phase, while metal ions do not appear in the metal phase, but in the equation for each species there would appear the rate of the reaction $M \rightleftharpoons M^{z_1 +} + z_1 e^-$, and, if this rate is eliminated, an equation for the metal species appears and this is what we write.

Dropping the superscripts denoting the solution and the metal, all variables henceforth, save c_M and ρ_M, refer to the solution and the equations where the solution contacts the electrode surfaces simplify to

$$\vec{n} \cdot \vec{N}_1 = [c_1 - c_M]\, \vec{n} \cdot \vec{u} - c_M\, \vec{n} \cdot \vec{U} \qquad (9.4)$$

$$\vec{n} \cdot \vec{N}_2 = [c_2]\vec{n} \cdot \vec{u} \qquad (9.5)$$

and

$$\vec{n} \cdot \vec{v} = \left[1 - \frac{\rho_M}{\rho}\right] \vec{n} \cdot \vec{u} - \frac{\rho_M}{\rho}\, \vec{n} \cdot \vec{U} \qquad (9.6)$$

and where the normal current density at each surface is then

$$\vec{n} \cdot \vec{i} = \sum Fz_i \vec{n} \cdot \vec{N}_i = \sum Fz_i c_i \vec{n} \cdot \vec{u} - Fz_1 c_M[\vec{n} \cdot \vec{u} + \vec{n} \cdot \vec{U}]$$

Notice that the normal to each surface and the speed at which it moves are present in these equations. They introduce one unknown at each surface, the function f which specifies the shape of that surface. In terms of f, $\vec{n}$ and u can be determined. But the two functions f must themselves be determined as part of the solution to these equations.

9.4   The Nonlinear Equations Simplified by Local Electroneutrality

Let the shapes of the two electrode surfaces be specified, as well as the function $\vec{v}$, and turn to equations (9.1) and (9.2) on the domain and equations (9.4) and (9.5) at its boundaries. Then, substituting

$$\vec{N}_1 = -D_1 \nabla c_1 - F \frac{D_1}{RT} z_1 c_1 \nabla \phi + c_1 \vec{v}$$

and

$$\vec{N}_2 = -D_2 \nabla c_2 - F \frac{D_2}{RT} z_2 c_2 \nabla \phi + c_2 \vec{v}$$

into these equations, two equations in the three variables c_1, c_2 and ϕ result, both on the domain and at its boundaries, and another equation is required. This third equation is obtained by making the approximation that the ionic solution remain everywhere electrically neutral. This approximation[6] requires

$$\sum z_i c_i = z_1 c_1 + z_2 c_2 = 0 \qquad (9.7)$$

---

[6]It is ordinarily thought to be reasonable outside small regions near the electrode surfaces, where it contradicts the charge separation that must obtain across these surfaces. These small regions will not be taken into account in this work.

and it simplifies the formula for I to

$$I = \int_S dA \, \vec{n} \cdot \vec{i}$$

Two linear combinations of $\vec{N}_1$ and $\vec{N}_2$ now become interesting. The first is

$$z_1\vec{N}_1 + z_2\vec{N}_2 = -D_1\nabla[z_1 c_1] + D_2\nabla[-z_2 c_2] + \left[-\frac{F}{RT}[D_1 z_1][z_1 c_1] \right.$$

$$\left. -\frac{F}{RT}[-D_2 z_2][-z_2 c_2] \right]\nabla\phi + [z_1 c_1 + z_2 c_2]\vec{v}$$

the second is

$$z_1 z_2 D_2\vec{N}_1 + z_1 z_2 D_1\vec{N}_2 = -z_2 D_1 D_2\nabla[z_1 c_1] + z_1 D_1 D_2\nabla[-z_2 c_2]$$

$$-\frac{F}{RT}D_1 D_2 z_1 z_2[z_1 c_1 + z_2 c_2]\nabla\phi + \left[z_2 D_2[z_1 c_1] \right.$$

$$\left. -z_1 D_1[-z_2 c_2] \right]\vec{v}$$

Then, using equation (9.7) and introducing the new variable c, where $c = z_1 c_1 = -z_2 c_2$, these equations simplify to

$$z_1\vec{N}_1 + z_2\vec{N}_2 = -[D_1 - D_2]\nabla c - \frac{F}{RT}[D_1 z_1 - D_2 z_2]c\nabla\phi$$

and

$$z_1 z_2 D_2\vec{N}_1 + z_1 z_2 D_1\vec{N}_2 = +D_1 D_2[z_1 - z_2]\nabla c - [z_1 D_1 - z_2 D_2]c\vec{v}$$

which can be used to reduce the three equations in c_1, c_2 and ϕ to two equations in c and ϕ. Turn, first to the equations on the domain. Multiply equation (9.1) by z_1, equation (9.2) by z_2 and add. Then, multiply equation (9.1) by $z_1 z_2 D_2$ and equation (9.2) by $z_1 z_2 D_1$ and add. By doing this and introducing D and P via

$$D = \frac{D_1 D_2}{z_1 D_1 - z_2 D_2}[z_1 - z_2]$$

and

$$P = \frac{z_1[D_1 - D_2]}{z_1 D_1 - z_2 D_2}$$

there obtains

$$\frac{\partial c}{\partial t} = D\nabla^2 c \quad \vec{v} \cdot \nabla c \tag{9.8}$$

and

$$P\nabla^2 c + \frac{z_1 F}{RT}\nabla \cdot [c\nabla\phi] = 0 \tag{9.9}$$

Now, turn to the equations at the boundary and multiply equation (9.4) by z_1, equation (9.5) by z_2 and add. Then, multiply equation (9.4) by $z_1 z_2 D_2$, equation (9.5) by $z_1 z_2 D_1$ and add. By doing this, there obtains, again in terms of D and P,

$$D\vec{n} \cdot \nabla c - c\vec{n} \cdot [\vec{v} - \vec{u}] = \frac{-z_1 z_2 D_2 c_M}{z_1 D_1 - z_2 D_2} [\vec{n} \cdot \vec{u} + \vec{n} \cdot \vec{U}] \qquad (9.10)$$

and

$$P\vec{n} \cdot \nabla c + \frac{z_1 F}{RT} c\vec{n} \cdot \nabla \phi = \frac{z_1^2 c_M}{z_1 D_1 - z_2 D_2} [\vec{n} \cdot \vec{u} + \vec{n} \cdot \vec{U}] \qquad (9.11)$$

These are the nonlinear equations that we will use. They are based on two approximations: The density of the electrolytic solution must be independent of the ion concentration and local electroneutrality must hold. The second discussion note ties these two approximations together.

If $\vec{v}$ were to be ignored, these equations would be enough to determine c and ϕ if their initial values were assigned throughout a solution of a specified geometry. But the geometry is not specified. To determine the shapes of the surfaces, the Butler–Volmer equation must be introduced and it must be satisfied at each point of the two electrode surfaces. It predicts $\vec{n} \cdot \vec{i}$ at these points in terms of the corresponding values of c and ϕ and the local curvature of the interface. Now, $\vec{n} \cdot \vec{i}$ is just $\vec{n} \cdot [F z_1 \vec{N}_1 + F z_2 \vec{N}_2]$, whence by equations (9.4), (9.5) and (9.7), there obtains

$$\vec{n} \cdot \vec{i} = -F z_1 c_M [\vec{n} \cdot \vec{u} + \vec{n} \cdot \vec{U}] \qquad (9.12)$$

where $[\vec{n} \cdot \vec{u} + \vec{n} \cdot \vec{U}]$ is present on the right-hand sides of equations (9.10) and (9.11), which hold at the two electrode surfaces. By this, the Butler–Volmer equation carries the local values of c and ϕ into the local speed of an electrode surface. This adjusts the local shape of the surface by

$$\vec{n} \cdot \vec{u} = u = \frac{\pm \dfrac{\partial f}{\partial t}}{\sqrt{1 + \left[\dfrac{\partial f}{\partial x}\right]^2}}$$

Outside of $\vec{v}$, it would appear that the nonlinear equations comprise a well-posed problem in c, ϕ and the geometry of the electrolytic solution.

To include $\vec{v}$ presents only a technical difficulty, but this does not reverse the above conclusion that the nonlinear equations, simplified by requiring local electroneutrality to hold, present a well-posed problem. However, we can go a long way by declaring only that $\vec{v}$ must satisfy

$$\nabla \cdot \vec{v} = 0$$

at the points of the solution and

$$\vec{n} \cdot \vec{v} = \left[1 - \frac{\rho_M}{\rho}\right] \vec{n} \cdot \vec{u} - \frac{\rho_M}{\rho} \vec{n} \cdot \vec{U}$$

at points along the interfaces which divide the electrolytic solution and the metal electrodes.

9.5 The Base Solution

Our first job is to find a steady one-dimensional solution to the nonlinear equations in which the electrode surfaces take the shape of parallel planes, a distance L_0 apart, at rest in the moving frame. To do this, affix the origin to the surface of the anode, whereupon all variables depend at most on y and all vectors lie parallel to $\vec{j}$. The superscripts, A or C, denote that a variable has been evaluated at the anode surface (i.e., at $y = 0$), or at the cathode surface (i.e., at $y = L_0$).

First, notice that a one-dimensional solution requires

$$f^A = 0, \quad \vec{n}^A = \vec{j}, \quad u^A = 0, \quad 2H^A = 0$$

and

$$f^C = L_0, \quad \vec{n}^C = -\vec{j}, \quad u^C = 0, \quad 2H^C = 0$$

Then, notice that v_y and i_y must satisfy

$$0 = \nabla \cdot \vec{v} = \frac{dv_y}{dy}$$

and

$$0 = \nabla \cdot \left[z_1 \vec{N}_1 + z_2 \vec{N}_2 \right] = \nabla \cdot \vec{i} = \frac{di_y}{dy}$$

whereupon both v_y and i_y remain constant across the electrolytic solution at values determined by what is going on at the electrode surfaces. Using equations (9.6) and (9.12), v_y and i_y are given by

$$v_y = -\frac{\rho_M}{\rho} U$$

and

$$i_y = -z_1 F c_M U$$

where $|U|$ is the speed at which a laboratory observer sees the two interfaces moving to the left. The two variables v_y and U, then, are just different ways of talking about the same thing, the electrical current density in the cell under steady, one-dimensional conditions.

It is of some interest to observe that a one-dimensional solution to the nonlinear equations would be ruled out at this point if the solution density were not the same at the two electrode surfaces.

Now, the two variables L_0 and $\bar{c}$ are input variables to the base solution. The remaining input variable is $V^A - V^C$, the voltage difference imposed across the cell. But the electrode voltages appear only in the Butler–Volmer equation and until this equation is written at the two electrode surfaces, it does little harm to think of U as an input variable for now.

To determine the base solution, we must solve

$$D\frac{d^2c}{dy^2} - v_y\frac{dc}{dy} = \frac{d}{dy}\left[D\frac{dc}{dy} - v_yc\right] = 0 \tag{9.13}$$

and

$$P\frac{d^2c}{dy^2} + \frac{d}{dy}\left[c\frac{d}{dy}\left[\frac{z_1F\phi}{RT}\right]\right] = \frac{d}{dy}\left[P\frac{dc}{dy} + c\frac{d}{dy}\left[\frac{z_1F\phi}{RT}\right]\right] = 0 \tag{9.14}$$

in the region $0 < y < L_0$ and satisfy

$$D\frac{dc}{dy} - v_yc = \frac{-z_1z_2D_2}{z_1D_1 - z_2D_2}\, c_MU$$

and

$$P\frac{dc}{dy} + c\frac{d}{dy}\left[\frac{z_1F\phi}{RT}\right] = \frac{z_1^2}{z_1D_1 - z_2D_2}\, c_MU$$

at $y = 0$ and at $y = L_0$.

To do this, integrate equations (9.13) and (9.14) one time to produce

$$D\frac{dc}{dy} - v_yc = \text{ const. } = \frac{-z_1z_2D_2}{z_1D_1 - z_2D_2}\, c_MU \tag{9.15}$$

and

$$P\frac{dc}{dy} + c\frac{d}{dy}\left[\frac{z_1F\phi}{RT}\right] = \text{ const. } = \frac{z_1^2}{z_1D_1 - z_2D_2}\, c_MU \tag{9.16}$$

and notice, by setting the two constants as indicated, that all four requirements at the two electrode surfaces can be satisfied. The first of these two equations can be solved to obtain c and the result substituted into the second which then can be integrated to obtain ϕ. The additional piece of information required to fix c is that its average value, $\bar{c}$, must be assigned, whereupon c must satisfy

$$\frac{1}{L_0}\int_0^{L_0} c\, dy = \bar{c}$$

Now, substitute $-\dfrac{\rho_M}{\rho}U$ for v_y and solve equation (9.15) to get

$$c = \frac{-z_1z_2D_2}{z_1D_1 - z_2D_2}\frac{\rho}{\rho_M}\, c_M + Ae^{-\frac{\rho_M}{\rho}\frac{U}{D}y}$$

where A must be taken to be

$$A = \frac{\bar{c} + \dfrac{z_1 z_2 D_2}{z_1 D_1 - z_2 D_2} \dfrac{\rho}{\rho_M} c_M}{\dfrac{1}{\dfrac{\rho_M}{\rho} \dfrac{U}{D} L_0} \left[1 - e^{-\frac{\rho_M}{\rho} \frac{U}{D} L_0}\right]}$$

to make the average value of c turn out to be $\bar{c}$. This formula determines c versus y and it indicates the way in which this function depends on $\bar{c}$, L_0 and U. In particular, it determines the values taken by c at the anode and at the cathode (i.e., the values c^A and c^C).

To determine ϕ versus y, and in particular the values ϕ^A and ϕ^C of ϕ at the anode and at the cathode, substitute the formula for c into equation (9.16), which then reads

$$\frac{d}{dy}\left[\frac{z_1 F \phi}{RT}\right] = \frac{1}{c}\left[\frac{z_1^2}{z_1 D_1 - z_2 D_2} c_M U - P \frac{dc}{dy}\right]$$

and integrate both sides. Doing this, one constant of integration appears which cannot, at this point, be determined, though the difference $\phi^A - \phi^C$ can be obtained.

Having c versus y and ϕ versus y, at least up to one constant of integration, it remains only to satisfy the Butler–Volmer equation at the two electrode surfaces. This requires, at the anode,

$$i_y^A = -z_1 F c_M U = i_0 \left[e^{[1-\beta]\frac{z_1 F}{RT}[\Delta\phi^A - \Delta\phi_{EQ}]} - \frac{c^A}{\bar{c}} e^{-\beta\frac{z_1 F}{RT}[\Delta\phi^A - \Delta\phi_{EQ}]}\right]$$

and, at the cathode,

$$-i_y^C = z_1 F c_M U = i_0 \left[e^{[1-\beta]\frac{z_1 F}{RT}[\Delta\phi^C - \Delta\phi_{EQ}]} - \frac{c^C}{\bar{c}} e^{-\beta\frac{z_1 F}{RT}[\Delta\phi^C - \Delta\phi_{EQ}]}\right]$$

where $\Delta\phi^A = V^A - \phi^A$ and $\Delta\phi^C = V^C - \phi^C$ and where, in these equations, $\Delta\phi_{EQ}$ and i_0 are constants, depending on $\bar{c}$.

Taking the values of $\bar{c}$ and L_0 to be set, these two equations amount to two equations in the four variables V^A, V^C, U and the constant of integration appearing in the formula for ϕ.

To establish a reference for the electrical potential, V^A will be set to zero, whence if V^C is set in the range $[0, -\infty)$, U and the constant of integration can be obtained and, hence, the steady one-dimensional base solution to the nonlinear equations. This solution is implicit.

The constant of integration can be eliminated by using the equation for ϕ versus y to obtain a formula for $\phi^A - \phi^C$, a formula in which U remains present. Then, the Butler–Volmer equation at the anode can be solved for $e^{\frac{z_1 F}{RT}[\Delta\phi^A - \Delta\phi_{EQ}]}$ in terms of U. Likewise, the Butler–Volmer equation at the cathode can be solved for $e^{\frac{z_1 F}{RT}[\Delta\phi^C - \Delta\phi_{EQ}]}$, again in terms of U. Dividing the first by the second leads to a formula for $\dfrac{z_1 F}{RT}[V^A - V^C - [\phi^A - \phi^C]]$ in terms of U and, hence, to a formula for $\dfrac{z_1 F}{RT}[V^A - V^C]$, also in terms of U. The current in the cell, then, depends on $V^A - V^C$, not on V^A and V^C independently and $\Delta\phi_{EQ}$ does not appear in this formula.

Now, there is a limiting value of the current (viz., a least value of U). It is determined by what is taking place at the cathode. As V^C decreases, and the potential difference across the cell increases, the rate of the reduction reaction at the cathode increases, but it does not become infinitely fast, being limited by a decreasing value of c^C. As V^C decreases to negative infinity, the value of c^C falls to zero and this determines the limiting value of U, denoted U_{lim}. By setting c^C to zero, a formula for U_{lim} appears. It is

$$\frac{xe^x}{1 + [x-1]e^x} = \frac{-z_1 z_2 D_2}{z_1 D_1 - z_2 D_2} \frac{c_M}{\bar{c}} \frac{\rho}{\rho_M}$$

where $x = \dfrac{\rho_M}{\rho} \dfrac{[-U_{lim}]}{D} L_0$. The function on the left-hand side decreases monotonically from positive infinity to one as x increases from zero. By this, as long as $\dfrac{-z_1 z_2 D_2}{z_1 D_1 - z_2 D_2} \dfrac{c_M}{\bar{c}} \dfrac{\rho}{\rho_M}$ is greater than one (i.e., so long as A is less than zero), there is one and only one value of U_{lim}. This value of U_{lim} depends on L_0 and $\bar{c}$ and also on the other parameters of the problem. But it does not, of course, depend on the value of V^C which must be negative infinity to produce the limiting current.

The signs of dc/dy and $d\phi/dy$ turn out to be important in what is to come and they can be determined. To do this, notice that

$$\frac{dc}{dy} = A\left[-\frac{\rho_M}{\rho} \frac{U}{D}\right] e^{-\frac{\rho_M}{\rho} \frac{U}{D} y}$$

whereupon the sign of dc/dy is the sign of A. Then, notice that the sign of A must be negative whenever $\bar{c}$ is less than $\dfrac{-z_1 z_2 D_2}{z_1 D_1 - z_2 D_2} \dfrac{\rho}{\rho_M} c_M$. This condition is ordinarily satisfied, typical values of $\dfrac{\rho}{\rho_M} c_M$ and $\bar{c}$ being 20 kg-mol/m^3 and 0.1 kg-mol/m^3.

Turning to $d\phi/dy$, substitute

$$\frac{dc}{dy} = \frac{1}{D}\left[\frac{-z_1 z_2 D_2}{z_1 D_1 - z_2 D_2} c_M U + v_y c\right]$$

into

$$\frac{d}{dy}\left[\frac{z_1 F}{RT}\phi\right] = \frac{1}{c}\left[\frac{z_1^2}{z_1 D_1 - z_2 D_2} c_M U - P\frac{dc}{dy}\right]$$

and use $v_y = -\dfrac{\rho_M}{\rho} U$ to obtain

$$\frac{d}{dy}\left[\frac{z_1 F}{RT}\phi\right] = \frac{z_1^2 c_M U}{c D_1 [z_1 - z_2]}\left[1 + \frac{D_1 - D_2}{D_2}\frac{\rho_M}{\rho}\frac{c/z_1}{c_M}\right]$$

Then, as $\dfrac{1}{z_1}\dfrac{\rho_M}{\rho}\dfrac{c}{c_M} = \dfrac{\rho_1}{\rho} < 1$, $\dfrac{d\phi}{dy}$ must always be negative.

In addition to the signs of dc/dy and $d\phi/dy$, the value of $e^{\frac{z_1 F}{RT}[\Delta\phi^C - \Delta\phi_{EQ}]}$ will also turn out to be of interest. It is determined, in terms of U, by the Butler–Volmer equation at the cathode, that is, by

$$\frac{z_1 F c_M U}{i_0} = \left[e^{\frac{z_1 F}{RT}[\Delta\phi^C - \Delta\phi_{EQ}]}\right]^{[1-\beta]} - \frac{c^C}{\overline{c}}\left[e^{\frac{z_1 F}{RT}[\Delta\phi^C - \Delta\phi_{EQ}]}\right]^{-\beta} \quad (9.17)$$

Now, due to the fact that the left-hand side of equation (9.17) must be negative, $\dfrac{c^C}{\overline{c}}\left[e^{\frac{z_1 F}{RT}[\Delta\phi^C - \Delta\phi_{EQ}]}\right]^{-\beta}$ must always exceed $\left[e^{\frac{z_1 F}{RT}[\Delta\phi^C - \Delta\phi_{EQ}]}\right]^{[1-\beta]}$ whence there obtains

$$1 > \frac{c^C}{\overline{c}} > e^{\frac{z_1 F}{RT}[\Delta\phi^C - \Delta\phi_{EQ}]}$$

By this, $\dfrac{z_1 F}{RT}[\Delta\phi^C - \Delta\phi_{EQ}]$ must be less than zero and it must decrease to negative infinity as c^C decreases to zero. In this limit, the first term on the right-hand side of equation (9.17) can be omitted and then, as c^C approaches zero and U approaches U_{lim}, $e^{\frac{z_1 F}{RT}[\Delta\phi^C - \Delta\phi_{EQ}]}$ can be approximated by

$$e^{\frac{z_1 F}{RT}[\Delta\phi^C - \Delta\phi_{EQ}]} = \left[\frac{c^C}{\overline{c}}\frac{i_0}{z_1 F c_M [-U]}\right]^{\frac{1}{\beta}} \quad (9.18)$$

This will prove to be useful later on.

9.6   The Perturbation Problem

The base solution will be denoted by affixing a subscript zero to the base variables. It is defined on the reference domain, namely on

$$0 < y_0 < L_0, \quad -\infty < x_0 < \infty$$

If the range of x_0 were finite instead of infinite, then homogeneous Neumann conditions would be specified along the side walls.

To determine the stability of the base solution, a small displacement will be imposed on the shapes of the electrode surfaces. This is indicated in Figure 9.6. The displaced domain, then, differs from the reference domain and it is taken to lie between the surfaces

$$y = Y^A(x, t, \epsilon)$$

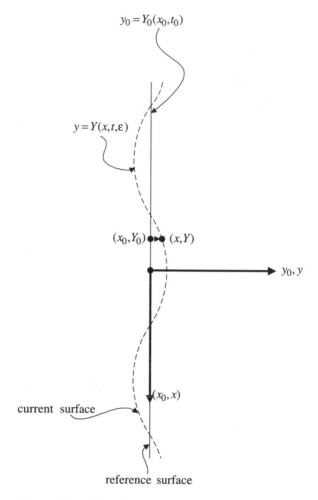

Figure 9.6. *A Displacement of the Reference Surface*

and

$$y = Y^C(x, \ t, \ \epsilon)$$

where ϵ denotes the amplitude of the displacement.

To account for a small displacement, the functions Y^A and Y^C will be expanded in powers of ϵ as

$$Y = Y_0 + \epsilon Y_1(x_0, \ t_0) + \cdots$$

where Y_0^A and Y_0^C identify the reference surfaces via $Y_0^A = 0$ and $Y_0^C = L_0$ and where Y_1^A and Y_1^C determine their displacement to order ϵ. Both Y_1^A and Y_1^C are part of the solution to the perturbation problem.

To obtain the perturbation equations at the electrode surfaces, the surface variables acquire their expansions via the expansion of Y while domain variables evaluated at the surface must be expanded in the form

$$c = c_0 + \epsilon \left[c_1 + Y_1 \frac{dc_0}{dy_0} \right] + \ldots$$

and

$$\phi = \phi_0 + \epsilon \left[\phi_1 + Y_1 \frac{d\phi_0}{dy_0} \right] + \ldots$$

where c_1 and ϕ_1 depend on x_0 and t_0 and where the subscript 1 no longer denotes species 1, except in the case z_1, but, instead, it denotes a term arising in the expansion of a variable in powers of ϵ.

The third discussion note introduces conditions that must be satisfied by Y_1^A, Y_1^C and c_1, due to our intention to impose perturbations on the base state that do not add mass to the system. In our case, where the density of the electrolytic solution is assumed to be independent of its ion concentration, these conditions apply only to zero wave number perturbations, where they take the simple form

$$Y_1^A = Y_1^C \tag{9.19}$$

and

$$\int_0^{L_0} c_1 \, dy_0 = c_0^A Y_1^A - c_0^C Y_1^C$$

9.6.1   The Total Current Drawn by the Cell

The current drawn by the cell cannot be a control variable. It is our aim to explain this before we present the perturbation equations.

To begin, let the width of the cell in the transverse direction be denoted by W_0 and notice that all variables are per unit depth of a two-dimensional cell.

Then, let I^A denote the total current passing through the cell at the anode. It is given by

$$I^A = \int_{S^A} dA \, [\vec{n} \cdot \vec{i}]^A$$

where in our two-dimensional problem dA is ds multiplied by the depth of the cell. Per unit depth, then, I^A can be written

$$I^A = \int_0^{W_0} \sqrt{1 + [Y_x^A]^2} \, dx \, \left[-z_1 F c_M [\vec{n}^A \cdot \vec{u}^A + \vec{n}^A \cdot \vec{U}] \right]$$

and, using $\vec{n}^A \cdot \vec{u}^A = u^A = \dfrac{Y_t^A}{\sqrt{1+[Y_x^A]^2}}$ and $\vec{n}^A \cdot \vec{U} = \dfrac{U}{\sqrt{1+[Y_x^A]^2}}$, this becomes

$$I^A = -z_1 F c_M \int_0^{W_0} dx_0 \, [Y_t^A + U]$$

where

$$I_0^A = -z_1 F c_M U \int_0^{W_0} dx_0$$

Then, expanding I^A as $I_0^A + \epsilon I_1^A + \cdots$, where $I_1^A = \dfrac{d}{d\epsilon} I^A(\epsilon = 0)$, and using $Y_t^A = \epsilon \dfrac{\partial Y_1^A}{\partial t_0} + \cdots$, there obtains

$$I_1^A = -z_1 F c_M \int_0^{W_0} \frac{\partial Y_1^A}{\partial t_0} \, dx_0$$

This tells us that a small, non-constant, periodic disturbance does not perturb the total current.

That the total current cannot be perturbed might be telling us that the total current must be a control variable in an electrodeposition experiment (i.e., a variable that must be held fixed). However, this conclusion cannot be entertained for two reasons; the first is technical, the second is physical.

First, the electrical potential at the cathode, V^C, could not then be an input variable. But, as an output variable, and hence as part of the solution to a problem, it, like c, ϕ, Y^A, Y^C, etc., would acquire a displacement away from its steady value when that solution has a small disturbance imposed upon it. Now, this displacement would be required to be uniform over the cathode, due to its infinite electrical conductivity, and this, then, would rule out non-constant periodic displacements being assigned to the other output variables. To determine periodic solutions to the perturbation equations, while maintaining the electrical conductivity of the metal electrodes infinitely large, requires the electrical potential of the cathode to be an input or control variable in order that its perturbation vanish. This is our plan henceforth. The electrical potential of the anode, V^A, is taken to be zero, to establish a reference, and the electrical potential of the cathode, V^C, $V^C < 0$, is what drives the problem. The control variables, then, are $\bar{c}$, L_0 and V^C.

Second, except in special cases, the variables V^C and I^A, when their values are assigned, impart different amounts of information to the nonlinear problem. If V^C is set, it is set everywhere along the metal side of the cathode surface, due to the perfect conductivity of the metal. Whereas, if I^A is set, only the sum total of what is taking place locally is set and this may leave the nonlinear problem underdetermined. By setting I^A, and not

$[\vec{n} \cdot \vec{i}]^A$, only the total rate of electron transfer is set, not the local rate, or only the average motion of the interface is set, not the local motion.

Again, taking V^C to be a control variable fixes its value at each point of the cathode surface and, then, the equation $V_1^C = 0$ at each such point leads to a closed set of perturbation equations. If, on the other hand, I^A could be a control variable, then V_1^C would need to be determined and the equation $I_1^A = 0$, holding on the average, not pointwise, would be insufficient to do this. In fact, I_1^A must be zero whether or not V^C is a control variable (i.e., whether or not $V_1^C = 0$). The perturbation equations would then be insufficient in number and the perturbation problem would require at least one more piece of information to determine V_1^C.

The input variables are then L_0, $\bar{c}$ and V^C. Their values determine the steady solution and its stability.

9.6.2 The Perturbation Equations at Fixed Values of L_0, $\bar{c}$ and V^C

The perturbation equations can be obtained by substituting into the nonlinear equations the expansions of the variables that appear therein. Collecting terms to order one produces the perturbation equations. The perturbation problem will be a linear initial-value problem in c_1, ϕ_1, Y_1^A and Y_1^C defined on the reference domain. To solve it, we turn it into an eigenvalue problem by taking the time dependence to be $e^{\sigma t_0}$. The variable σ is then the eigenvalue and it determines whether a perturbation in the form of the corresponding eigenfunction grows or dies.

To obtain the eigenvalue problem, derive the perturbation equations in the usual way[7] and then substitute

$$c_1 = \hat{c}_1(y_0)e^{\sigma t_0}e^{ikx_0}$$

$$\phi_1 = \hat{\phi}_1(y_0)e^{\sigma t_0}e^{ikx_0}$$

$$Y_1^A = \hat{Y}_1^A e^{\sigma t_0}e^{ikx_0}$$

and

$$Y_1^C = \hat{Y}_1^C e^{\sigma t_0}e^{ikx_0}$$

The result, taking into account that c_0 and ϕ_0 depend at most on y_0, that Y_0^A and Y_0^C do not depend on x_0 or t_0 and that $\vec{U}$ is fixed and lies in the $\vec{j}$ direction, is as follows:

---

[7] This is left to the reader. But some help can be found in the last endnote

First, on the reference domain (viz., on $0 < y_0 < L_0$), $\hat{c}_1$ and $\hat{\phi}_1$ must satisfy

$$D\left[\frac{d^2}{dy_0^2} - k^2\right]\hat{c}_1 - \hat{v}_{y_1}\frac{dc_0}{dy_0} - v_{y_0}\frac{d\hat{c}_1}{dy_0} = \sigma\hat{c}_1 \tag{9.20}$$

and

$$P\left[\frac{d^2}{dy_0^2} - k^2\right]\hat{c}_1 + \frac{dc_0}{dy_0}\left[\frac{z_1 F}{RT}\right]\frac{d\hat{\phi}_1}{dy_0} + c_0\left[\frac{z_1 F}{RT}\right]\left[\frac{d^2}{dy_0^2} - k^2\right]\hat{\phi}_1$$

$$+ \left[\frac{z_1 F}{RT}\right]\frac{d\phi_0}{dy_0}\frac{d\hat{c}_1}{dy_0} + \left[\frac{z_1 F}{RT}\right]\frac{d^2\phi_0}{dy_0^2}\hat{c}_1 = 0 \tag{9.21}$$

Then at the anode surface, four equations must be satisfied. The first three are

$$D\frac{d\hat{c}_1}{dy_0} - v_{y_0}\hat{c}_1 - c_0\left[\hat{v}_{y_1} - \sigma\hat{Y}_1^A\right] = -\frac{z_1 z_2 D_2}{z_1 D_1 - z_2 D_2}c_M\sigma\hat{Y}_1^A \tag{9.22}$$

$$P\frac{d\hat{c}_1}{dy_0} + \hat{c}_1\left[\frac{z_1 F}{RT}\right]\frac{d\phi_0}{dy_0} + c_0\left[\frac{z_1 F}{RT}\right]\frac{d\hat{\phi}_1}{dy_0} = \frac{z_1^2}{z_1 D_1 - z_2 D_2}c_M\sigma\hat{Y}_1^A \tag{9.23}$$

and

$$\hat{v}_{y_1} = \left[1 - \frac{\rho_M}{\rho}\right]\sigma\hat{Y}_1^A \tag{9.24}$$

The fourth equation at $y_0 = 0$ comes from the expansion of the Butler–Volmer equation. It takes its form from the expansion of the variables c^A, ϕ^A, $2H^A$, etc. appearing therein and it is

$$-\frac{z_1 F c_M}{i_0}\sigma\hat{Y}_1^A = \left[[1 - \beta]e^{[1-\beta]\frac{z_1 F}{RT}[\Delta\phi_0^A - \Delta\phi_{EQ}]} + \frac{c_0^A}{\bar{c}}\beta e^{-\beta\frac{z_1 F}{RT}[\Delta\phi_0^A - \Delta\phi_{EQ}]}\right]$$

$$\times\left[\frac{z_1 F}{RT}\right]\left[\Delta\hat{\phi}_1^A + \hat{Y}_1^A\frac{d}{dy_0}\Delta\phi_0^A + \frac{\gamma}{z_1 F c_M}k^2\hat{Y}_1^A\right]$$

$$-e^{-\beta\frac{z_1 F}{RT}[\Delta\phi_0^A - \Delta\phi_{EQ}]}\frac{[\hat{c}_1^A + \hat{Y}_1^A\frac{dc_0^A}{dy_0}]}{\bar{c}} \tag{9.25}$$

where $\Delta\hat{\phi}_1^A = -\hat{\phi}_1^A$.

At the cathode surface, a set of equations must hold that are much like those holding at the anode surface. The first three can be taken over directly. They are

$$D\frac{d\hat{c}_1}{dy_0} - v_{y_0}\hat{c}_1 - c_0\left[\hat{v}_{y_1} - \sigma\hat{Y}_1^C\right] = \frac{-z_1 z_2 D_2}{z_1 D_1 - z_2 D_2}c_M\sigma\hat{Y}_1^C \tag{9.26}$$

$$P\frac{d\hat{c}_1}{dy_0} + \hat{c}_1\left[\frac{z_1 F}{RT}\right]\frac{d\phi_0}{dy_0} + c_0\left[\frac{z_1 F}{RT}\right]\frac{d\hat{\phi}_1}{dy_0} = \frac{z_1^2}{z_1 D_1 - z_2 D_2}\, c_M \sigma \hat{Y}_1^C \quad (9.27)$$

and

$$\hat{v}_{y_1} = \left[1 - \frac{\rho_M}{\rho}\right]\sigma \hat{Y}_1^C \quad (9.28)$$

The fourth equation differs by two sign changes from the corresponding equation at the anode surface. It is

$$+\frac{z_1 F c_M}{i_0}\,\sigma \hat{Y}_1^C = \left[[1 - \beta]e^{[1-\beta]\frac{z_1 F}{RT}[\Delta\phi_0^C - \Delta\phi_{EQ}]} + \frac{c_0^C}{\bar{c}}\beta e^{-\beta\frac{z_1 F}{RT}[\Delta\phi_0^C - \Delta\phi_{EQ}]}\right]$$

$$\times \left[\frac{z_1 F}{RT}\right]\left[\Delta\hat{\phi}_1^C + \hat{Y}_1^C\frac{d}{dy_0}\Delta\phi_0^C - \frac{\gamma}{z_1 F c_M}k^2\hat{Y}_1^C\right]$$

$$-e^{-\beta\frac{z_1 F}{RT}[\Delta\phi_0^C - \Delta\phi_{EQ}]}\frac{[\hat{c}_1^C + \hat{Y}_1^C\frac{dc_0^C}{dy_0}]}{\bar{c}} \quad (9.29)$$

where $\Delta\hat{\phi}_1^C = -\hat{\phi}_1^C$.

It remains to solve these equations.

9.7 The Stability of the Equilibrium Base Solution

The base solution obtained by setting V^C to zero is not only a steady solution, it is an equilibrium solution. In this case, U and v_{y_0} vanish while c_0 and ϕ_0 take the constant values $\bar{c}$ and $-\Delta\phi_{EQ}$. This is the simplest base solution. The electrode surfaces are at rest in the laboratory frame, while the ion concentration and the electrical potential remain uniform throughout the electrolytic solution. The cell draws no current.

The perturbation equations then reduce to

$$\left[\frac{d^2}{dy_0^2} - k^2\right]\hat{c}_1 - \frac{\sigma}{D}\hat{c}_1 \qquad\qquad (9.30)$$

$$\left[\frac{d^2}{dy_0^2} - k^2\right]\left[P\hat{c}_1 + \bar{c}\frac{z_1 F}{RT}\hat{\phi}_1\right] = 0 \qquad (9.31)$$

$$0 \le y_0 \le L_0$$

$$\frac{d\hat{c}_1}{dy_0} = \left[-\frac{\rho_M}{\rho}\bar{c} - \frac{z_1 z_2 D_2 c_M}{z_1 D_1 - z_2 D_2}\right]\frac{\sigma}{D}\hat{Y}_1^A \tag{9.32}$$

$$\frac{d}{dy_0}\left[P\hat{c}_1 + \bar{c}\frac{z_1 F}{RT}\hat{\phi}_1\right] = \frac{z_1^2}{z_1 D_1 - z_2 D_2}c_M \sigma \hat{Y}_1^A \tag{9.33}$$

$$y_0 = 0$$

$$\hat{v}_{y_1} = \left[1 - \frac{\rho_M}{\rho}\right]\sigma\hat{Y}_1^A$$

$$-\left[\frac{z_1 F c_M}{i_0}\sigma + \frac{\gamma}{c_M RT}k^2\right]\hat{Y}_1^A = -\frac{z_1 F}{RT}\hat{\phi}_1 - \frac{\hat{c}_1}{\bar{c}} \tag{9.34}$$

and

$$\frac{d\hat{c}_1}{dy_0} = \left[-\frac{\rho_M}{\rho}\bar{c} - \frac{z_1 z_2 D_2 c_M}{z_1 D_1 - z_2 D_2}\right]\frac{\sigma}{D}\hat{Y}_1^C \tag{9.35}$$

$$\frac{d}{dy_0}\left[P\hat{c}_1 + \bar{c}\frac{z_1 F}{RT}\hat{\phi}_1\right] = \frac{z_1^2}{z_1 D_1 - z_2 D_2}c_M \sigma \hat{Y}_1^C \tag{9.36}$$

$$y_0 = L_0$$

$$\hat{v}_{y_1} = \left[1 - \frac{\rho_M}{\rho}\right]\sigma\hat{Y}_1^C$$

$$\left[\frac{z_1 F c_M}{i_0}\sigma + \frac{\gamma}{c_M RT}k^2\right]\hat{Y}_1^C = -\frac{z_1 F}{RT}\hat{\phi}_1 - \frac{\hat{c}_1}{\bar{c}} \tag{9.37}$$

This eigenvalue problem determines, as its solution, $\hat{c}_1$, $\hat{\phi}_1$, $\hat{Y}_1^A$ and $\hat{Y}_1^C$ and it has solutions other than $\hat{c}_1 = 0$, $\hat{\phi}_1 = 0$, $\hat{Y}_1^A = 0$ and $\hat{Y}_1^C = 0$ only for special values of σ. Determining these values of σ as a function of k^2, will tell us whether a disturbance of wavelength $2\pi/k$, imposed on the equilibrium state, will grow or die out.

To solve for $\hat{c}_1$ and then for $P\hat{c}_1 + \bar{c}\frac{z_1 F}{RT}\hat{\phi}_1$ in terms of $\hat{Y}_1^A$ and $\hat{Y}_1^C$, notice that equation (9.30) for $\hat{c}_1$ can be satisfied by writing

$$\hat{c}_1 = A_1 e^{my_0} + B_1 e^{-my_0}$$

where $m = \sqrt{\frac{\sigma}{D} + k^2}$ and where A_1 and B_1 must be set to satisfy equations (9.32) and (9.35) which lead to

$$A_1 - B_1 = \left[-\frac{\rho_M}{\rho}\bar{c} - \frac{z_1 z_2 D_2}{z_1 D_1 - z_2 D_2}c_M\right]\frac{\sigma}{mD}\hat{Y}_1^A$$

and

$$A_1 e^{mL_0} - B_1 e^{-mL_0} = \left[-\frac{\rho_M}{\rho}\bar{c} - \frac{z_1 z_2 D_2}{z_1 D_1 - z_2 D_2}c_M\right]\frac{\sigma}{mD}\hat{Y}_1^C$$

whereupon the value of $\hat{c}_1$ at the anode turns out to be

$$\hat{c}_1^A = A_1 + B_1$$

$$= \left[\frac{\rho_M}{\rho}\,\bar{c} + \frac{z_1 z_2 D_2}{z_1 D_1 - z_2 D_2} c_M\right]\frac{\sigma}{mD}\left[\hat{Y}_1^A \coth mL_0 - \hat{Y}_1^C \operatorname{csch} mL_0\right] \quad (9.38)$$

while at the cathode, it turns out to be

$$\hat{c}_1^C = A_1 e^{mL_0} + B_1 e^{-mL_0}$$

$$= \left[\frac{\rho_M}{\rho}\,\bar{c} + \frac{z_1 z_2 D_2}{z_1 D_1 - z_2 D_2} c_M\right]\frac{\sigma}{mD}\left[\hat{Y}_1^A \operatorname{csch} mL_0 - \hat{Y}_1^C \coth mL_0\right] \quad (9.39)$$

Likewise, equation (9.31) for $P\hat{c}_1 + \bar{c}\dfrac{z_1 F}{RT}\hat{\phi}_1$ can be satisfied by writing

$$P\hat{c}_1 + \bar{c}\frac{z_1 F}{RT}\hat{\phi}_1 = A_2 e^{ky_0} + B_2 e^{-ky_0}$$

where k is taken to be non negative and where A_2 and B_2 must be set to satisfy equations (9.33) and (9.36) whence there obtains

$$A_2 - B_2 = \frac{z_1^2}{z_1 D_1 - z_2 D_2}\,c_M\,\frac{\sigma}{k}\,\hat{Y}_1^A$$

and

$$A_2 e^{kL_0} - B_2 e^{-kL_0} = \frac{z_1^2}{z_1 D_1 - z_2 D_2}\,c_M\,\frac{\sigma}{k}\,\hat{Y}_1^C$$

and the value of $P\hat{c}_1 + \bar{c}\dfrac{z_1 F}{RT}\hat{\phi}_1$ at the anode turns out to be

$$P\hat{c}_1^A + \bar{c}\frac{z_1 F}{RT}\hat{\phi}_1^A = A_2 + B_2$$

$$= -\left[\frac{z_1^2}{z_1 D_1 - z_2 D_2}\right]c_M\frac{\sigma}{k}\left[\hat{Y}_1^A \coth kL_0 - \hat{Y}_1^C \operatorname{csch} kL_0\right] \quad (9.40)$$

while at the cathode, it is

$$P\hat{c}_1^C + \bar{c}\frac{z_1 F}{RT}\hat{\phi}_1^C = A_2 e^{kL_0} + B_2 e^{-kL_0}$$

$$= -\left[\frac{z_1^2}{z_1 D_1 - z_2 D_2}\right]c_M\frac{\sigma}{k}\left[\hat{Y}_1^A \operatorname{csch} kL_0 - \hat{Y}_1^C \coth kL_0\right] \quad (9.41)$$

By using equations (9.38) through (9.41), the variables $\hat{c}_1$ and $\hat{\phi}_1$ can be eliminated from the perturbed Butler–Volmer equations, first at the anode and then at the cathode. These are equations (9.34) and (9.37). This leads to two equations in $\hat{Y}_1^A$ and $\hat{Y}_1^C$. They are

$$-\left[\frac{z_1 F c_M}{i_0}\,\sigma + \frac{\gamma}{c_M RT}\,k^2\right]\hat{Y}_1^A = \frac{P-1}{\bar{c}}\,[A_1 + B_1] - \frac{1}{\bar{c}}\,[A_2 + B_2]$$

and

$$\left[\frac{z_1 F c_M}{i_0} \sigma + \frac{\gamma}{c_M RT} k^2 \right] \hat{Y}_1^C$$

$$= \frac{P-1}{\bar{c}} \left[A_1 e^{mL_0} + B_1 e^{-mL_0} \right] - \frac{1}{\bar{c}} \left[A_2 e^{kL_0} + B_2 e^{-kL_0} \right]$$

and they are linear, homogeneous, algebraic equations in $\hat{Y}_1^A$ and $\hat{Y}_1^C$. These equations can be written

$$\begin{pmatrix} A & B \\ B & A \end{pmatrix} \begin{pmatrix} \hat{Y}_1^A \\ \hat{Y}_1^C \end{pmatrix} = \begin{pmatrix} 0 \\ 0 \end{pmatrix}$$

where

$$A = -\frac{P-1}{\bar{c}} \left[-\frac{\rho_M}{\rho} \bar{c} - \frac{z_1 z_2 D_2}{z_1 D_1 - z_2 D_2} c_M \right] \frac{\sigma}{mD} \coth mL_0$$

$$+ \frac{1}{\bar{c}} \left[\frac{z_1^2}{z_1 D_1 - z_2 D_2} c_M \right] \frac{\sigma}{k} \coth kL_0 + \left[\frac{z_1 F c_M}{i_0} \right] \sigma + \left[\frac{\gamma}{c_M RT} \right] k^2$$

and

$$B = \frac{P-1}{\bar{c}} \left[-\frac{\rho_M}{\rho} \bar{c} - \frac{z_1 z_2 D_2}{z_1 D_1 - z_2 D_2} c_M \right] \frac{\sigma}{mD} \operatorname{csch} mL_0$$

$$- \frac{1}{\bar{c}} \left[\frac{z_1^2}{z_1 D_1 - z_2 D_2} c_M \right] \frac{\sigma}{k} \operatorname{csch} kL_0$$

Now, a solution other than $\hat{Y}_1^A = 0 = \hat{Y}_1^C$ can be found if and only if

$$\det \begin{pmatrix} A & B \\ B & A \end{pmatrix} = A^2 - B^2 = 0$$

and this requires

$$A = \pm B$$

where each of the two possibilities establishes a formula for σ as a function of k^2. The two formulas, one coming from $A = B$, the other from $A = -B$, are implicit in σ. They are

$$-\frac{P-1}{\bar{c}}\left[-\frac{\rho_M}{\rho}\,\bar{c}-\frac{z_1 z_2 D_2}{z_1 D_1 - z_2 D_2}\,c_M\right]\frac{\sigma}{mD}\frac{1+\cosh mL_0}{\sinh mL_0}$$

$$+\frac{1}{\bar{c}}\left[\frac{z_1^2}{z_1 D_1 - z_2 D_2}\,c_M\right]\frac{\sigma}{k}\frac{1+\cosh kL_0}{\sinh kL_0}+\left[\frac{z_1 F c_M}{i_0}\right]\sigma$$

$$+\left[\frac{\gamma}{c_M RT}\right]k^2 = 0$$

and

$$-\frac{P-1}{\bar{c}}\left[-\frac{\rho_M}{\rho}\,\bar{c}-\frac{z_1 z_2 D_2}{z_1 D_1 - z_2 D_2}\,c_M\right]\frac{\sigma}{mD}\frac{1-\cosh mL_0}{\sinh mL_0}$$

$$+\frac{1}{\bar{c}}\left[\frac{z_1^2}{z_1 D_1 - z_2 D_2}\,c_M\right]\frac{\sigma}{k}\frac{1-\cosh kL_0}{\sinh kL_0}-\left[\frac{z_1 F c_M}{i_0}\right]\sigma$$

$$-\left[\frac{\gamma}{c_M RT}\right]k^2 = 0$$

For each value of k^2, each formula produces at least one value of σ, the first formula corresponding to eigenfunctions where $\hat{Y}_1^A = +\hat{Y}_1^C$, the second corresponding to eigenfunctions where $\hat{Y}_1^A = -\hat{Y}_1^C$.

The roots of these two equations will be taken to be real. There is every reason to imagine this to be so by virtue of the base state being a true equilibrium state, although a direct algebraic proof does not appear to be a simple matter. If σ is real, then, it cannot be positive, at least not if $\bar{c}$ is sufficiently small. To see this, notice that $P - 1$ and z_2 cannot be positive and that m, where $m = \sqrt{k^2 + \dfrac{\sigma}{D}}$, must be positive if σ is positive. Hence, if σ is positive, all terms on the left-hand sides of both equations must be positive and cannot add to zero. Due to this, σ must be negative for all positive values of k^2 and a cell at equilibrium must be stable to all small displacements.

Now, small wave numbers, those where the stabilizing effect of surface tension is the weakest, are of most interest as we prepare to turn our attention to nonequilibrium cells. At small values of k^2, the eigenvalues nearest zero, corresponding first to $Y_1^A = Y_1^C$, and then to $Y_1^A = -Y_1^C$, can be estimated via

$$\frac{\sigma}{Dk^2} \simeq -1 + [P-1]\left[-\frac{z_2 D_2}{z_1 D}\right]$$

and

$$\frac{\sigma}{Dk^2} \simeq \frac{-\dfrac{\bar{c}}{c_M}\dfrac{\gamma}{c_M RT}\dfrac{2}{L_0}}{\left[\dfrac{z_1^2 D}{z_1 D_1 - z_2 D_2}-[P-1]\dfrac{-z_1 z_2 D_2}{z_1 D_1 - z_2 D_2}+\dfrac{z_1 F\bar{c}}{i_0}\dfrac{2D}{L_0}\right]}$$

where, in these formulas, the small term $\dfrac{\rho_M}{\rho}\,\bar{c}$ has been dropped vis-à-vis

$-\dfrac{z_1 z_2 D_2}{z_1 D_1 - z_2 D_2}\, c_M.$

The two formulas would seem to tell us that a double root, $\sigma = 0$, at $k^2 = 0$ splits into two roots as k^2 increases. One branch is independent of $\bar{c}$, L_0, γ and i_0, the other is not. One formula depends only on what is going on in the solution. The other depends on what is going on at the electrode surfaces as well.

An order of magnitude approximation to the first formula is

$$\frac{\sigma}{Dk^2} = -2$$

And an order of magnitude approximation to the second formula is

$$\frac{\sigma}{Dk^2} = -\frac{\bar{c}}{c_M}\,\frac{\gamma/c_M RT}{\frac{1}{2}L_0}$$

which takes into account the fact that the denominator on the right-hand side of the second formula is ordinarily near one. By this, as long as the capillary length scale does not exceed half the distance between the electrodes by more than an order of magnitude, the graph of the second formula lies above the first, and this is ordinarily the case. It might be thought, then, that if a cell becomes unstable when it draws current, it would be the eigenvalue corresponding to this second formula that turns positive.

To see, in fact, whether or not $\sigma = 0$ can be a solution to the eigenvalue problem, set σ to zero, whereupon equations (9.30) through (9.37) reduce to

$$\left[\frac{d^2}{dy_0^2} - k^2\right]\hat{c}_1 = 0$$

$$\left[\frac{d^2}{dy_0^2} - k^2\right]\left[\frac{z_1 F}{RT}\,\hat{\phi}_1\right] = 0$$

$\quad\quad\quad\quad\quad 0 \le y_0 \le L_0$

$$\frac{d\hat{c}_1}{dy_0} = 0$$

$$\frac{d}{dy_0}\left[\frac{z_1 F}{RT}\,\hat{\phi}_1\right] = 0$$

$$\hat{v}_{y_1} = 0$$

$$-\frac{\gamma}{c_M RT}\,k^2 \hat{Y}_1^A = -\frac{z_1 F}{RT}\,\hat{\phi}_1 - \frac{\hat{c}_1}{\bar{c}}$$

$\quad\quad\quad\quad\quad y_0 = 0$

and

$$\left.\begin{array}{c} \dfrac{d\hat{c}_1}{dy_0} = 0 \\[2ex] \dfrac{d}{dy_0}\left[\dfrac{z_1 F}{RT}\,\hat{\phi}_1\right] = 0 \\[2ex] \hat{v}_{y_1} = 0 \\[2ex] \dfrac{\gamma}{c_M RT}\,k^2 \hat{Y}_1^C = -\dfrac{z_1 F}{RT}\,\hat{\phi}_1 - \dfrac{\hat{c}_1}{\bar{c}} \end{array}\right\} \qquad y_0 = L_0$$

This is an eigenvalue problem in k^2. It has no solution other than $\hat{c}_1 = 0$, $\hat{\phi}_1 = 0$, $\hat{Y}_1^A = 0$, $\hat{Y}_1^C = 0$ as long as $k^2 > 0$. However, it has a solution when $k^2 = 0$, a solution wherein $\hat{c}_1$ and $\hat{\phi}_1$ take constant values satisfying $\dfrac{z_1 F}{RT}\,\hat{\phi}_1 + \dfrac{\hat{c}_1}{\bar{c}} = 0$ and where $\hat{Y}_1^A$ and $\hat{Y}_1^C$ can take any values. Hence, $\sigma = 0$ cannot be a solution to the perturbation eigenvalue problem if k^2 is positive. But, in case k^2 is zero, $\sigma = 0$ is a solution. Equation (9.19) then comes into play, with the result that the eigenfunction must take the form $\hat{Y}_1^A = \hat{Y}_1^C$, $\hat{c}_1 = 0$ and, hence, $\hat{\phi}_1 = 0$. Only one of the two branches of the dispersion formula that tend to $\sigma = 0$ as k^2 goes to zero, in fact arrives at $k^2 = 0$. It is the branch that is independent of the input variables. The other branch, then, must determine $k^2_{critical}$ as the cell begins to draw current.

In the first discussion note, $\sigma = 0$ is found to be a solution to the perturbation eigenvalue problem when $k^2 = 0$, whether or not the cell draws current. This solution always corresponds to $\hat{Y}_1^A = \hat{Y}_1^C$, not to $\hat{Y}_1^A = -\hat{Y}_1^C$. It is a single root, not a double root. It might then be thought that the equilibrium branch of the dispersion formula corresponding to $\hat{Y}_1^A = \hat{Y}_1^C$ is left behind as the cell begins to draw current and that it might block other branches from crossing the line $\sigma = 0$. If this is so, only one positive value of $k^2_{critical}$ ought to be found away from equilibrium, and this expectation will be borne out.

The stability result in the case of an equilibrium base state is the following: The equilibrium state is stable to any small displacement so long as its wavelength is finite, but it is neutral to an infinite wavelength displacement.

At equilibrium, then, there is a critical point at $k^2 = 0$. This critical point remains at $k^2 = 0$ as the cell draws current. However, a second critical point ought to appear and it ought to shift to the right as the cell moves away from equilibrium. Our job is to determine this second critical point.

9.8   The Stability of the Base Solution When Current Passes Through the Cell

In case the cell draws current, the base variables, c_0 and ϕ_0, are no longer uniform across the cell and this makes the determination of the eigenvalues corresponding to each wave number now a little more difficult than it was at equilibrium. Yet, determining a formula for the critical value of k^2 remains within the reach of pencil-and-paper work.

To do this, set σ to zero in the perturbation eigenvalue problem, turning it into an eigenvalue problem in k^2. The values of k^2 at which this problem has a solution other than $\hat{c}_1$, $\hat{\phi}_1$, $\hat{Y}_1^A$ and $\hat{Y}_1^C$ all zero are then critical values of k^2. There are two values of k^2 corresponding to $\sigma = 0$. One is again zero; it is the other that is of interest.

First, put σ to zero in equations (9.20) through (9.29), whence $\hat{c}_1$ and $\hat{\phi}_1$ satisfy

$$D \left[\frac{d^2}{dy_0^2} - k^2 \right] \hat{c}_1 - v_{y_0} \frac{d\hat{c}_1}{dy_0} - \hat{v}_{y_1} \frac{dc_0}{dy_0} = 0 \qquad (9.42)$$

and

$$P \left[\frac{d^2}{dy_0^2} - k^2 \right] \hat{c}_1 + \hat{c}_1 \left[\frac{z_1 F}{RT} \right] \frac{d^2\phi_0}{dy_0^2} + \frac{d\hat{c}_1}{dy_0} \left[\frac{z_1 F}{RT} \right] \frac{d\phi_0}{dy_0}$$

$$+ c_0 \left[\frac{z_1 F}{RT} \right] \left[\frac{d^2}{dy_0^2} - k^2 \right] \hat{\phi}_1 + \frac{dc_0}{dy_0} \left[\frac{z_1 F}{RT} \right] \frac{d\hat{\phi}_1}{dy_0} = 0 \qquad (9.43)$$

on $0 < y_0 < L_0$, while

$$D \frac{d\hat{c}_1}{dy_0} - v_{y_0} \hat{c}_1 = 0 \qquad (9.44)$$

$$P \frac{d\hat{c}_1}{dy_0} + \hat{c}_1 \left[\frac{z_1 F}{RT} \right] \frac{d\phi_0}{dy_0} + c_0 \left[\frac{z_1 F}{RT} \right] \frac{d\hat{\phi}_1}{dy_0} = 0 \qquad (9.45)$$

and

$$\hat{v}_{y_1} = 0$$

must hold at $y_0 = 0$ and at $y_0 = L_0$.

Only $\hat{c}_1$, $\hat{\phi}_1$ and $\hat{v}_{y_1}$ are present here, as $\hat{Y}_1^A$ and $\hat{Y}_1^C$ do not come into play until the perturbed Butler–Volmer equations at $y_0 = 0$ and at $y_0 = L_0$ are taken into account.

Now, $\hat{v}_{y_1}$ must be zero under neutral conditions. To see why this is so, notice that at neutral conditions the perturbation velocity $\vec{v}_1$ must satisfy

$$\rho \vec{v}_0 \cdot \nabla \vec{v}_1 = -\nabla p_1 + \mu \nabla^2 \vec{v}_1$$

and

$$\nabla \cdot \vec{v}_1 = 0$$

throughout the ionic solution and

$$\vec{v}_1 = \vec{0}$$

at its boundaries. The term $\partial \vec{v}_1 / \partial t$ is omitted at neutral conditions and $\vec{v}_1 \cdot \nabla \vec{v}_0$ is zero, as $\vec{v}_0$ is spatially uniform. Then multiply the perturbed Navier-Stokes equation by $\vec{v}_1 \cdot$ and integrate the product over V_0 to get

$$\rho \vec{v}_0 \cdot \iiint_{V_0} \nabla \frac{1}{2} |\vec{v}_1|^2 \, dV$$

$$= - \iiint_{V_0} \nabla \cdot [\vec{v}_1 p_1] \, dV + \mu \iiint_{V_0} \nabla \cdot \left[\nabla \frac{1}{2} |\vec{v}_1|^2 \right] \, dV$$

$$- \mu \iiint_{V_0} \nabla \vec{v}_1 : \nabla \vec{v}_1^T \, dV$$

which simplifies to

$$\rho \vec{v}_0 \cdot \iint_{S_0} dA \, \vec{n} \frac{1}{2} |\vec{v}_1|^2$$

$$= - \iint_{S_0} dA \, \vec{n} \cdot \vec{v}_1 p_1 + \mu \iint_{S_0} dA \, \vec{n} \cdot \nabla \frac{1}{2} |\vec{v}_1|^2$$

$$- \mu \iiint_{V_0} \nabla \vec{v}_1 : \nabla \vec{v}_1^T \, dV$$

where $\nabla \cdot \vec{v}_1 = 0$ has been used, where S_0 denotes the boundary of V_0 and where $\vec{n} \cdot \nabla \frac{1}{2} |\vec{v}_1|^2 = |\vec{v}_1| \frac{d}{dn} |\vec{v}_1|$. Now, as $\vec{v}_1$ must vanish at the boundary, this result requires

$$\iiint_{V_0} \nabla \vec{v}_1 : \nabla \vec{v}_1^T \, dV = 0$$

whence $\vec{v}_1$ can be, at most, a constant; hence, it must vanish everywhere.

Due to the fact that $\hat{v}_{y_1}$ must be zero, $\hat{c}_1$ must now satisfy

$$D \left[\frac{d^2}{dy_0^2} - k^2 \right] \hat{c}_1 - v_{y_0} \frac{d\hat{c}_1}{dy_0} = 0 \tag{9.46}$$

on $0 < y_0 < L_0$ and equation (9.44) at both $y_0 = 0$ and $y_0 = L_0$. These equations can be satisfied by writing

$$\hat{c}_1 = A e^{m_+ y_0} + B e^{m_- y_0}$$

where

$$2m_{\pm} = \frac{v_{y_0}}{D} \pm \sqrt{\frac{v_{y_0}^2}{D^2} + 4k^2}$$

and where

$$\left[m_+ - \frac{v_{y_0}}{D}\right] A + \left[m_- - \frac{v_{y_0}}{D}\right] B = 0$$

and

$$e^{m_+ L_0} \left[m_+ - \frac{v_{y_0}}{D}\right] A + e^{m_- L_0} \left[m_- - \frac{v_{y_0}}{D}\right] B = 0$$

must hold, whereupon A and B must be zero because

$$\det \begin{pmatrix} \left[m_+ - \frac{v_{y_0}}{D}\right] & \left[m_- - \frac{v_{y_0}}{D}\right] \\ e^{m_+ L_0} \left[m_+ - \frac{v_{y_0}}{D}\right] & e^{m_- L_0} \left[m_- - \frac{v_{y_0}}{D}\right] \end{pmatrix}$$

$$= -k^2 [e^{m_- L_0} - e^{m_+ L_0}]$$

and this is not zero so long as k^2 is not zero.

Due to this, $\hat{c}_1$ must be zero, and substituting $\hat{c}_1 = 0$ into equations (9.43) and (9.45), $\hat{\phi}_1$ must satisfy

$$\frac{d}{dy_0} \left[c_0 \frac{d\hat{\phi}_1}{dy_0}\right] - k^2 c_0 \hat{\phi}_1 = 0$$

on $0 < y_0 < L_0$ and

$$c_0 \frac{d\hat{\phi}_1}{dy_0} = 0$$

at both $y_0 = 0$ and $y_0 = L_0$. Now, c_0 is positive for all values of y_0 whence this Sturm–Liouville problem has only the solution $\hat{\phi}_1 = 0$.

Our first conclusion, then, is this: At neutral conditions $\hat{v}_{y_1}$, $\hat{c}_1$ and $\hat{\phi}_1$ must all vanish and, hence, the critical value of k^2 must come out of the perturbed Butler–Volmer equations.

Turning to the anode, the perturbed Butler–Volmer equation [i.e., equation (9.25)], comes down to

$$0 = \left[[1-\beta]e^{[1-\beta]\frac{z_1 F}{RT}[-\phi_0^A - \Delta\phi_{EQ}]} - \frac{c_0^A}{\bar{c}}[-\beta]e^{-\beta\frac{z_1 F}{RT}[-\phi_0^A - \Delta\phi_{EQ}]}\right]$$

$$\times \left[-\frac{z_1 F}{RT}\frac{d\phi_0^A}{dy_0} + \frac{\gamma}{c_M RT} k^2\right]\hat{Y}_1^A - e^{-\beta\frac{z_1 F}{RT}[-\phi_0^A - \Delta\phi_{EQ}]}\left[\frac{1}{\bar{c}}\frac{dc_0^A}{dy_0}\right]\hat{Y}_1^A$$

on setting σ, $\hat{c}_1^A$ and $\hat{\phi}_1^A$ to zero. This is a linear, homogeneous equation in $\hat{Y}_1^A$ and, as $d\phi_0^A/dy_0$ and dc_0^A/dy_0 must both be negative, the factor multiplying $\hat{Y}_1^A$ cannot vanish; hence, the only solution is $\hat{Y}_1^A = 0$.

It remains to turn to the Butler–Volmer equation at the cathode, equation (9.29), and to try to find conditions under which $\hat{Y}_1^C$ need not vanish. On setting σ, $\hat{c}_1^C$ and $\hat{\phi}_1^C$ to zero, this turns out to be a linear, homogeneous equation in $\hat{Y}_1^C$. It is

$$0 = \left[[1-\beta]e^{[1-\beta]\frac{z_1 F}{RT}[V^C-\phi_0^C-\Delta\phi_{EQ}]} + \frac{c_0^C}{\bar{c}}\beta e^{-\beta\frac{z_1 F}{RT}[V^C-\phi_0^C-\Delta\phi_{EQ}]} \right]$$

$$\times \left[-\frac{z_1 F}{RT}\frac{d\phi_0^C}{dy_0} - \frac{\gamma}{c_M RT}k^2 \right]\hat{Y}_1^C - e^{-\beta\frac{z_1 F}{RT}[V^C-\phi_0^C-\Delta\phi_{EQ}]}\left[\frac{1}{\bar{c}}\frac{dc_0^C}{dy_0} \right]\hat{Y}_1^C$$

and it has a solution other than $\hat{Y}_1^C = 0$ if and only if the factor multiplying $\hat{Y}_1^C$ can be made to vanish. That this is possible at the cathode, whereas it was not at the anode, is due to the sign by which the term $\dfrac{\gamma}{c_M RT}k^2$ comes into the two equations, as the derivatives, $d\phi_0/dy_0$ and dc_0/dy_0, come into the problem with the same signs at both electrodes.

It is the cathode, then, that determines the critical wave number, not the anode. The different physics at the two electrode surfaces comes down to the different signs by which the surface tension term comes into the Butler–Volmer equation at the two interfaces.

If γ were set to zero, it would appear that $\hat{Y}_1^C$ must then be zero. However, if γ becomes small, k^2 must become large and the unstable range of wave numbers must increase. It is possible only to say that the product γk^2 is what this equation predicts.

Setting the factor multiplying $\hat{Y}_1^C$ to zero leads to a formula for the critical wave number. It is

$$\frac{\gamma}{c_M RT}k^2_{critical}$$

$$= -\frac{z_1 F}{RT}\frac{d\phi_0^C}{dy_0} - \frac{1}{[1-\beta]e^{\frac{z_1 F}{RT}[V^C-\phi_0^C-\Delta\phi_{EQ}]} + \beta\frac{c_0^C}{\bar{c}}}\frac{1}{\bar{c}}\frac{dc_0^C}{dy_0} \qquad (9.47)$$

and the fact that the instability comes out of the cathode is a prediction that first came out of the pictures drawn in the early paragraphs of this essay.

This formula for $k^2_{critical}$ is based on two approximations: The first is constant and uniform solution density, the second is local electrical neutrality. The formula, itself, does not depend on the specific form taken by the base solution, so long as it is one dimensional. However, the base solution is needed to evaluate the terms on the right-hand side, and different additional approximations will lead to different estimates of $k^2_{critical}$.

Now, $d\phi_0^C/dy_0$, dc_0^C/dy_0 and c_0^C can be evaluated directly in terms of U, $\bar{c}$ and L_0. But ϕ_0^C presents a problem. To resolve it, integrate the ϕ equation across the ionic solution to determine $\phi_0^C - \phi_0^A$. This and the two Butler–Volmer equations, one at the anode and one at the cathode, can then be

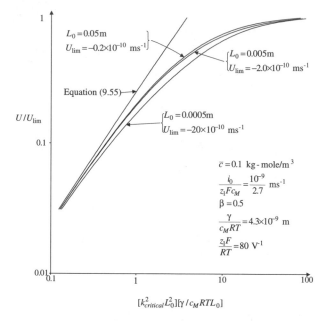

Figure 9.7. *A Graph of $k^2_{critical}$ versus U from Equation (9.47)*

solved for either V^C, ϕ_0^C and ϕ_0^A in terms of U, or U, ϕ_0^C and ϕ_0^A in terms of V^C. By doing this, $k^2_{critical}$ can be determined as it depends either on U or on V^C at fixed values of $\bar{c}$ and L_0. Two such graphs are drawn as Figures 9.7 and 9.8 using data corresponding to a copper, copper-sulphate system. These graphs present information about the stability of the base solution; hence, on both graphs, V^C is the control variable, not U.

The graphs indicate that our problem has some structure. To see what accounts for this, we evaluate the terms in the formula for $k^2_{critical}$ under the ordinary condition that the density of the metal ions in the solution is far smaller than the density of the metal electrodes.

To do this, return to equations (9.15) and (9.16), that is, to

$$\frac{dc_0}{dy_0} = -\frac{z_1 z_2 D_2}{z_1 D_1 - z_2 D_2} \frac{c_M U}{D} + \frac{v_{y_0}}{D} c_0$$

$$= -\frac{z_1 z_2 D_2}{z_1 D_1 - z_2 D_2} \frac{c_M U}{D} - \frac{c_M U}{D} \frac{\rho_M}{\rho} \frac{c_0}{c_M}$$

and

$$\frac{d}{dy_0}\left[\frac{z_1 F}{RT}\phi_0\right] = \frac{1}{c_0}\left[\frac{z_1^2}{z_1 D_1 - z_2 D_2} c_M U - P\frac{dc_0}{dy_0}\right]$$

$$= \frac{z_1^2 c_M U}{c_0 D_1 [z_1 - z_2]}\left[1 + \frac{D_1 - D_2}{z_1 D_2} \frac{\rho_M}{\rho} \frac{c_0}{c_M}\right]$$

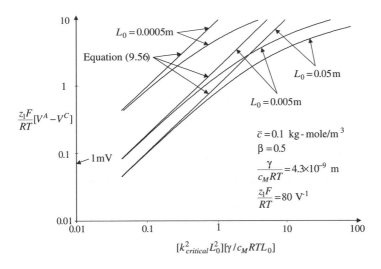

Figure 9.8. *A Graph of $k^2_{critical}$ versus $V^A - V^C$ from Equation (9.47)*

and omit the second term on the right-hand side of both equations. Then, solve the resulting equations to obtain

$$-\frac{1}{\bar{c}}\frac{dc_0^C}{dy_0} = \frac{z_1 z_2 D_2}{z_1 D_1 - z_2 D_2}\frac{c_M}{\bar{c}}\frac{U}{D} = \frac{2}{L_0}\frac{U}{U_{lim}} \tag{9.48}$$

$$\frac{c_0^C}{\bar{c}} = 1 - \frac{z_1 z_2 D_2}{z_1 D_1 - z_2 D_2}\frac{c_M}{\bar{c}}\frac{U}{D}\frac{1}{2}L_0 = 1 - \frac{U}{U_{lim}} \tag{9.49}$$

$$-\frac{d}{dy_0}\left[\frac{z_1 F}{RT}\phi_0\right]^C = \frac{-z_1^2 c_M U}{D_1[z_1 - z_2]}\frac{1}{c_0^C} = \frac{z_1}{-z_2}\frac{2}{L_0}\frac{U/U_{lim}}{1 - U/U_{lim}} \tag{9.50}$$

and

$$\frac{z_1 F}{RT}\left[\phi_0^A - \phi_0^C\right] = \frac{z_1}{-z_2}\ln\frac{\left[1 + \dfrac{U}{U_{lim}}\right]}{\left[1 - \dfrac{U}{U_{lim}}\right]} \tag{9.51}$$

where U_{lim}, the value of U at which c_0^C vanishes, is now

$$U_{lim} = \frac{2}{L_0}\frac{z_1 D_1 - z_2 D_2}{z_1 z_2 D_2}D\frac{\bar{c}}{c_M}$$

Equations (9.48) through (9.51) supply most of what is needed on the right-hand side of equation (9.47). It remains only to evaluate the term $[1 - \beta]e^{\frac{z_1 F}{RT}[\Delta\phi_0^C - \Delta\phi_{EQ}]}$ and this can be done when U is near U_{lim} and when U is near zero.

9.8.1   The Case Where U Is Near U_{lim}

As U approaches U_{lim}, $e^{\frac{z_1 F}{RT}[\Delta\phi_0^C - \Delta\phi_{EQ}]}$ approaches

$$\left[\frac{1 - \dfrac{U}{U_{lim}}}{\dfrac{U}{U_{lim}}} \frac{i_0}{z_1 F c_M [-U_{lim}]}\right]^{1/\beta}$$

[cf. equation (9.18)], and in this limit, $k_{critical}^2$ can be estimated by substituting this limit and equations (9.48), (9.49) and (9.50) into equation (9.47) to obtain

$$\frac{\gamma}{c_M RT} k_{critical}^2 = \frac{2}{L_0} \frac{z_1}{-z_2} \frac{\dfrac{U}{U_{lim}}}{1 - \dfrac{U}{U_{lim}}}$$

$$+ \frac{\dfrac{2}{L_0} \dfrac{U}{U_{lim}}}{[1-\beta]\left[\dfrac{1 - \dfrac{U}{U_{lim}}}{\dfrac{U}{U_{lim}}} \dfrac{i_0}{z_1 F c_M [-U_{lim}]}\right]^{\frac{1}{\beta}} + \beta\left[1 - \dfrac{U}{U_{lim}}\right]} \quad (9.52)$$

In this formula, which explains the shape of the curves[8] in Figure 9.7 as U approaches U_{lim}, the estimate of $e^{\frac{z_1 F}{RT}[\Delta\phi_0^C - \Delta\phi_{EQ}]}$ is not a good one for all values of U. It is good as U goes to U_{lim}. It is better the smaller the value of $\dfrac{-i_0}{z_1 F c_M U_{lim}}$. Under the conditions of Figure 9.7, if $L_0 = 0.0005$ m, the approximation is accurate to 1 percent at U/U_{lim} as low as 0.8, whereas if $L_0 = 0.05$ m, it is accurate to 25 percent at $U/U_{lim} = 0.999$.

There is no corresponding estimate of $e^{\frac{z_1 F}{RT}[\Delta\phi_0^A - \Delta\phi_{EQ}]}$ as U goes to U_{lim} and, hence, it is not a simple matter to obtain a formula for $k_{critical}^2$ versus V^C as U approaches U_{lim} and V^C approaches $-\infty$. Due to this, the dependence of $k_{critical}^2$ on V^C remains a matter of calculation. This is not so at values of U near zero, to which we now turn.

9.8.2   The Case Where U Is Near Zero

In case U is near zero and the cell is near equilibrium, both $\Delta\phi_0^A - \Delta\phi_{EQ}$ and $\Delta\phi_0^C - \Delta\phi_{EQ}$ must be near zero and the exponential functions in the Butler–Volmer equation at the two electrodes can be

---

[8]Equations (9.48) through (9.51) are accurate at the conditions of Figures 9.7 and 9.8 as the contribution due to v_{y0} is negligible there.

approximated by the first two terms in their Taylor series expansions. Introducing these approximations, substituting $1 + \dfrac{U}{U_{lim}}$ for $\dfrac{c_0^A}{\bar{c}}$ and $1 - \dfrac{U}{U_{lim}}$ for $\dfrac{c_0^C}{\bar{c}}$ and then solving the Butler–Volmer equation at the anode for $\Delta\phi_0^A$ and the Butler–Volmer equation at the cathode for $\Delta\phi_0^C$ leads to

$$\frac{z_1 F}{RT}\left[\Delta\phi_0^A - \Delta\phi_{EQ}\right] = \frac{\dfrac{-z_1 F c_M U_{lim}}{i_0} + 1}{1 + \beta\dfrac{U}{U_{lim}}}\frac{U}{U_{lim}} \tag{9.53}$$

and

$$\frac{z_1 F}{RT}\left[\Delta\phi_0^C - \Delta\phi_{EQ}\right] = \frac{\dfrac{-z_1 F c_M U_{lim}}{i_0} + 1}{-1 + \beta\dfrac{U}{U_{lim}}}\frac{U}{U_{lim}} \tag{9.54}$$

Equation (9.54) can be used to derive a low U estimate of $k_{critical}^2$ from equation (9.47). It is

$$\frac{\gamma}{c_M RT}k_{critical}^2 = \frac{2}{L_0}\left[\frac{z_1}{-z_2} + 1\right]\frac{U}{U_{lim}} \tag{9.55}$$

Then equations (9.51), (9.53) and (9.54) can be used, again at low values of U, to express $V^A - V^C$ in terms of $\dfrac{U}{U_{lim}}$ and to eliminate $\dfrac{U}{U_{lim}}$ from equation (9.55). The result is

$$\frac{\gamma}{c_M RT}k_{critical}^2 = \frac{1}{L_0}\frac{\dfrac{z_1 F}{RT}[V^A - V^C]}{1 + \dfrac{z_1 z_2}{z_1 - z_2}\dfrac{F c_M U_{lim}}{i_0}} \tag{9.56}$$

The predictions of equations (9.55) and (9.56) appear in Figures 9.7 and 9.8.

These figures illustrate an important fact: The near-equilibrium formulas can be used over practical ranges of cell operation. This stems from the fact that a cell cannot be run in a stable mode unless the displacements to which it is unstable (i.e., the displacements corresponding to wave numbers less than $k_{critical}$), can be prevented from being imposed on the cell. The simplest way to do this is to control the width of the cell. Let W_0 denote this width. Then, W_0 determines the admissible wave numbers and in the case of homogeneous Neumann side wall conditions the smallest,[9] and hence the

---

[9]The wave number $k = 0$ cannot be ruled out by homogeneous Neumann end conditions. But this causes no problem. At $k = 0$, there are eigenvalues corresponding to $\hat{Y}_1^A = \hat{Y}_1^C$ and eigenvalues corresponding to $\hat{Y}_1^A = -\hat{Y}_1^C$. Those corresponding to $\hat{Y}_1^A = \hat{Y}_1^C$ cannot be positive. Those corresponding to $\hat{Y}_1^A = -\hat{Y}_1^C$ may be positive for reasons that led to positive eigenvalues in the solidification problem (cf. the discussion note in the seventh essay). In electrodeposition, these dangerous eigenvalues are ruled out by Equation (9.19).

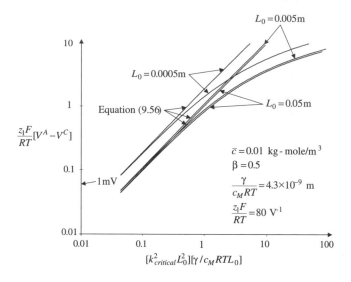

Figure 9.9. *Figure 9.8 Redrawn with $\bar{c} = 0.01$ kg-mole/m^3*

most dangerous, admissible wave number is π/W_0. By this, at $L_0 = 0.5$ cm, a value of $k_{critical}$ as large as 1.4×10^5 m^{-1}, which is near the end of the near-equilibrium range, requires a cell width of about 0.022 mm in order to open up a range of stable operation. All cells of greater width require lower values of $k_{critical}$, and to these, the near-equilibrium formulas apply.

To introduce a control variable, let the width of the cell be fixed at some finite value. This fixes the admissible wave numbers, the smallest or most dangerous lying to the right of zero. Then, start with the cell at equilibrium and begin to increase the potential difference across the cell. This moves the critical wave number upward from zero, but as long as it lies to the left of the most dangerous wave number, the cell remains stable. However, as the potential difference applied across the cell continues its increase, sooner or later it will reach a value at which the critical wave number will just become equal to the least admissible wave number. This will be the critical value of $V^A - V^C$. Increasing $V^A - V^C$ beyond this value will produce at least one displacement to which the cell is unstable. In this way, the potential difference across the cell can be viewed as a control variable in just the same way that the length of a liquid bridge was viewed as a control variable earlier. In both cases, control is achieved by screening out the long wavelength displacements to which the cell or the bridge is unstable.

Equation (9.47) is our main result. It tells us the inverse surface tension dependence of $k_{critical}^2$. To get an idea of the dependence of $k_{critical}^2$ on L_0 and $\bar{c}$, one can do no better than to examine the near-equilibrium formula, equation (9.56). What it predicts depends on the value of the term

$\dfrac{z_1 z_2}{z_1 - z_2} \dfrac{F c_M U_{lim}}{i_0}$, where, in case $\beta = \dfrac{1}{2}$, $\dfrac{U_{lim}}{i_0}$ is a multiple of $\dfrac{\sqrt{\bar{c}}}{L_0}$. If this term is small compared to one, then $k^2_{critical}$ is independent of $\bar{c}$, but it is a multiple of L_0^{-1}. If, on the other hand, it is large compared to one, then $k^2_{critical}$ is independent of L_0, but it is a multiple of $\bar{c}^{-1/2}$.

Under the conditions of Figure 9.8, the value of $\dfrac{z_1 z_2}{z_1 - z_2} \dfrac{F c_M U_{lim}}{i_0}$ is near one at $L_0 = 0.005$ m, hence increasing L_0 decreases its value to zero, leaving $\dfrac{\gamma}{c_M R T L_0} k^2_{critical} L_0^2$ insensitive to L_0 and to $\bar{c}$, as long as $\bar{c}$ remains near 0.1 kg-mole/m^3. It is by decreasing L_0 that sensitivity to L_0 and to $\bar{c}$ can be found and this is illustrated in the figure.

Figure 9.8 is redrawn as Figure 9.9 with $\bar{c}$ reduced from 0.1 to 0.01 kg-mole/m^3. This figure illustrates the fact that lowering $\bar{c}$ weakens the dependence on L_0.

Afterword

It is worth noticing that the eigenfunction corresponding to $\sigma = 0$ turns out to be $\hat{c}_1 = 0$, $\hat{\phi}_1 = 0$, $\hat{Y}_1^A = 0$ and $\hat{Y}_1^C \neq 0$. This is a familiar result in these essays. In two-surface problems, one surface is not displaced at critical conditions. This was discovered in both the two-fluid jet and the three-fluid Rayleigh–Taylor problems. In cases where an instability can arise due to what is taking place at one or the other of two surfaces, one surface will determine the outcome.

9.9 ENDNOTES

9.9.1 Putting the Butler–Volmer Equation into a Little More Useful Form

It is of some advantage to eliminate the constants $\overleftarrow{k}\,c_M$ and $\overrightarrow{k}$ in the Butler–Volmer equation and to introduce the average metal ion concentration, denoted by $\bar{c}_{M^+}$, as a reference concentration.

To do this, let the reaction transfer one electron and start with the Butler–Volmer equation in the form[10]

$$\vec{n}\cdot\vec{i} = F\overleftarrow{k}\,c_M\; e^{[1-\beta]\frac{F}{RT}\left[\Delta\phi-\frac{\gamma}{Fc_M}2H\right]}\; - \; F\overrightarrow{k}\,c_{M^+}\; e^{-\beta\frac{F}{RT}\left[\Delta\phi-\frac{\gamma}{Fc_M}2H\right]}$$

Then, hold both c_{M^+} and $2H$ fixed and determine the value of $\Delta\phi$ required to bring the reaction to equilibrium. This value of $\Delta\phi$, denoted $\Delta\phi_{EQ}$, is the solution to the equation

$$0 = F\overleftarrow{k}\,c_M\,e^{[1-\beta]\frac{F}{RT}\left[\Delta\phi_{EQ}-\frac{\gamma}{Fc_M}2H\right]}\; - \; F\overrightarrow{k}\,c_{M^+}e^{-\beta\frac{F}{RT}\left[\Delta\phi_{EQ}-\frac{\gamma}{Fc_M}2H\right]}$$

At equilibrium the rates of the oxidation and the reduction reactions on the right-hand side, being equal, can both be denoted i_0, where i_0 is called the exchange current density. It, like $\Delta\phi_{EQ}$, depends on c_{M^+} and on $2H$.

Now, arrange the equilibrium equation into the form

$$\frac{c_{M^+}}{\overleftarrow{k}\,c_M\,/\,\overrightarrow{k}} = e^{\frac{F}{RT}\left[\Delta\phi_{EQ}-\frac{\gamma}{Fc_M}2H\right]}$$

and let $\overline{\Delta\phi}_{EQ}$ denote the equilibrium potential difference appearing across the surface when the ion concentration at the surface is $\bar{c}_{M^+}$ instead of c_{M^+}. Then, we have

$$\frac{\bar{c}_{M^+}}{\overleftarrow{k}\,c_M\,/\,\overrightarrow{k}} = e^{\frac{F}{RT}[\overline{\Delta\phi}_{EQ}-\frac{\gamma}{Fc_M}2H]}$$

whereupon $c_{M^+}/\bar{c}_{M^+}$ is given by

$$\frac{c_{M^+}}{\bar{c}_{M^+}} = e^{\frac{F}{RT}[\Delta\phi_{EQ}-\overline{\Delta\phi}_{EQ}]}$$

and although $\Delta\phi_{EQ}$ and $\overline{\Delta\phi}_{EQ}$ both depend on the curvature of the interface, their difference does not.

Then, take for reference a flat interface and denote $\Delta\phi$ by $\Delta\phi°$. In this case, there obtains

$$\frac{c_{M^+}}{\bar{c}_{M^+}} = e^{\frac{F}{RT}\Delta\phi°_{EQ}}\; e^{-\frac{F}{RT}\overline{\Delta\phi}°_{EQ}}$$

---

[10]In Bockris and Reddy's book, (cf. Preface), Chapter 8, the reader can find an account of this equation which emphasizes its physics.

and, hence,

$$e^{\frac{F}{RT}\Delta\phi_{EQ}^{\circ}} = e^{\frac{F}{RT}\overline{\Delta\phi}_{EQ}^{\circ}} \left[\frac{c_{M^+}}{\bar{c}_{M^+}}\right]$$

Now, turn to i_0 and write

$$i_0 = F\overleftarrow{k}\, c_M e^{[1-\beta]\frac{F}{RT}[\Delta\phi_{EQ} - \frac{\gamma}{Fc_M}2H]}$$

If the interface is flat, denote i_0 by i_0°, and if the ionic concentration at the interface is $\bar{c}_{M^+}$, instead of c_{M^+}, denote i_0 by $\bar{i}_0$. Then, there obtains

$$i_0^{\circ} = F\overleftarrow{k}\, c_M e^{[1-\beta]\frac{F}{RT}[\Delta\phi_{EQ}^{\circ}]}$$

and

$$\bar{i}_0^{\,\circ} = F\overleftarrow{k}\, c_M e^{[1-\beta]\frac{F}{RT}[\overline{\Delta\phi}_{EQ}^{\circ}]}$$

which lead to

$$i_0^{\circ} = \bar{i}_0^{\,\circ} e^{[1-\beta]\frac{F}{RT}[\Delta\phi_{EQ}^{\circ} - \overline{\Delta\phi}_{EQ}^{\circ}]}$$

and, hence, to

$$i_0^{\circ} = \bar{i}_0^{\,\circ} \left[\frac{c_{M^+}}{\bar{c}_{M^+}}\right]^{[1-\beta]}$$

Now, go back to the Butler–Volmer equation and substitute

$$F\overleftarrow{k}\, c_M = i_0^{\circ} e^{-[1-\beta]\frac{F}{RT}[\Delta\phi_{EQ}^{\circ}]}$$

and

$$F\overrightarrow{k}\, c_{M^+} = i_0^{\circ} e^{\beta\frac{F}{RT}[\Delta\phi_{EQ}^{\circ}]}$$

to get

$$\vec{n}\cdot\vec{i} = i_0^{\circ}\left[e^{[1-\beta]\frac{F}{RT}[\Delta\phi - \Delta\phi_{EQ}^{\circ} - \frac{\gamma}{Fc_M}2H]} - e^{-\beta\frac{F}{RT}[\Delta\phi - \Delta\phi_{EQ}^{\circ} - \frac{\gamma}{Fc_M}2H]}\right]$$

Then, using

$$i_0^{\,\circ} = \bar{i}_0^{\,\circ}\left[\frac{c_{M^+}}{\bar{c}_{M^+}}\right]^{[1-\beta]}$$

and

$$e^{\frac{F}{RT}\Delta\phi_{EQ}^{\circ}} = e^{\frac{F}{RT}\overline{\Delta\phi}_{EQ}^{\circ}}\left[\frac{c_{M^+}}{\bar{c}_{M^+}}\right]$$

we obtain the Butler–Volmer equation in the form[11] it is used in the essay, namely

---

[11]Twenty years or so ago, in order to generalize the Butler–Volmer equation from a single-electron oxidation-reduction step to a z_1-electron step, the factor z_1 would have

$$\vec{n} \cdot \vec{i} = \bar{i}_0{}^\circ \left[e^{[1-\beta]\frac{F}{RT}[\Delta\phi - \overline{\Delta\phi}^\circ_{EQ} - \frac{\gamma}{Fc_M}2H]} - \left[\frac{c_{M^+}}{\bar{c}_{M^+}}\right] e^{-\beta\frac{F}{RT}[\Delta\phi - \overline{\Delta\phi}^\circ_{EQ} - \frac{\gamma}{Fc_M}2H]} \right]$$

The exchange current density, $\bar{i}_0{}^\circ$, and the equilibrium potential difference, $\overline{\Delta\phi}^\circ_{EQ}$, refer to a plane electrode surface contacting an electrolytic solution whose metal ion concentration is $\bar{c}_{M^+}$, and their values depend on the value of $\bar{c}_{M^+}$. In the essay, they will be denoted simply by i_0 and $\Delta\phi_{EQ}$.

The derivation in this endnote could have been carried out in terms of any reference concentration, denoted, say, $c_{M^+\text{ref}}$. Our choice for $c_{M^+\text{ref}}$ is $\bar{c}_{M^+}$.

9.9.2   The Perturbation Equations at the Boundary of the Reference Domain

Due to the fact that the reference electrode surfaces are parallel planes (i.e., that Y_0 is constant), independent of x_0 and t_0, the perturbation equations at the boundary of the reference domain take fairly simple forms.

First, the expansions of $\vec{n}$, $u = \vec{n} \cdot \vec{u}$ and $2H$ come from the expansion of Y, namely from

$$Y = Y_0 + \epsilon Y_1 + \cdots$$

With Y_0 constant, they are, to first order, and at the anode,

$$\vec{n} = \vec{j} - \epsilon \frac{\partial Y_1}{\partial x_0}\vec{i}$$

$$u = \epsilon \frac{\partial Y_1}{\partial t_0}$$

and

$$2H = \epsilon \frac{\partial^2 Y_1}{\partial x_0^2}$$

Then, to see how equation (9.23), that is,

$$P\frac{d\hat{c}_1}{dy_0} + c_0 \left[\frac{z_1 F}{RT}\right]\frac{d\hat{\phi}_1}{dy_0} + \left[\frac{z_1 F}{RT}\right]\frac{d\phi_0}{dy_0}\hat{c}_1 = \frac{z_1^2}{z_1 D_1 - z_2 D_2}c_M\sigma\hat{Y}_1$$

---

been introduced in three places as we have done in the essay. But it is now thought to be unlikely that z_1 electrons can tunnel simultaneously through an interface. Presently (cf. A. J. Bard and L. R. Faulkner, Electrochemical Methods-Fundamentals and Applications, second edition, J. Wiley, N.Y., [2001]), multiple electron transfer is thought to take place as a sequence of single electron transfer steps, one step being rate limiting. In the Butler–Volmer equation for the rate limiting step the factor z_1 would not be introduced into the two exponents. For example, in the case of a copper electrode and a copper sulphate solution, it would be required to account for both cuperous and cuperic ions in the solution. We do not take this complication into account.

comes from equation (9.11), that is,

$$P\vec{n} \cdot \nabla c + c\left[\frac{z_1 F}{RT}\right]\vec{n} \cdot \nabla \phi = \frac{z_1^2}{z_1 D_1 - z_2 D_2} c_M [\vec{n} \cdot \vec{u} + \vec{n} \cdot \vec{U}]$$

write

$$\nabla c = \nabla_0 c_0 + \epsilon\left[\nabla_0 c_1 + Y_1 \frac{d}{dy_0}\nabla_0 c_0\right]$$

and

$$\nabla \phi = \nabla_0 \phi_0 + \epsilon\left[\nabla_0 \phi_1 + Y_1 \frac{d}{dy_0}\nabla_0 \phi_0\right]$$

and notice that $\vec{U}$, $\nabla_0 c_0$, and $\nabla_0 \phi_0$ all lie in the $\vec{j}$ direction. Due to this,

$$\vec{n} \cdot \nabla c = \frac{dc_0}{dy_0} + \epsilon\left[\frac{\partial c_1}{\partial y_0} + Y_1 \frac{d^2 c_0}{dy_0^2}\right]$$

and

$$c\vec{n} \cdot \nabla \phi = c_0\frac{d\phi_0}{dy_0} + \epsilon\left[\left[c_1 + Y_1 \frac{dc_0}{dy_0}\right]\frac{d\phi_0}{dy_0} + c_0\left[\frac{\partial \phi_1}{\partial y_0} + Y_1 \frac{d^2 \phi_0}{dy_0^2}\right]\right]$$

obtain. The perturbation equation results on substituting these expansions into the nonlinear equation and using the base equations to eliminate the term

$$Y_1\left[P\frac{d^2 c_0}{dy_0^2} + \frac{dc_0}{dy_0}\left[\frac{z_1 F}{RT}\right]\frac{d\phi_0}{dy_0} + c_0\left[\frac{z_1 F}{RT}\right]\frac{d^2 \phi_0}{dy_0}\right]$$

which would otherwise appear on the left-hand side. By this, there is no term in Y_1 on the left-hand side of equation (9.23).

In equation (9.22), which is

$$D\frac{d\hat{c}_1}{dy_0} - v_{y_0}\hat{c}_1 - c_0\left[\hat{v}_{y_1} - \sigma \hat{Y}_1\right] = -\frac{z_1 z_2 D_2}{z_1 D_1 - z_2 D_2} c_M \sigma \hat{Y}_1$$

the last three terms on the left-hand side come from the term, $c\vec{n} \cdot [\vec{v} - \vec{u}]$ in equation (9.10) on using

$$c = c_0 + \epsilon\left[c_1 + Y_1 \frac{dc_0}{dy_0}\right]$$

and

$$\vec{v} = \vec{v}_0 + \epsilon \vec{v}_1$$

9.10   DISCUSSION NOTES

9.10.1   The Special Case k^2 Equal to Zero

Let k^2 be zero, then the solution to the perturbation eigenvalue problem must meet two requirements which are not needed otherwise. They are

$$\hat{Y}_1^A - \hat{Y}_1^C = 0$$

and

$$\int_0^{L_0} \hat{c}_1 \, dy_0 = [c_0^A - c_0^C]\hat{Y}_1$$

where $\hat{Y}_1$ denotes the common value of $\hat{Y}_1^A$ and $\hat{Y}_1^C$. These conditions stem from limitations placed on the class of admissible perturbations. They are derived in the third discussion note.

Our job is to determine whether $\sigma = 0$ can be an eigenvalue in the case where k^2 is zero. To do this, set $k^2 = 0 = \sigma$ in the perturbation eigenvalue problem and look for a solution other than $\hat{c}_1$, $\hat{\phi}_1$, $\hat{Y}_1^A$ and $\hat{Y}_1^C$ all zero. Then, $\hat{c}_1$ must satisfy

$$\frac{d}{dy_0}\left[D\frac{d\hat{c}_1}{dy_0} - v_{y_0}\hat{c}_1\right] = 0$$

on $0 < y_0 < L_0$,

$$D\frac{d\hat{c}_1}{dy_0} - v_{y_0}\hat{c}_1 = 0$$

at both $y_0 = 0$ and $y_0 = L_0$ and

$$\int_0^{L_0} \hat{c}_1 \, dy_0 = [c_0^A - c_0^C]\hat{Y}_1$$

This problem ought to remind the reader of the c_0 problem:

$$\frac{d}{dy_0}\left[D\frac{dc_0}{dy_0} - v_{y_0}c_0\right] = 0$$

on $0 < y_0 < L_0$,

$$D\frac{dc_0}{dy_0} - v_{y_0}c_0 = \frac{-z_1 z_2 D_2}{z_1 D_1 - z_2 D_2} c_M U$$

at both $y_0 = 0$ and $y_0 = L_0$ and

$$\int_0^{L_0} c_0 \, dy_0 = \bar{c}$$

Due to this, $\hat{c}_1$ can be written

$$\hat{c}_1 = -\frac{dc_0}{dy_0}\,\hat{Y}_1$$

Turn now to $\hat{\phi}_1$. In the case where $k^2 = 0 = \sigma$, $\hat{\phi}_1$ must satisfy

$$\frac{d}{dy_0}\left[P\frac{d\hat{c}_1}{dy_0} + \hat{c}_1\left[\frac{z_1 F}{RT}\right]\frac{d\phi_0}{dy_0} + c_0\left[\frac{z_1 F}{RT}\right]\frac{d\hat{\phi}_1}{dy_0}\right] = 0$$

on $0 < y_0 < L_0$ and

$$P\frac{d\hat{c}_1}{dy_0} + \hat{c}_1\left[\frac{z_1 F}{RT}\right]\frac{d\phi_0}{dy_0} + c_0\left[\frac{z_1 F}{RT}\right]\frac{d\hat{\phi}_1}{dy_0} = 0$$

at both $y_0 = 0$ and $y_0 = L_0$. By this,

$$P\frac{d\hat{c}_1}{dy_0} + \hat{c}_1\left[\frac{z_1 F}{RT}\right]\frac{d\phi_0}{dy_0} + c_0\left[\frac{z_1 F}{RT}\right]\frac{d\hat{\phi}_1}{dy_0}$$

must vanish everywhere. Then, by using $\hat{c}_1 = -\dfrac{dc_0}{dy_0}\,\hat{Y}_1$, $\hat{\phi}_1$ can be determined by integrating

$$\left[\frac{z_1 F}{RT}\right]\frac{d\hat{\phi}_1}{dy_0} = \left[\frac{1}{c_0}\frac{dc_0}{dy_0}\left[\frac{z_1 F}{RT}\right]\frac{d\phi_0}{dy_0} + P\frac{1}{c_0}\frac{d^2 c_0}{dy_0^2}\right]\hat{Y}_1 = -\left[\frac{z_1 F}{RT}\right]\frac{d^2\phi_0}{dy_0^2}\,\hat{Y}_1$$

where the far-right-hand side obtains on use of the base equations. The result is that

$$\left[\frac{z_1 F}{RT}\right]\hat{\phi}_1 = -\left[\frac{z_1 F}{RT}\right]\frac{d\phi_0}{dy_0}\,\hat{Y}_1 + C$$

where C denotes a constant of integration.

Now, turn to the Butler–Volmer equations at $y_0 = 0$ and at $y_0 = L_0$, set σ to zero and notice that the factor $\hat{c}_1 + \dfrac{dc_0}{dy_0}\,\hat{Y}_1$ vanishes while the factor

$$-\left[\frac{z_1 F}{RT}\right]\hat{\phi}_1 - \left[\frac{z_1 F}{RT}\right]\frac{d\phi_0}{dy_0}\,\hat{Y}_1 \text{ is just } -C. \text{ Due to this, these two equations}$$

can be satisfied only by setting C to zero, leaving $\hat{Y}_1$ free to be set to any value we please, with the result that $\sigma = 0$ is always a solution to the eigenvalue problem at $k^2 = 0$, whether or not the cell draws current.

9.10.2   Uniform Solution Density and Local Electrical Neutrality

There are two approximations made in the essay: uniform and constant solution density and local electroneutrality. Each, by itself, leads to the volume of the solution, as well as the average concentration of the ions in the solution, remaining constant in time. This discussion note presents the proof. An electrolytic cell is indicated in Figure 9.10. The electrodes are at rest in the laboratory frame and two fixed planes, $A - A$ and $C - C$, lie inside the electrodes away from their moving surfaces. Let W_0 denote the width of the cell and take the cell to be two dimensional, the third dimension being into the plane of the paper.

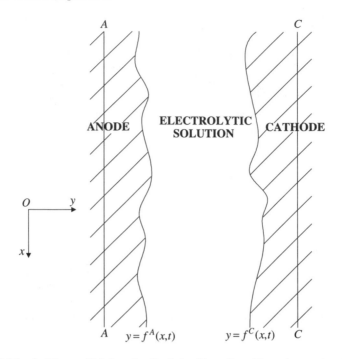

Figure 9.10. *A Figure Helping to Explain How Our Two Approximations Fit Together*

Uniform Solution Density

Let the density of the solution be constant and use the fact that the total mass of the material lying between planes $A - A$ and $C - C$, given by

$$\int_0^{W_0} \int_A^{f^A(x,t)} \rho_M \, dy \, dx + \int_0^{W_0} \int_{f^A(x,t)}^{f^C(x,t)} \rho \, dy \, dx + \int_0^{W_0} \int_{f^C(x,t)}^{C} \rho_M \, dy \, dx$$

must remain constant in time. Then, by setting its derivative with respect to time to zero, we have

$$[\rho_M - \rho] \int_0^{W_0} dx \left[\frac{\partial f^A}{\partial t} - \frac{\partial f^C}{\partial t} \right] = 0$$

and, hence,

$$\frac{d}{dt} \int_0^{W_0} \int_{f^A(x,t)}^{f^C(x,t)} dy \, dx = 0$$

whence the volume of the electrolytic solution must remain constant.

In addition, both the number of moles of the metal, in both forms, and the number of moles of the non metallic ion, lying between planes $A - A$

and $C - C$, must remain constant in time. Hence, we have

$$\frac{d}{dt}\left[\int_0^{W_0}\int_A^{f^A(x,t)} c_M \, dy \, dx + \int_0^{W_0}\int_{f^A(x,t)}^{f^C(x,t)} c_1 \, dy \, dx \right.$$

$$\left. + \int_0^{W_0}\int_{f^C(x,t)}^{C} c_M \, dy \, dx \right] = 0$$

and

$$\frac{d}{dt}\int_0^{W_0}\int_{f^A(x,t)}^{f^C(x,t)} c_2 \, dy \, dx = 0$$

whence $\bar{c}_2$ must be constant and constant solution volume implies that $\bar{c}_1$ must be constant.

Local Electrical Neutrality

Require local electroneutrality to hold and use $\vec{i} = \sum F z_i \vec{N}_i$ to obtain

$$\nabla \cdot \vec{i} = \sum F z_i \nabla \cdot \vec{N}_i = -\sum F z_i \frac{\partial c_i}{\partial t} = 0$$

Then, integrate $\nabla \cdot \vec{i}$ over the volume of the solution, taking the side walls to be electrical insulators, to get

$$\int_{S^A} dA \, \vec{n} \cdot \vec{i} + \int_{S^C} dA \, \vec{n} \cdot \vec{i} = 0$$

where S^A and S^C denote the anode and cathode surfaces and where

$$dA = \left[1 + [f_x]^2\right]^{1/2} dx$$

Now, at each electrode $\vec{n} \cdot \vec{i}$ is a multiple of $\vec{n} \cdot \vec{u}$ and, as the proportionality factor is common to the two electrodes, we have

$$\int_0^{W_0} \vec{n} \cdot \vec{u}^A \left[1 + [f_x^A]^2\right]^{1/2} dx + \int_0^{W_0} \vec{n} \cdot \vec{u}^C \left[1 + [f_x^C]^2\right]^{1/2} dx = 0$$

whence, using

$$\vec{n} \cdot \vec{u}^A = \frac{\dfrac{\partial f^A}{\partial t}}{\left[1 + [f_x^A]^2\right]^{1/2}}$$

and

$$\vec{n} \cdot \vec{u}^C = \frac{-\dfrac{\partial f^C}{\partial t}}{\left[1 + [f_x^C]^2\right]^{1/2}}$$

we get

$$\int_0^{W_0} \frac{\partial f^A}{\partial t} \, dx - \int_0^{W_0} \frac{\partial f^C}{\partial t} \, dx = 0$$

and again we see that

$$\int_0^{W_0} \int_{f^A(x,\, t)}^{f^C(x,\, t)} dy \, dx$$

which is the volume of the solution, remains constant in time. Once constant volume is established, constant average ion concentrations in the solution follow.

Both of our approximations, then, lead to a common set of time invariants. In fact, local electroneutrality goes a little further. It implies constant average solution density. This can be seen by writing

$$0 = \frac{d}{dt} \left[\int_0^{W_0} \int_A^{f^A(x,\, t)} \rho_M \, dy \, dx + \int_0^{W_0} \int_{f^A(x,\, t)}^{f^C(x,\, t)} \rho \, dy \, dx \right.$$

$$\left. + \int_0^{W_0} \int_{f^C(x,\, t)}^C \rho_M \, dy \, dx \right]$$

as

$$\left[-\rho_M \frac{d}{dt} \int_0^{W_0} \int_{f^A(x,\, t)}^{f^C(x,\, t)} dy \, dx + \frac{d}{dt} \int_0^{W_0} \int_{f^A(x,\, t)}^{f^C(x,\, t)} \rho \, dy \, dx \right] = 0$$

and noticing that the first term on the left-hand side vanishes if local electroneutrality is assumed to hold.

9.10.3 Limitations on the Admissible Class of Perturbations

Attention is again drawn to the figure in the second discussion note. But now our interest is in what does or does not happen as we impose a disturbance on the base state. In particular, our interest is in integrals of the form

$$I = \int_{V(\epsilon)} f(\epsilon) \, dV$$

where ϵ denotes the amplitude of the disturbance and where only the ϵ dependence of f is indicated.

Let us begin by writing the formula

$$\frac{d}{d\epsilon} I = \int_V \frac{\partial f}{\partial \epsilon} \, dV + \int_S dA \, \vec{n} \cdot \frac{d\vec{R}}{d\epsilon} f$$

where the surface S encloses the region V, but only the electrode surfaces can be displaced, where $\vec{n}$ denotes the normal to S pointing out of V and where $d\vec{R}/d\epsilon$ comes from the mapping of the reference domain into the present domain.

At $\epsilon = 0$ we have

$$I_1 = \frac{dI}{d\epsilon}(\epsilon = 0) = \int_{V_0} f_1 \, dV_0 + \int_{S_0} dA_0 \, \vec{n}_0 \cdot \vec{R}_1 f_0$$

where, to two terms, the expansion of I is

$$I = I_0 + \epsilon I_1$$

First, take I to be V, the volume of the electrolytic solution, and denote the electrode surfaces[12] by $y = Y(x_0, t_0, \epsilon)$. Then, using $\vec{n}_0^A = \vec{j} = -\vec{n}_0^C$ and $\vec{R}_1 = Y_1 \vec{j}$ we have

$$V_1 = -\int_{S_0^A} Y_1^A \, dx_0 + \int_{S_0^C} Y_1^C \, dx_0$$

whence V_1 must be zero for all non constant periodic displacements, while for zero wave number, or uniform displacements, it is given by

$$W_0[Y_1^C - Y_1^A]$$

where W_0 denotes the width of the cell.

Second, take I to be M, the mass of all of the material lying between the planes $A - A$ and $C - C$. Then, with the density ρ_M of the electrodes and the density ρ of the solution both constant and uniform, we have

$$M_1 = [\rho_M - \rho]\left[\int_{S_0^A} Y_1^A \, dx_0 - \int_{S_0^C} Y_1^C \, dx_0\right]$$

Now, so long as a perturbation of the base state does not add mass to the system, M must equal M_0 and M_1 must be zero. While M_1 must surely be zero for all non constant periodic displacements, to make it zero for zero wave number displacements introduces a condition on Y_1^A and Y_1^C. It is

$$Y_1^A = Y_1^C$$

and due to this, V_1 must also be zero, whence as long as the base state is perturbed with no mass being added to the system, the volume of the electrolytic solution must remain at its base state value.

Third, take I to be N, the number of moles of metal in both forms lying between the plane $A - A$ and $C - C$. Then, with c_0^A and c_0^C constant over the electrode surfaces, we have

---

[12]The problem is taken to be two dimensional, the third dimension being into the plane of the figure.

$$N_1 = \left[c_M - \frac{c_0^A}{z_1}\right] \int_{S_0^A} Y_1^A \, dx_0$$

$$- \left[c_M - \frac{c_0^C}{z_1}\right] \int_{S_0^C} Y_1^C \, dx_0 + \int_{V_0} \frac{c_1}{z_1} \, dx_0 \, dy_0$$

As with M, so too, with N: So long as a perturbation of the base state does not add metal to the system, N must be equal to N_0 and hence N_1 must be zero. Like M_1, N_1 is always zero for non constant periodic displacements, but to make it zero for zero wave number displacements requires, with $Y_1^A = Y_1^C$,

$$\int_0^{L_0} c_1 \, dy_0 - c_0^A Y_1^A + c_0^C Y_1^C = 0$$

Our conclusion then is this: By limiting the admissible perturbations to those which add neither mass nor metal to the system, two conditions must be satisfied. Both apply only to zero wave number displacements and they are

$$Y_1^A = Y_1^C$$

and

$$\int_0^{L_0} c_1 \, dy_0 = c_0^A Y_1^A - c_0^C Y_1^C$$

10
The Last Word

Go back to the Guide for the Reader and look over the themes presented there. One of these themes can be called our main aim. It is this: to present problems that can be solved using only pencil and paper. These should serve teachers in need of lectures and students in need of problems to work through.

Keeping to the class of problems where deflecting surfaces are important, we can add to the references listed in the Guide for the Reader several more. Working through these ought to be enjoyable.

- S. Tomotika. On the Stability of a Cylindrical Thread of a Viscous Liquid Surrounded by Another Viscous Fluid. *Proc. Roy. Soc. Lond. A*, 150:322, 1935.
- J.B. Keller, S.I. Rubinow and Y.O. Tu. Spatial Instability of a Jet. *Phys. Fluids*, 16:2052, 1957.
- S.L. Goren. The Instability of an Annular Thread of Fluid. *J. Fluid Mech.*, 12:309, 1962.
- T.S. Chow and J.J. Hermans. Stability of a Cylindrical Viscous Jet Surrounded by a Flowing Gas. *Phys. Fluids*, 14(2):244, 1971.
- C. Hickox. Instability Due to a Viscosity and Density Stratification in Axisymmetric Pipe Flow. *Phys. Fluids*, 14:251, 1971.
- N.R. Lebovitz. Perturbation Expansions on Perturbed Domains. *SIAM Rev.*, 24(4):381, 1982.
- C.-S. Yih and J.F.C. Kingman. Instability of a Rotating Liquid Film with a Free Surface. *Proc. Roy. Soc. Lond. A*, 258(1292):63, 1960.
- J. Gillis and B. Kaufman. The Stability of a Rotating Viscous Jet. *Quart. Appl. Math.*, XIX(4):301, 1962.

- J. Gillis and K.S. Suh. Stability of a Rotating Liquid Jet. *Phys. Fluids*, 5(10):1149, 1962.
- W.W. Mullins and R.F. Sekerka. Stability of a Planar Interface During Solidification of a Dilute Binary Alloy. *J. Appl. Phys.*, 35:444, 1964.
- D.P. Woodruff. The Stability of a Planar Interface During the Melting of a Binary Alloy. *Phil. Mag.*, 17:283, 1968.
- D.J. Wollkind and R.N. Maurer. The Influence on a Stability Analysis Involving a Prototype Solidification Problem of Including an Interfacial Surface Entropy Effect in the Heat Balance Relationship at the Interface. *J. Crystal Growth*, 42:24, 1977.
- S.R. Coriell and R.L. Parker. Stability of the Shape of a Solid Cylinder Growth in a Diffusion Field. *J. Appl. Phys.*, 16(2):632, 1965.
- C.A. Miller. Stability of Moving Surfaces in Fluid Systems with Heat and Mass Transport. *AIChE J.*, 19(5):909, 1973.
- L.G. Sundström and F.H. Bark. On Morphological Instability During Electrodeposition with a Stagnant Binary Electrolyte. *Electrochim. Acta*, 40(5), 599, 1995.
- J. Elezgaray, C. Léger and F. Argoul. Linear Stability Analysis of Unsteady Galvanostatic Electrodeposition in the Two-Dimensional Diffusion-Limited Regime. *J. Electrochem. Soc.*, 145(6):2016, 1998.

Appendix A
The Application of the Expansion Formulas of the Second Essay to Some Physical Problems

A.1   The Shape of the Surface of a Spinning Beaker

Our first problem in this appendix is one whose solution can be obtained directly. We turn it into a perturbation problem only for the reader's benefit. Indeed, it may be a little too simple even for this.

Let a cylindrical beaker of radius R be filled with a liquid to a depth d and then be set into steady rotation about its axis of symmetry at an angular speed Ω. Our job is to determine the domain occupied by the liquid as a function of the angular speed at which the beaker turns, assuming that the liquid attains a state of rigid body rotation.

To do this, the position of the free surface of the liquid needs to be determined, but the cylindrical wall and the bottom of the beaker, not being displaced, do not come into the calculation.

Let the origin, O, lie on the axis of symmetry where it crosses the bottom of the beaker and let the axis Oz point upward. Figure A.1 indicates the important notation.

Cylindrical coordinates, r, θ and z will be used to identify the points of the liquid in its present configuration and its free surface will be denoted

$$z = Z(r,\ \Omega^2)$$

where the square of the angular speed will be used as the expansion variable in a perturbation calculation (viz., $\epsilon = \Omega^2$).

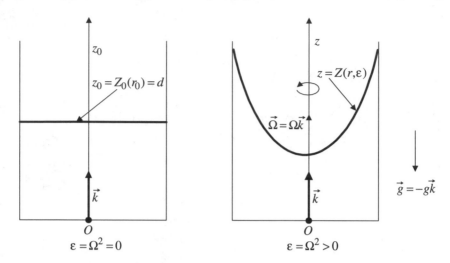

Figure A.1. *Some Notation*

To solve this problem, we must find the pressure throughout the liquid and the domain occupied by the liquid. The pressure must satisfy

$$\rho \vec{\Omega} \times \vec{\Omega} \times \vec{r} = -\nabla p + \rho \vec{g} \tag{A.1}$$

on the domain; at the free surface, it must satisfy

$$p = p_{ambient} \tag{A.2}$$

where $p_{ambient}$ is specified and held constant.

There is the additional demand that the volume of the liquid remain at its zero spin value for all $\Omega^2 > 0$, and this requires $Z(r, \Omega^2)$ to satisfy

$$\int_0^R 2\pi r Z(r, \Omega^2) dr = \pi R^2 d \tag{A.3}$$

Now, equation (A.1) can be written

$$\left. \begin{aligned} \frac{\partial p}{\partial r} &= \rho \Omega^2 r \\[2em] \frac{\partial p}{\partial z} &= -\rho g \end{aligned} \right\} \quad 0 < r < R, \ \ 0 < z < Z(r, \Omega^2)$$

and, before going on, observe that its solution is

$$p = \frac{1}{2}\rho \Omega^2 r^2 - \rho g z + C$$

where C remains to be determined. The requirement that the free surface be at constant pressure then determines its shape via

$$Z(r, \ \Omega^2) = \frac{1}{2}\frac{\Omega^2}{g}r^2 + D$$

where $D = \dfrac{C - p_{ambient}}{\rho g}$ and where, to satisfy the constant-volume requirement, D must be

$$d - \frac{1}{4}\frac{\Omega^2}{g}R^2$$

Now, return to the problem as it was originally stated and view it as a family of problems for increasing values of Ω^2. The simplest problem corresponds to $\Omega^2 = 0$ and this defines the reference problem. Its solution determines the reference domain. In terms of reference domain variables, this problem is

$$\left.\begin{array}{l} \dfrac{\partial p_0}{\partial r_0} = 0 \\[2em] \dfrac{\partial p_0}{\partial z_0} = -\rho g \end{array}\right\} \qquad 0 < r_0 < R, \ \ 0 < z_0 < Z_0(r_0)$$

where at the free surface, $z_0 = Z_0(r_0), 0 < r_0 < R$, p_0 must satisfy

$$p_0 = p_{ambient}$$

and where, to maintain constant volume, Z_0 must satisfy

$$\int_0^R 2\pi r_0 Z_0(r_0) \ dr_0 = \pi R^2 d$$

The solution to this problem is

$$p_0 = -\rho g[z_0 - d] + p_{ambient}$$

and

$$Z_0(r_0) = d$$

whereupon the reference domain is the set of points (r_0, z_0) defined by $0 < r_0 < R$ and $0 < z_0 < d$.

To derive the equations holding on the reference domain and at its boundary, to orders zero, one, two, etc. in ϵ, introduce the mapping

$$r = r_0$$

and

$$z = z_0 + \epsilon z_1(r_0, \ z_0) + \frac{1}{2}\epsilon^2 z_2(r_0, \ z_0) + \cdots$$

which ties the present domain to the reference domain. Observe that at the boundary this is written

$$r = r_0$$

and

$$z = Z(r, \ \epsilon) = Z_0 + \epsilon Z_1(r_0) + \frac{1}{2}\epsilon^2 Z_2(r_0) + \ \cdots$$

Now, introduce the expansions of p, $\dfrac{\partial p}{\partial r}$ and $\dfrac{\partial p}{\partial z}$ along the mapping, namely

$$p(r, \ z, \ \epsilon) = p_0 + \epsilon \left[p_1 + z_1 \frac{\partial p_0}{\partial z_0} \right]$$

$$+ \frac{1}{2}\epsilon^2 \left[p_2 + 2z_1 \frac{\partial p_1}{\partial z_0} + z_1^2 \frac{\partial^2 p_0}{\partial z_0^2} + z_2 \frac{\partial p_0}{\partial z_0} \right] + \ \cdots$$

$$\frac{\partial p}{\partial r}(r, \ z, \ \epsilon) = \frac{\partial p_0}{\partial r_0} + \epsilon \left[\frac{\partial p_1}{\partial r_0} + z_1 \frac{\partial^2 p_0}{\partial r_0 \partial z_0} \right]$$

$$+ \frac{1}{2}\epsilon^2 \left[\frac{\partial p_2}{\partial r_0} + 2z_1 \frac{\partial^2 p_1}{\partial r_0 \partial z_0} + z_1^2 \frac{\partial^3 p_0}{\partial r_0 \partial z_0^2} + z_2 \frac{\partial^2 p_0}{\partial r_0 \partial z_0} \right] + \cdots$$

and

$$\frac{\partial p}{\partial z}(r, \ z, \ \epsilon) = \frac{\partial p_0}{\partial z_0} + \epsilon \left[\frac{\partial p_1}{\partial z_0} + z_1 \frac{\partial^2 p_0}{\partial z_0^2} \right]$$

$$+ \frac{1}{2}\epsilon^2 \left[\frac{\partial p_2}{\partial z_0} + 2z_1 \frac{\partial^2 p_1}{\partial z_0^2} + z_1^2 \frac{\partial^3 p_0}{\partial z_0^3} + z_2 \frac{\partial^2 p_0}{\partial z_0^2} \right] + \ \cdots$$

and notice that all the variables on the right-hand side depend at most on r_0 and z_0. Then, substitute these expansions into equations (A.1) and (A.2) to get the problems satisfied by p_0 at zeroth order, p_1 at first order, etc.

At zeroth order, there obtains

$$\left. \begin{aligned} \frac{\partial p_0}{\partial r_0} &= 0 \\[2em] \frac{\partial p_0}{\partial z_0} &= -\rho g \end{aligned} \right\} \qquad 0 < r_0 < R, \ \ 0 < z_0 < d \qquad \text{(A.4)}$$

and

$$p_0 = p_{ambient}, \ \ 0 < r_0 < R, \ \ z_0 = d \qquad \text{(A.5)}$$

At first order, there obtains

$$\left.\begin{array}{l} \dfrac{\partial p_1}{\partial r_0} = \rho r_0 \\[3mm] \dfrac{\partial p_1}{\partial z_0} = 0 \end{array}\right\} \qquad 0 < r_0 < R, \ \ 0 < z_0 < d \qquad \text{(A.6)}$$

and

$$p_1 + Z_1 \frac{\partial p_0}{\partial z_0} = 0, \ \ 0 < r_0 < R, \ \ z_0 = d \qquad \text{(A.7)}$$

At second order, there obtains

$$\left.\begin{array}{l} \dfrac{\partial p_2}{\partial r_0} = 0 \\[3mm] \dfrac{\partial p_2}{\partial z_0} = 0 \end{array}\right\} \qquad \begin{array}{l} 0 < r_0 < R \\[3mm] 0 < z_0 < d \end{array} \qquad \text{(A.8)}$$

and

$$p_2 + 2Z_1 \frac{\partial p_1}{\partial z_0} + Z_1^2 \frac{\partial^2 p_0}{\partial z_0^2} + Z_2 \frac{\partial p_0}{\partial z_0} = 0, \ \ 0 < r_0 < R, \ \ z_0 = d \qquad \text{(A.9)}$$

etc.

Notice that on the reference domain, but not on its boundary, the zeroth-order equations eliminate the mapping in the first-order equations, the zeroth- and first-order equations eliminate the mapping at second order, etc.

Now, equations (A.4), (A.6), (A.8), $\cdots$ can be used to evaluate the derivatives appearing in equations (A.7), (A.9), $\cdots$ and doing this simplifies these equations to

$$\left.\begin{array}{l} p_0 = p_{ambient} \\[4mm] p_1 - Z_1 \rho g = 0 \\[4mm] p_2 - Z_2 \rho g = 0 \\[4mm] \text{etc.} \end{array}\right\} \qquad 0 < r_0 < R, \ \ z_0 = d$$

where Z_1, Z_2, $\cdots$ depend on r_0.

Then, equations (A.4), (A.6), (A.8), $\cdots$ on the domain can be satisfied by writing

$$p_0 = -\rho g z_0 + C_0$$

$$p_1 = \frac{1}{2}\rho r_0^2 + C_1$$

$$p_2 = C_2$$

etc.

where the constants C_0, C_1, C_2, $\cdots$ along with the functions Z_1, Z_2, $\cdots$ remain to be determined by using equations (A.5), (A.7), (A.9), $\cdots$. These require

$$C_0 = p_{ambient} + \rho g d$$

$$Z_1 = \frac{1}{2g}r_0^2 + \frac{C_1}{\rho g}$$

$$Z_2 = \frac{C_2}{\rho g}$$

etc.

and it remains only to satisfy the requirement that the volume of the liquid remain fixed as its surface deflects.

Substituting the expansion of the surface shape, namely

$$Z(r,\ \epsilon) = Z_0(r_0) + \epsilon Z_1(r_0) + \frac{1}{2}\epsilon^2 Z_2(r_0) + \ \cdots$$

into

$$\pi R^2 d = \int_0^R 2\pi r Z(r,\ \epsilon)dr$$

and using $Z_0 = d$ produces the conditions

$$\int_0^R r_0 Z_1(r_0)dr_0 = 0$$

$$\int_0^R r_0 Z_2(r_0)dr_0 = 0$$

etc.

which can be used to determine the constants C_1, C_2, $\cdots$. The result is

$$C_1 = -\frac{1}{4}\rho R^2$$

$$C_2 = 0$$

etc.

whence

$$p_0 = -\rho g[z_0 - d] + p_{ambient}, \quad Z_0 = d$$

$$p_1 = \frac{1}{2}\rho r_0^2 - \frac{1}{4}\rho R^2, \quad Z_1 = \frac{1}{2g}r_0^2 - \frac{1}{4g}R^2$$

$$p_2 = 0, \quad Z_2 = 0$$

etc.

The reader can go on and discover that p_3 and Z_3 must satisfy

$$\left.\begin{array}{l} \dfrac{\partial p_3}{\partial r_0} = 0 \\[1.2em] \dfrac{\partial p_3}{\partial z_0} = 0 \end{array}\right\} \qquad 0 < r_0 < R, \quad 0 < z_0 < d$$

$$p_3 - Z_3\rho g = 0, \quad 0 < r_0 < R, \quad z_0 = d$$

and

$$\int_0^R r_0 Z_3 dr_0 = 0$$

whence $p_3 = 0 = Z_3$. This result obtains (viz., $p_i = 0 = Z_i$), for all $i = 2, 3, \cdots$.

The present domain can now be determined. The shape of its free surface is given by

$$Z(r, \ \Omega^2) = Z_0(r_0) + \epsilon Z_1(r_0) + \frac{1}{2}\epsilon^2 Z_2(r_0) + \cdots$$

$$= Z_0(r_0) + \Omega^2 Z_1(r_0) = d + \Omega^2 \left[\frac{r_0^2}{2g} - \frac{R^2}{4g}\right]$$

and this turns out to be the correct formula for Z, as obtained earlier.

It remains to determine the pressure throughout this domain [viz., throughout $0 < r < R, \ 0 < z < Z(r, \ \Omega^2)$]. The problem we face in doing this is indicated in Figure A.2. It is this: While some of the points of the present domain are also points of the reference domain, others are not. The first set of points is shaded in the figure, and at these points, p can be estimated by carrying z straight back to z_0, as explained in the essay. By doing this, p is obtained as

$$p(r, \ z, \ \epsilon) = p_0(r_0, \ z) + \epsilon p_1(r_0, \ z) + \cdots$$

where $r = r_0$, whence at these points the value of p is found to be

$$p(r, \ z, \ \Omega^2) - p_{ambient} = \rho g[z - d] + \Omega^2 \left[\frac{1}{2}\rho r^2 - \frac{1}{4}\rho R^2\right]$$

Again, this turns out to be correct, as $p_3, \ p_4, \cdots$ all vanish. It also turns out to be correct if it is used at the unshaded points, but this is only because it is correct on the shaded points. Indeed, if the correct result can

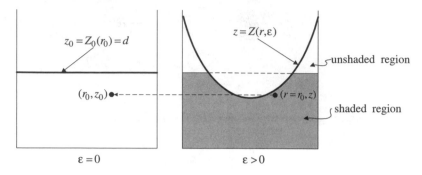

Figure A.2. *Points That Can Be Carried Straight Back to the Reference Domain and Points That Cannot*

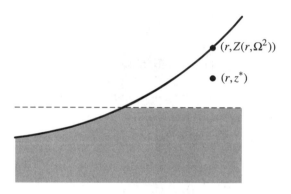

Figure A.3. *A Point Near the Boundary That Cannot Be Carried Straight Back to the Reference Domain*

be obtained on a part of a domain, it extends to the entire domain by analytic continuation. The weakness of this example is that it is so simple that its solution, rather than an estimate of its solution, is obtained after just a few terms in the expansion.

Nonetheless, the values of p at points of the unshaded region can be estimated, as they would have been if this lucky accident had not occurred, by observing that p and all its derivatives can be evaluated at points of the boundary of the present domain and by observing that these points must be neighbors of the points in the unshaded region. This is illustrated in Figure A.3.

To determine the value of p at the point $(r,\ z^*)$, substitute into

$$p(r, \ z^*, \ \epsilon) = p(r, Z, \epsilon) + \frac{\partial p}{\partial z}(r, \ Z, \ \epsilon)[z^* - Z]$$

$$+ \frac{1}{2}\frac{\partial^2 p}{\partial z^2}(r, \ Z, \ \epsilon)[z^* - Z]^2 + \ \cdots$$

p and its derivatives determined according to

$$p(r, \ Z, \ \epsilon) = p_0(r_0, Z_0) + \epsilon \left[p_1(r_0, \ Z_0) + Z_1(r_0)\frac{\partial p_0}{\partial z_0}(r_0, \ Z_0) \right] + \cdots$$

$$\frac{\partial p}{\partial z}(r, \ Z, \ \epsilon) = \frac{\partial p_0}{\partial z_0}(r_0, Z_0) + \epsilon \left[\frac{\partial p_1}{\partial z_0}(r_0, \ Z_0) + Z_1(r_0)\frac{\partial^2 p_0}{\partial z_0^2}(r_0, \ Z_0) \right] + \cdots$$

etc.

Doing this and using equations (A.4), (A.5), (A.6) and (A.7) to write

$$p_0 = p_{ambient}$$

$$\frac{\partial p_0}{\partial z_0} = -\rho g$$

$$p_1 + Z_1\frac{\partial p_0}{\partial z_0} = 0$$

and

$$\frac{\partial p_1}{\partial z_0} + Z_1\frac{\partial^2 p_0}{\partial z_0^2} = 0$$

at points $(r_0, \ Z_0)$ along the boundary of the reference domain, produces

$$p(r, \ z^*, \ \Omega^2) = p_{ambient} - \rho g[z^* - Z]$$

where $r = r_0$ and where $Z(r, \ \Omega^2) = d + \Omega^2 \left[\frac{r_0^2}{2g} - \frac{R^2}{4g} \right]$. Again, this is the correct result.

It might be worthwhile to explain a little more about this last calculation. It requires the use of several series. Going back to the notation in the second essay, the problem is to calculate u at a point $(x, \ y^*)$ lying near a boundary point $(x, \ Y(x, \ \epsilon))$. To do this, the series

$$u(x, \ y^*, \ \epsilon) = u(x, \ Y, \ \epsilon) + \frac{\partial u}{\partial y}(x, \ Y, \ \epsilon)[y^* - Y]$$

$$+ \frac{1}{2}\frac{\partial^2 u}{\partial y^2}(x, \ Y, \ \epsilon)[y^* - Y]^2 + \ \cdots \qquad (A.10)$$

must be used, into which the several series, namely

$$Y = Y_0 + \epsilon Y_1 + \frac{1}{2}\epsilon^2 Y_2 + \ \cdots \qquad (A.11)$$

$$u(x,\, Y,\, \epsilon) = u_0 + \epsilon \left[u_1 + Y_1 \frac{\partial u_0}{\partial y_0} \right] + \cdots \tag{A.12}$$

$$\frac{\partial u}{\partial y}(x,\, Y,\, \epsilon) = \frac{\partial u_0}{\partial y_0} + \epsilon \left[\frac{\partial u_1}{\partial y_0} + Y_1 \frac{\partial^2 u_0}{\partial y_0^2} \right] + \cdots \tag{A.13}$$

etc.

must be substituted, where all the variables on the right-hand sides of these series are evaluated at x_0 and $y_0 = Y_0(x_0)$, and where, on the left-hand side, $x = x_0$.

Then, to obtain $u(x,\, y^*,\, \epsilon)$ to a certain order in ϵ, all of the terms in the first series are required, each to the assigned order in ϵ. To illustrate this, let u be required to first order in ϵ; then, substituting equations (A.11), (A.12), (A.13), $\cdots$ into equation (A.10), there obtains

$$u(x,\, y^*,\, \epsilon) = \left[u_0 + \frac{\partial u_0}{\partial y_0}[y^* - Y_0] + \frac{1}{2}\frac{\partial^2 u_0}{\partial y_0^2}[y^* - Y_0]^2 + \cdots \right]$$

$$+ \epsilon \left[u_1 + \frac{\partial u_1}{\partial y_0}[y^* - Y_0] + \frac{1}{2}\frac{\partial^2 u_1}{\partial y_0^2}[y^* - Y_0]^2 + \cdots \right]$$

Hence, to get $u(x,\, y^*,\, \epsilon)$ to order ϵ requires only that the sums of these two series be estimated with sufficient accuracy. Presumably, y^* lies close enough to Y_0 that this does not present a new difficulty. Indeed, if $y^* - Y_0$ is of order ϵ, then $u(x,\, y^*,\, \epsilon)$ can be estimated to order ϵ by

$$u_0 + \frac{\partial u_0}{\partial y_0}[y^* - Y] + \epsilon u_1$$

A.2   The Position of a Melting Front

Turning now to a second example, which is physically and mathematically more interesting, we try to determine the configuration of a heat conducting body where the shape of the body depends upon the rate at which it is losing heat. To define the problem, let ice at its melting point occupy the half-space $x > 0$. Then, introduce heat across the wall bounding the ice at $x = 0$. As a result of doing this, the ice originally lying between $x = 0$ and $x = d$ is melted and the temperature of the water formed is raised above its freezing point. At time $t = 0$, the heating is stopped and the wall at $x = 0$ is thereafter insulated. The system is turned over to us at $t = 0$. At that time, it is made up of a layer of water occupying the region $0 < x < d$, exhibiting a positive distribution of temperature, and the unmelted ice, at zero temperature, occupying the region $x > d$. As time begins to run, the water begins to cool, and in so doing, it begins to melt the ice. Our job is to determine the position of the water–ice interface, as well as the temperature of the water, as they depend on time.

The temperature of the ice remains always at zero and so it is only the water that is of interest. To take the simplest possible case, the water is bounded by an insulated plane wall on the left and by the water–ice coexistence plane on the right. The heat conduction is one dimensional.

At any time, t, let u denote the temperature, as it depends on position, at the points presently occupied by water and let X denote the position of the water–ice interface. The speed at which this interface moves is then given by $\dfrac{dX}{dt}$.

Let $f(x), 0 < x < d$, indicate the way in which the temperature is initially distributed throughout the water. Then, express the initial condition of the water by

$$u(t = 0) = \epsilon f(x), \quad 0 < x < d$$

where ϵ is a measure of the amount of heat, over that just required to melt the ice, that the water receives while the initial state is being created. This is the amount of sensible heat on the water side that is available to melt the ice once the system is allowed to seek its equilibrium configuration.

By introducing ϵ in this way, a family of problems is created and the family member corresponding to $\epsilon = 0$ is especially simple. Each member of the family is defined by a specific value of ϵ. Its domain the current domain, current in terms of ϵ, not t, is not the spatial domain occupied by the water at some time t but the entire history of these spatial domains corresponding to a specified value of ϵ. The current domain is then defined by three boundaries:

(i) the insulated wall at $x = 0$ for all $t \geq 0$
(ii) the initial spatial domain of the water at $t = 0$ for all x, $0 < x < d$
(iii) the water–ice coexistence plane at $x = X(t,\ \epsilon)$ for all $t \geq 0$, where the function $X(t,\ \epsilon)$ determining this part of the boundary of the current domain is part of the solution of the problem

This is illustrated in Figure A.4.

Then, u and X must be determined. On the current domain and on its boundary, $u(t,\ x,\ \epsilon)$ and $X(t,\ \epsilon)$ must satisfy[1]

$$\frac{\partial u}{\partial t} = \frac{\partial^2 u}{\partial x^2}, \quad t > 0, \quad 0 < x < X(t,\ \epsilon)$$

---

[1] It may be observed that as long as the right-hand phase is maintained at the phase change temperature, the fact that the density of the two phases differ is of no account. The flow produced by this takes place in the right-hand phase, but it is at uniform temperature.

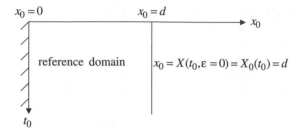

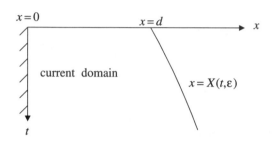

Figure A.4. *The Current and the Reference Domains*

$$\frac{\partial u}{\partial x} = 0, \quad t > 0, \quad x = 0$$

$$u = 0$$

$$\left. \frac{\partial u}{\partial x} = -\frac{\partial X}{\partial t} \right\} \quad t > 0, \quad x = X(t, \ \epsilon)$$

and

$$u = \epsilon f, \quad t = 0, \quad 0 < x < d$$

where $X(t = 0, \ \epsilon) = d$.

   This is the nonlinear problem and its solution at $\epsilon = 0$ defines the reference domain. In terms of reference domain variables, u_0, X_0, t_0 and x_0, this solution is

$$u_0 = 0$$

and

$$X_0 = d$$

The reference domain is then the simple domain $0 \leq x_0 \leq d$, $t_0 \geq 0$.

   To introduce the mapping that ties the reference domain to the current domain, write

$$t = t_0$$

and

$$x = g(t_0, \ x_0, \ \epsilon) = x_0 + \epsilon x_1(t_0, \ x_0) + \frac{1}{2}\epsilon^2 x_2(t_0, \ x_0) + \ \cdots$$

and observe that all parts of the boundary of the reference domain must be carried by this mapping into the corresponding parts of the boundary of the present domain. This introduces certain requirements that the mapping must meet. At the insulated wall, the points $(t_0, \ x_0 = 0)$ must be carried into the points $(t, \ x = 0)$ whence

$$x_1(t_0, \ x_0 = 0) = 0$$

$$x_2(t_0, \ x_0 = 0) = 0$$

etc.

Then, requiring the points $(t_0 = 0, \ x_0)$ of the initial spatial domain of the reference configuration to be carried into the corresponding points $(t = 0, \ x = x_0)$ of the initial spatial domain of the present configuration leads to

$$x_1(t_0 = 0, \ x_0) = 0$$

$$x_2(t_0 = 0, \ x_0) = 0$$

etc.

Turn now to the water–ice interface. In the reference configuration, it is denoted $x_0 = X_0(t_0) = d$, while in the present configuration, it is denoted $x = X(t, \ \epsilon)$. The expansion of $X(t, \ \epsilon)$ along the mapping is written

$$X(t, \ \epsilon) = X_0(t_0) + \epsilon X_1(t_0) + \frac{1}{2}\epsilon^2 X_2(t_0) + \ \cdots$$

and $X(t = 0, \epsilon) = d = X_0(t_0 = 0)$ requires

$$X_1(t_0 = 0) = 0$$

$$X_2(t_0 = 0) = 0$$

etc.

where

$$X_1(t_0) = x_1(t_0, \ x_0 = X_0(t_0) = d)$$

$$X_2(t_0) = x_2(t_0, \ x_0 = X_0(t_0) = d)$$

etc.

To produce the equations satisfied by u_0, u_1, u_2, etc. on the reference domain, substitute into $\dfrac{\partial u}{\partial t} - \dfrac{\partial^2 u}{\partial x^2} = 0$ the expansions

$$\frac{\partial^2 u}{\partial x^2}(t,\ x,\ \epsilon) = \frac{\partial^2 u_0}{\partial x_0^2} + \epsilon \left[\frac{\partial^2 u_1}{\partial x_0^2} + x_1 \frac{\partial^3 u_0}{\partial x_0^3} \right] + \cdots$$

and

$$\frac{\partial u}{\partial t}(t,\ x,\ \epsilon) = \frac{\partial u_0}{\partial t_0} + \epsilon \left[\frac{\partial u_1}{\partial t_0} + x_1 \frac{\partial^2 u_0}{\partial t_0 \partial x_0} \right] + \cdots$$

and set the coefficient of each power of ϵ to zero. The result is

$$\left. \begin{array}{c} \dfrac{\partial u_0}{\partial t_0} - \dfrac{\partial^2 u_0}{\partial x_0^2} = 0 \\[3mm] \dfrac{\partial u_1}{\partial t_0} - \dfrac{\partial^2 u_1}{\partial x_0^2} = 0 \\[3mm] \text{etc.} \end{array} \right\} \qquad t_0 > 0,\ \ 0 < x_0 < d$$

where the first equation has been used to eliminate the mapping in the second, etc.

Likewise, the equations at the boundary of the reference domain obtain via substitution of the expansions of u and its derivatives. Along the insulated wall, substitution of

$$\frac{\partial u}{\partial x}(t,\ x,\ \epsilon) = \frac{\partial u_0}{\partial x_0} + \epsilon \left[\frac{\partial u_1}{\partial x_0} + x_1 \frac{\partial^2 u_0}{\partial x_0^2} \right] + \cdots$$

into

$$\frac{\partial u}{\partial x}(t,\ x = 0,\ \epsilon) = 0$$

and use of $x_1(t_0,\ x_0 = 0) = 0$, $x_2(t_0,\ x_0 = 0) = 0$, etc. leads to

$$\left. \begin{array}{c} \dfrac{\partial u_0}{\partial x_0} = 0 \\[3mm] \dfrac{\partial u_1}{\partial x_0} = 0 \\[3mm] \text{etc.} \end{array} \right\} \qquad t_0 > 0,\ \ x_0 = 0$$

while on the initial spatial domain, $t = 0$, $0 < x < d$, substitution of

$$u(t,\ x,\ \epsilon) = u_0 + \epsilon \left[u_1 + x_1 \frac{\partial u_0}{\partial x_0} \right] + \cdots$$

into

$$u(t = 0,\ x,\ \epsilon) = \epsilon f(x)$$

and use of $x_1(t_0 = 0,\ x_0) = 0,\ x_2(t_0 = 0,\ x_0) = 0$, etc. produces

$$\left.\begin{array}{c} u_0 = 0 \\[2ex] u_1 = f \\[2ex] \text{etc.} \end{array}\right\} \quad t_0 = 0,\ \ 0 < x_0 < d$$

At the water–ice interface, two equations must be satisfied in the present configuration, one to make certain that the heat conduction problem in the water is properly posed, the other to determine the speed at which ice is turned into water. Substitution of the expansion of $u(t,\ x,\ \epsilon)$ into

$$u(t,\ X,\ \epsilon) = 0$$

and use of $x_1(t_0,\ x_0 = X_0) = X_1,\ x_2(t_0,\ x_0 = X_0) = X_2$, etc. requires

$$\left.\begin{array}{c} u_0 = 0 \\[2ex] u_1 = -X_1\dfrac{\partial u_0}{\partial x_0} \\[2ex] \text{etc.} \end{array}\right\} \quad t_0 > 0,\ \ x_0 = d$$

The substitution of the expansions of $\dfrac{\partial u}{\partial x}(t,\ x,\ \epsilon)$ and $X(t,\ \epsilon)$ into

$$\frac{\partial u}{\partial x}(t,\ X,\ \epsilon) = -\frac{\partial X}{\partial t}(t,\ \epsilon)$$

and, again, the use of $x_1(t_0,\ x_0 = X_0) = X_1, x_2(t_0,\ x_0 = X_0) = X_2$, etc. requires

$$\left.\begin{array}{c} \dfrac{\partial u_0}{\partial x_0} = -\dfrac{dX_0}{dt_0} \\[2ex] \dfrac{\partial u_1}{\partial x_0} + X_1\dfrac{\partial^2 u_0}{\partial x_0^2} = -\dfrac{dX_1}{dt} \\[2ex] \text{etc.} \end{array}\right\} \quad t_0 > 0,\ \ x_0 = d$$

The zeroth-order problem on the reference domain is then

$$\frac{\partial u_0}{\partial t_0} = \frac{\partial^2 u_0}{\partial x_0^2}, \quad t_0 > 0,\ \ 0 < x_0 < X_0 = d$$

$$u_0(t_0 = 0) = 0$$

$$\frac{\partial u_0}{\partial x_0}(x_0 = 0) = 0$$

$$u_0(x_0 = d) = 0$$

and

$$\frac{\partial u_0}{\partial x_0}(x_0 = d) = -\frac{dX_0}{dt_0}$$

where

$$X_0(t_0 = 0) = d$$

Its solution is

$$u_0 = 0 \quad \text{and} \quad X_0 = d$$

Again, the zeroth-order problem is the original problem. This is always the case. The solution to the zeroth-order problem is always a solution to the original nonlinear problem.

The first-order problem begins to yield some information. It is

$$\frac{\partial u_1}{\partial t_0} = \frac{\partial^2 u_1}{\partial x_0^2}, \quad t_0 > 0, \ 0 < x_0 < d$$

$$u_1(t_0 = 0) = f$$

$$\frac{\partial u_1}{\partial x_0}(x_0 = 0) = 0$$

$$u_1(x_0 = d) = -X_1 \frac{\partial u_0}{\partial x_0}(x_0 = d)$$

and

$$\frac{\partial u_1}{\partial x_0}(x_0 = d) = -X_1 \frac{\partial^2 u_0}{\partial x_0^2}(x_0 = d) - \frac{dX_1}{dt_0}$$

where

$$X_1(t_0 = 0) = 0$$

This problem must be solved on the reference domain (viz., $0 < x_0 < d$, $t_0 > 0$), and u_1 and X_1 must be determined together. They depend on u_0 and X_0. Then, at the next order, u_2 and X_2 must be determined together and they depend on u_1 and X_1, as well as on u_0 and X_0. This is the way the calculation proceeds and it is the main purpose of this example to make just this point. The position of the water–ice interface then becomes known as the terms in the series

$$X(t, \ \epsilon) = X_0(t_0) + \epsilon X_1(t_0) + \frac{1}{2}\epsilon^2 X_2(t_0) + \cdots$$

become known.

It might be worth observing that, in this particular case, u_0 is so simple that the first four equations at first order can be used to determine u_1 and then the last two can be used to determine X_1. Because u_0 and all its derivatives vanish, this uncoupling obtains at all orders. The readers can satisfy themselves that u_2 satisfies a simple heat conduction problem on the reference domain and that, once it is determined, X_2 can be obtained by an integration. The only difference between u_1 and u_2 is that u_1 is driven by a source at $t_0 = 0$, whereas u_2 is driven by a source at $x_0 = d$. Like u_2, all higher-order u's are driven by sources at $x_0 = d$ that depend on what is discovered at lower orders.

It remains to explain how it is that a point of the current domain can be evaluated in terms of u_0, u_1, u_2, $\cdots$ known only at points of the reference domain. But, as this ought to present little difficulty, we go on and rework this problem from another point of view.

A.3   Carrying the Melting Front Forward in Time

When this problem was first stated, it did not come equipped with a ready-made small parameter. It was simply a problem defined on a domain where the domain was at first not known. We, ourselves, introduced the parameter ϵ; it lies wholly outside the original statement of the problem. By doing this, we traded a problem on an unknown domain for many problems on a known and simple domain. In many ways, this is a fair trade, but it raises the question: Just what is ϵ? There is no one answer to this question. Sometimes, the problem itself determines what ϵ must be. Other times, we invent ϵ.

Even in a definite problem, many ways of introducing the expansion parameter will ordinarily present themselves. The definition ought to depend on what we are trying to discover. To illustrate this, we rework the foregoing problem, taking ϵ to be a difference in time.

To do this, let the solution be available to us at some reference time, denoted t_0. Then, our job is simply to advance it to a later time t. Looking at it in this way, the reference domain becomes the spatial domain of the water at the time t_0, described as $0 \leq x_0 \leq X_0 = X(t_0)$, where X_0 is assigned. The current domain becomes the spatial domain of the water at the later time t of interest, described as $0 \leq x \leq X(t)$, where $X(t)$ must be determined in the course of turning $u(x_0, t_0)$ on the reference domain into $u(x, t)$ on the present domain. The expansion parameter ϵ is now $t - t_0$ and the corresponding family of problems is now just what the problem itself was as originally stated. All this is illustrated in Figure A.5.

Our goal, now, is not, as it was before, to determine the reference domain and then to determine the equations satisfied by a sequence of functions

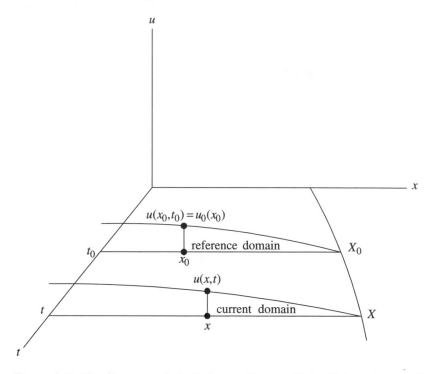

Figure A.5. *The Current and the Reference Domain If the Melting Front Is Being Carried Forward in Time*

u_0, u_1, u_2, etc. on the reference domain. Instead, it is to use u_0, which is now assigned, to determine u_1, u_2, etc., as well as, X_1, X_2, etc., and thereby to advance the solution in time from the reference domain to the present domain.

To do this, we introduce a mapping of the reference domain into the present domain and then introduce the expansions of u and its derivatives along the mapping.

The mapping can be written

$$x = f(x_0, \; \epsilon) = x_0 + \epsilon x_1(x_0) + \frac{1}{2}\epsilon^2 x_2(x_0) + \; \cdots$$

where at the water–ice interface

$$x = X(\epsilon) = X_0 + \epsilon X_1 + \frac{1}{2}\epsilon^2 X_2 + \; \cdots$$

whence

$$x_1(x_0 = X_0) = X_1$$

$$x_2(x_0 = X_0) = X_2$$

etc.

while at the insulated wall

$$x = 0$$

whence

$$x_1(x_0 = 0) = 0$$

$$x_2(x_0 = 0) = 0$$

etc.

Again, denote by $u_0(x_0)$ the assigned values of $u(x_0, t_0)$ on the reference domain $0 \leq x_0 \leq X_0$. The goal, then, is to determine the values of $u(x, t)$, denoted by $u(x, \epsilon)$, on the present domain $0 \leq x \leq X(\epsilon)$, where $X(\epsilon)$ must also be determined. To do this, expand u and its derivatives along the mapping.

The expansion of u is

$$u(x, \epsilon) = u(\epsilon = 0) + \epsilon \frac{du}{d\epsilon}(\epsilon = 0) + \frac{1}{2}\epsilon^2 \frac{d^2 u}{d\epsilon^2}(\epsilon = 0) + \cdots$$

where $d/d\epsilon$, which is also d/dt, denotes a total derivative holding x_0 fixed. On carrying out the calculation, much as before, this reduces to

$$u(x, \epsilon) = u_0 + \epsilon \left[u_1 + x_1 \frac{du_0}{dx_0} \right]$$

$$+ \frac{1}{2}\epsilon^2 \left[u_2 + 2x_1 \frac{du_1}{dx_0} + x_1^2 \frac{d^2 u_0}{dx_0^2} + x_2 \frac{du_0}{dx_0} \right] + \cdots$$

where all variables on the right-hand side depend only on x_0, namely

$$u_0(x_0) = u(x_0, \epsilon = 0)$$

$$u_1(x_0) = \frac{\partial u}{\partial t}(x_0, \epsilon = 0)$$

$$u_2(x_0) = \frac{\partial^2 u}{\partial t^2}(x_0, \epsilon = 0)$$

etc.

The expansions of $\partial u/\partial x$ and $\partial^2 u/\partial x^2$ likewise take their familiar forms

$$\frac{\partial u}{\partial x}(x, \epsilon) = \frac{du_0}{dx_0} + \epsilon \left[\frac{du_1}{dx_0} + x_1 \frac{d^2 u_0}{dx_0^2} \right] + \cdots$$

and

$$\frac{\partial^2 u}{\partial x^2}(x, \epsilon) = \frac{d^2 u_0}{dx_0^2} + \epsilon \left[\frac{d^2 u_1}{dx_0^2} + x_1 \frac{d^3 u_0}{dx_0^3} \right] + \cdots$$

but the expansion of $\dfrac{\partial u}{\partial t}(x,\ \epsilon)$, which is also $\dfrac{\partial u}{\partial \epsilon}(x,\ \epsilon)$, might seem to be new and so we work it out. The main idea is simply to calculate $\partial/\partial\epsilon$ of both sides of the expansion of $u(x,\ \epsilon)$, where x must be held fixed. This presents the usual problem on the right-hand side: x_0 cannot remain fixed if x is fixed, and in carrying out the differentiation, x_0 must be viewed as a function of ϵ at fixed x. This function, x_0 versus ϵ at fixed x, is specified by the mapping, and differentiating the mapping leads to a formula for $\partial x_0/\partial\epsilon$. Calling this $\partial x_0/\partial t$, the formula is

$$0 = \frac{\partial x_0}{\partial t} + x_1 + \epsilon\frac{\partial x_1}{\partial t} + \epsilon x_2 + \frac{1}{2}\epsilon^2\frac{\partial x_2}{\partial t} + \cdots$$

Then, differentiating the expansion of u with respect to t produces

$$\frac{\partial u}{\partial t}(x,\ \epsilon) = \frac{du_0}{dx_0}\frac{\partial x_0}{\partial t} + \left[u_1 + x_1\frac{du_0}{dx_0}\right]$$

$$+\epsilon\left[\frac{du_1}{dx_0}\frac{\partial x_0}{\partial t} + x_1\frac{d^2u_0}{dx_0^2}\frac{\partial x_0}{\partial t} + \frac{\partial x_1}{\partial t}\frac{du_0}{dx_0}\right]$$

$$+\epsilon\left[u_2 + 2x_1\frac{du_1}{dx_0} + x_1^2\frac{d^2u_0}{dx_0^2} + x_2\frac{du_0}{dx_0}\right] + \text{terms of order } \epsilon^2$$

whence, eliminating $\dfrac{\partial x_0}{\partial t}$ via

$$\frac{\partial x_0}{\partial t} = -x_1 - \epsilon\frac{\partial x_1}{\partial t} - \epsilon x_2 - \frac{1}{2}\epsilon^2\frac{\partial x_2}{\partial t}$$

there obtains

$$\frac{\partial u}{\partial t}(x,\ \epsilon) = u_1 + \epsilon\left[u_2 + x_1\frac{du_1}{dx_0}\right] + \cdots$$

The plan, now, is to substitute our expansions into the equations satisfied by u on the present domain and on its boundary to discover how u_1, u_2, $\cdots$ can be obtained from u_0.

Taking the domain equation, $\dfrac{\partial u}{\partial t} = \dfrac{\partial^2 u}{\partial x^2}$, first, we get

$$u_1 = \frac{d^2u_0}{dx_0^2}$$

$$u_2 = \frac{d^2u_1}{dx_0^2}$$

etc.

where the first equation has been used to eliminate the mapping in the second. This determines all of the functions u_1, u_2, $\cdots$ present in the expansion of u.

Then, going to the water–ice interface, the equation $u(x = X, t) = 0$ requires

$$u_0(x_0 = X_0) = 0$$

$$u_1(x_0 = X_0) + X_1 \frac{du_0}{dx_0}(x_0 = X_0) = 0$$

etc.

These formulas can be used to determine X_1, X_2, $\cdots$, but they make X_1, for instance, depend on $\dfrac{d^2 u_0}{dx_0^2}(x_0 = X_0)$. The remaining equation at the water–ice interface, namely

$$\frac{\partial u}{\partial x}(x = X, t) = -\frac{dX}{dl}(t)$$

requires

$$\frac{du_0}{dx_0}(x_0 = X_0) = -X_1$$

$$\frac{du_1}{dx_0}(x_0 = X_0) + X_1 \frac{d^2 u_0}{dx_0^2}(x_0 = X_0) = -X_2$$

etc.

which also determine X_1, X_2, $\cdots$. By this, the region occupied by the water is determined to order ϵ^2 by an order ϵ calculation. But as X_1, X_2, $\cdots$ can all be determined in two ways, a consistency condition on the data, u_0, presents itself.

Due to the fact that it is at the boundary of the reference domain, and only there, that the mapping into current domain can be determined, by continuing the calculations, u and all its derivatives, $\partial u/\partial x$, $\partial^2 u/\partial x^2$, $\cdots$, can be estimated at the end points of the present domain (viz., at $x = 0$ and $x = X$). But this is sufficient, as indicated in the essay, to determine u everywhere on the present domain.

Now, the reader may be skeptical upon learning that a well-posed nonlinear problem turns out to be overdetermined when its solution is advanced in time, as above. What, then, is going on? How can two formulas be produced, where both can be used to predict X_1, but where each comes from one or the other of two independent equations which the solution to the problem must satisfy? Is it possible that the two formulas predict the same value for X_1? Our view is this: If u_0 were assigned arbitrarily at $t_0 = 0$, as it could be if it were an initial condition, then the two formulas would, most likely, not be compatible. But if, instead, u_0 is determined by taking it to be the solution at some time t_0 after an arbitrary initial condition has already been assigned, then the two formulas must be compatible. Notice that the two formulas for X_1 come, on the one hand, out of the phase-equilibrium equation at the interface, while, on the other hand, out of the heat balance equation across the interface. Then, notice that u_0 denotes u after time t_0, where u, the solution to the problem, has been satisfying both equations over a finite interval of time. By this, predictions beyond t_0 based on u_0 using one equation ought to be consistent with predictions using the other.

To illustrate this, take a nonlinear problem whose solution can be found, namely

$$\frac{\partial u}{\partial t} = \frac{\partial^2 u}{\partial x^2}, \quad 0 < x < X(t), \quad t > 0$$

$$u(x = 0) = -1, \quad t > 0$$

$$u(x = X) = 0, \quad t > 0$$

and

$$\frac{\partial u}{\partial x}(x = X) = \frac{dX}{dt}, \quad t > 0$$

where

$$X(t = 0) = 0$$

This corresponds to water, held at $u = 0$ at $t = 0$, being frozen by reducing its temperature at $x = 0$ to $u = -1$ and holding it there. The new sign in the heat balance equation across the interface is accounted for by the fact that now ice lies to the left, water to the right, instead of water to the left, ice to the right as before.[2] Also, the adiabatic boundary condition at $x = 0$ has been replaced by an isothermal boundary condition, but this does not alter the two formulas for X_1.

This problem can be solved by taking u and X to be

$$u = -1 + \frac{\operatorname{erf}\left(\frac{x}{\sqrt{4t}}\right)}{\operatorname{erf} c}$$

and

$$X = c\sqrt{4t}$$

where c must satisfy

$$c = \frac{1}{\sqrt{\pi}} \frac{e^{-c^2}}{\operatorname{erf} c}$$

but the value of c is not important in what is to come.

---

[2]Putting ice in place of water, water in place of ice, all else remaining the same, changes the sign of the term which accounts for the latent heat.

Now, let $t_0 > 0$ be assigned; then, u_0 and X_0 turn out to be

$$u_0 = -1 + \frac{\text{erf}\left(\frac{x_0}{\sqrt{4t_0}}\right)}{\text{erf } c}$$

and

$$X_0 = c\sqrt{4t_0}$$

whereupon our two formulas for X_1, namely

$$X_1 = -\frac{u_1}{\dfrac{\partial u_0}{\partial x_0}} = -\frac{\dfrac{\partial^2 u_0}{\partial x_0^2}}{\dfrac{\partial u_0}{\partial x_0}}$$

and

$$X_1 = \frac{\partial u_0}{\partial x_0}$$

predict

$$X_1 = \frac{\dfrac{1}{\sqrt{\pi}} \dfrac{e^{-c^2}}{\text{erf } c} c \dfrac{1}{\sqrt{t_0}} \dfrac{1}{\sqrt{t_0}}}{\dfrac{1}{\sqrt{\pi}} \dfrac{e^{-c^2}}{\text{erf } c} \dfrac{1}{\sqrt{t_0}}}$$

and

$$X_1 = \frac{1}{\sqrt{\pi}} \frac{e^{-c^2}}{\text{erf } c} \frac{1}{\sqrt{t_0}}$$

That these are the same, illustrates that the two formulas for X_1 are consistent. Again, this consistency is thought to reside in the fact that u_0 satisfies the nonlinear equations and is the current value of a function, u, that has been doing so for some time.

Still, why did two formulas turn up? The reason seems to be this: To estimate u to order 1 via

$$u(x_0, \ t_0 + \epsilon) = u_0(x_0) + \epsilon u_1(x_0)$$

by carrying points of the present domain straight back to the reference domain requires only that u_1 be determined. Surprisingly, this can be obtained at order 0, i.e., before order 1, via

$$u_1 = \frac{d^2 u_0}{dx_0^2}$$

The calculation may then be said to be getting ahead of itself when ϵ is a time difference, thereby requiring fewer equations for its completion.

Another way to think about this is to notice that the phase-equilibrium equation, namely

$$u(X, t) = 0$$

can be differentiated to produce

$$\frac{\partial u}{\partial x}\frac{dX}{dt} + \frac{\partial u}{\partial t} = 0$$

whereupon there obtains

$$\frac{dX}{dt} = -\frac{\dfrac{\partial u}{\partial t}}{\dfrac{\partial u}{\partial x}} = -\frac{\dfrac{\partial^2 u}{\partial x^2}}{\dfrac{\partial u}{\partial x}}$$

and this is just the formula for X_1 that comes out of the heat equation across the interface.

A.4   Laminar Flow Through an Off-centered Annulus

To introduce our fourth example, let a rod of radius R_0 be placed inside a pipe of radius κR_0, $\kappa > 1$. Let a fixed pressure gradient, denoted $\dfrac{dp}{dz}$, be imposed on the fluid lying between the rod and the pipe, causing it to flow at a volumetric rate Q. Our job is to find out by how much Q differs from its base value Q_0, in case the axis of the rod is displaced a small distance, parallel to itself, from its on-centered position in the reference configuration. This is not a problem where the domain must be discovered. In fact, the domain is specified in advance, but it is not a symmetric domain and this leads us to a domain perturbation, just as surely as do the problems presented earlier in this appendix and in the essays.

The important geometric variables are indicated in Figure A.6 which illustrates a cross section perpendicular to the z direction, the direction in which the fluid is flowing. The axis of the pipe lies at $x = 0$, $y = 0$, while the axis of the rod is displaced to $x = \epsilon$, $y = 0$.

The pipe presents no problem; it is the rod whose surface is displaced. Let this surface be denoted

$$r = R(\theta, \epsilon)$$

Then, if X and Y denote the Cartesian coordinates of the point (R, θ) on the surface of the rod, there obtains

$$[X - \epsilon]^2 + Y^2 = R_0^2$$

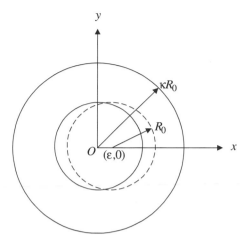

Figure A.6. *The Cross Section of an Off-centered Annulus.*

whereupon R, as a function of θ and ϵ, must satisfy

$$R^2 - \epsilon 2R \cos\theta = R_0^2 - \epsilon^2$$

This fourth problem differs in an important way from the three earlier problems. Here, the present configuration is known in advance and so too the reference configuration which we can take to be the on-centered configuration. Due to this, the mapping of the displaced surface can be determined at the outset. It is not part of the solution. Hence, by writing the mapping of the surface of the rod in the form

$$\theta = \theta_0$$

and

$$R = R(\theta, \epsilon) = R_0 + \epsilon R_1(\theta_0) + \frac{1}{2}\epsilon^2 R_2(\theta_0) + \cdots$$

there obtains

$$R_1 = \frac{dR}{d\epsilon}(\epsilon = 0) = \cos\theta_0$$

and

$$R_2 = \frac{d^2 R}{d\epsilon^2}(\epsilon = 0) = -\frac{\sin^2\theta_0}{R_0}$$

Now, the volumetric flow rate is determined by the speed of the fluid via the formula

$$Q = \int_0^{2\pi} \int_{R(\theta,\,\epsilon)}^{\kappa R_0} v_z(r,\,\theta,\,\epsilon) r\, dr\, d\theta$$

where the integral is over the cross section of the displaced annulus. Carrying out the integration over the cross section of the reference annulus turns this formula into

$$Q = \int_0^{2\pi} \int_{R_0}^{\kappa R_0} v_z r \, \frac{\partial(r, \theta)}{\partial(r_0, \theta_0)} \, dr_0 \, d\theta_0$$

whereupon substituting

$$v_z = v_{z_0}(r_0) + \epsilon \left[v_{z_1}(r_0, \theta_0) + r_1(r_0, \theta_0) \frac{dv_{z_0}}{dr_0}(r_0) \right] + \cdots$$

$$r = r_0 + \epsilon r_1(r_0, \theta_0) + \cdots$$

and

$$\theta = \theta_0$$

leads to

$$Q = Q_0 + \epsilon Q_1 + \frac{1}{2} \epsilon^2 Q_2 + \cdots$$

The readers can work out the formulas for Q_1 and Q_2 and thereby assure themselves that only the mapping of the surface of the rod appears in these formulas, and the surface of the rod is specified. The second endnote to the second essay will help, but it does not go far enough.

Our aim is to carry the calculation of Q to second order in ϵ and this requires v_z to second order in ϵ. The problem to be solved for $v_z(r, \theta)$ on the present domain is

$$\frac{1}{\mu} \frac{dp}{dz} = \nabla^2 v_z, \quad R(\theta, \, \epsilon) < r < \kappa R_0, \quad 0 < \theta < 2\pi$$

where v_z must vanish at the surface of the rod [i.e., at $r = R(\theta, \, \epsilon)$, $0 < \theta < 2\pi$] and at the wall of the pipe [i.e., at $r = \kappa R_0$, $0 < \theta < 2\pi$].

The base problem on the reference domain is then

$$\frac{1}{\mu} \frac{dp}{dz} = \frac{d^2 v_{z_0}}{dr_0^2} + \frac{1}{r_0} \frac{dv_{z_0}}{dr_0}, \quad R_0 < r_0 < \kappa R_0$$

where v_{z_0} must vanish at $r = R_0$ and at $r = \kappa R_0$. Its solution is

$$v_{z_0} = \frac{1}{2\mu} \frac{dp}{dz} \left[\frac{1}{2} r_0^2 - \frac{1}{2} R_0^2 \frac{\kappa^2 - 1}{\ln \kappa} \ln \frac{r_0}{R_0} - \frac{1}{2} R_0^2 \right]$$

and this can be used to find Q_0. To find v_{z_1} and v_{z_2} and, hence, Q_1 and Q_2, two derivatives of v_{z_0} will be needed. They are given by

$$\frac{dv_{z_0}}{dr_0}(r_0 = R_0) = \frac{1}{2\mu} \frac{dp}{dz} R_0 \left[1 - \frac{1}{2} \frac{\kappa^2 - 1}{\ln \kappa} \right]$$

and

$$\frac{d^2 v_{z_0}}{dr_0^2}(r_0 = R_0) = \frac{1}{2\mu} \frac{dp}{dz} \left[1 + \frac{1}{2} \frac{\kappa^2 - 1}{\ln \kappa} \right]$$

The problem to be solved for v_{z_1}, on the reference domain, is

$$0 = \frac{\partial^2 v_{z_1}}{\partial r_0^2} + \frac{1}{r_0}\frac{\partial v_{z_1}}{\partial r_0} + \frac{1}{r_0^2}\frac{\partial^2 v_{z_1}}{\partial \theta_0^2}, \quad R_0 < r_0 < \kappa R_0, \ 0 < \theta_0 < 2\pi \quad \text{(A.14)}$$

where

$$v_{z_1}(r_0 = \kappa R_0) = 0 \quad \text{(A.15)}$$

and

$$v_{z_1}(r_0 = R_0) = -R_1 \frac{dv_{z_0}}{dr_0}(r_0 = R_0) \quad \text{(A.16)}$$

while the problem for v_{z_2}, again on the reference domain, is

$$0 = \frac{\partial^2 v_{z_2}}{\partial r_0^2} + \frac{1}{r_0}\frac{\partial v_{z_2}}{\partial r_0} + \frac{1}{r_0^2}\frac{\partial^2 v_{z_2}}{\partial \theta_0^2}, \quad R_0 < r_0 < \kappa R_0, \ 0 < \theta_0 < 2\pi \quad \text{(A.17)}$$

where

$$v_{z_2}(r_0 = \kappa R_0) = 0 \quad \text{(A.18)}$$

and

$$v_{z_2}(r_0 = R_0) = -2R_1 \frac{\partial v_{z_1}}{\partial r_0}(r_0 = R_0) - R_1^2 \frac{d^2 v_{z_0}}{dr_0^2}(r_0 = R_0)$$

$$- R_2 \frac{dv_{z_0}}{dr_0}(r_0 = R_0) \quad \text{(A.19)}$$

Equations (A.16) and (A.19), satisfied by v_{z_1} and v_{z_2} at $r_0 = R_0$, are determined by the expansion of v_z at $r = R(\theta, \ \epsilon)$, namely by

$$v_z(R, \ \theta, \ \epsilon) = v_{z_0}(R_0, \ \theta_0) + \epsilon\left[v_{z_1}(R_0, \ \theta_0) + R_1(\theta_0)\frac{dv_{z_0}}{dr_0}(R_0)\right]$$

$$+ \frac{1}{2}\epsilon^2\left[v_{z_2}(R_0, \ \theta_0) + 2R_1(\theta_0)\frac{\partial v_{z_1}}{\partial r_0}(R_0, \ \theta_0)\right.$$

$$\left. + R_1^2(\theta_0)\frac{d^2 v_{z_0}}{dr_0^2}(R_0) + R_2(\theta_0)\frac{dv_{z_0}}{dr_0}(R_0)\right] + \cdots$$

and by the fact that $v_z(R, \ \theta, \ \epsilon)$ must vanish for all values of ϵ.
The solution to the v_{z_1} problem can be written

$$v_{z_1} = \left[A_1 r_0 + \frac{B_1}{r_0}\right]\cos\theta_0$$

where A_1 and B_1 must satisfy

$$A_1 \kappa R_0 + \frac{B_1}{\kappa R_0} = 0$$

and

$$A_1 R_0 + \frac{B_1}{R_0} = -\frac{dv_{z0}}{dr_0}(r_0 = R_0)$$

whence v_{z_1} is given by

$$v_{z_1} = -\frac{dv_{z0}}{dr_0}(r_0 = R_0)\,\frac{r_0 - \kappa^2 R_0^2/r_0}{R_0[1 - \kappa^2]}\cos\theta_0$$

and by this, there obtains

$$\frac{\partial v_{z_1}}{\partial r_0}(r_0 = R_0) = \frac{dv_{z0}}{dr_0}(r_0 = R_0)\frac{1}{R_0}\frac{\kappa^2+1}{\kappa^2-1}\cos\theta_0$$

which is needed in the v_{z_2} problem.

To solve the v_{z_2} problem, expand the right-hand side of equation (A.19) in terms of the functions 1, $\cos\theta_0$, $\cos 2\theta_0$, etc. All the information required to do this is available and there obtains

$$v_{z_2}(r_0 = R_0)$$

$$= \frac{1}{2\mu}\frac{dp}{dz}\left[\left[\frac{1+\kappa^2}{1-\kappa^2}+\frac{1}{\ln\kappa}\right]+\left[\frac{2\kappa^2}{1-\kappa^2}+\frac{1}{2}\frac{1+\kappa^2}{\ln\kappa}\right]\cos 2\theta_0\right] \quad (A.20)$$

In view of this, the solution to the v_{z_2} problem can be written

$$v_{z_2} = A_0 + B_0\ln\frac{r_0}{R_0} + \left[A_2 r_0^2 + \frac{B_2}{r_0^2}\right]\cos 2\theta_0$$

where the constants A_0, B_0, A_2 and B_2 must satisfy

$$A_0 + B_0\ln\kappa = 0 \quad (A.21)$$

$$A_0 = \frac{1}{2\mu}\frac{dp}{dz}\left[\frac{1+\kappa^2}{1-\kappa^2}+\frac{1}{\ln\kappa}\right] \quad (A.22)$$

$$A_2\kappa^2 R_0^2 + \frac{B_2}{\kappa^2 R_0^2} = 0 \quad (A.23)$$

and

$$A_2 R_0^2 + \frac{B_2}{R_0^2} = \frac{1}{2\mu}\frac{dp}{dz}\left[\frac{2\kappa^2}{1-\kappa^2}+\frac{1}{2}\frac{1+\kappa^2}{\ln\kappa}\right] \quad (A.24)$$

where equations (A.21) and (A.23) come from equation (A.18), and where equations (A.22) and (A.24) come from equation (A.19), rewritten as equation (A.20).

Now, A_2 and B_2 are not required in order to obtain Q_2. Our solution, then, dropping the terms in A_2 and B_2, is

$$v_{z_2} = \frac{1}{2\mu}\frac{dp}{dz}\left[\frac{1+\kappa^2}{1-\kappa^2}+\frac{1}{\ln\kappa}\right]\left[1-\frac{\ln(r_0/R_0)}{\ln\kappa}\right]$$

Again, this is not all of v_{z_2}.

Let us turn now to the estimation of Q to second order in ϵ. The formulas for Q_1 and Q_2 will be needed. Hopefully, the reader has derived these formulas and found them to be[3]

$$Q_1 = \int_0^{2\pi} \int_{R_0}^{\kappa R_0} v_{z_1} r_0 \, dr_0 \, d\theta_0 - \int_0^{2\pi} R_1 v_{z_0}(r_0 = R_0) R_0 \, d\theta_0$$

and

$$Q_2 = \int_0^{2\pi} \int_{R_0}^{\kappa R_0} v_{z_2} r_0 \, dr_0 \, d\theta_0 - \int_0^{2\pi} R_2 v_{z_0}(r_0 = R_0) R_0 \, d\theta_0$$

$$-2 \int_0^{2\pi} R_1 v_{z_1}(r_0 = R_0) R_0 \, d\theta_0 - \int_0^{2\pi} R_1^2 v_{z_0}(r_0 = R_0) \, d\theta_0$$

$$-\int_0^{2\pi} R_1^2 \frac{dv_{z_0}}{dr_0}(r_0 = R_0) R_0 \, d\theta_0$$

These formulas include all of the terms due to the displacement of the surface of the rod, even those that turn out to be zero in this problem. Then, as v_{z_1} and R_1 are multiples of $\cos\theta_0$ and $v_{z_0}(r_0 = R_0)$ is zero, Q_1 turns out to be zero and Q_2 is simply given by

$$\int_0^{2\pi} \int_{R_0}^{\kappa R_0} v_{z_2} r_0 \, dr_0 \, d\theta_0 - \int_0^{2\pi} 2R_1 v_{z_1}(r_0 = R_0) R_0 \, d\theta_0$$

$$-\int_0^{2\pi} R_1^2 \frac{dv_{z_0}}{dr_0}(r_0 = R_0) R_0 \, d\theta_0$$

whereupon Q_2 turns out to be

$$\left[\frac{1}{2\mu}\frac{dp}{dz}\pi R_0^2\right]\left[1 - \frac{1}{2}\frac{\kappa^2 - 1}{\ln\kappa}\right]\left[\frac{2\kappa^2}{\kappa^2 - 1} - \frac{1}{\ln\kappa}\right]$$

The first and second factors are negative for all $\kappa > 1$ and the last factor is positive, whence Q_2 is positive.

---

[3]Should the rule have been used, Q_1 and Q_2 would have been

$$Q_1 = \int_0^{2\pi} \int_{R_0}^{\kappa R_0} v_{z_1} r_0 \, dr_0 \, d\theta_0$$

and

$$Q_2 = \int_0^{2\pi} \int_{R_0}^{\kappa R_0} v_{z_2} r_0 \, dr_0 \, d\theta_0$$

but these formulas account only in part for the displacement of the rod. Of course, nobody would use the rule to obtain Q_1 and Q_2!

Two results [viz., $Q_1 = 0$ and $Q_2 > 0$], could have been anticipated and the readers might wish to satisfy themselves about this. Then, the reader might redo this calculation setting dp/dz to zero and requiring, instead, $v_z(r = R) = V$, where V is specified and held fixed. Again, Q_1 must be zero.

Three additional calculations can be suggested where some of the above may be helpful.

In the first, a fluid lies between two circular cylinders, spinning at certain specified angular speeds. Take the case where the inner cylinder is spinning while the outer cylinder is held fixed, and where it is required to determine the torque required to spin the inner cylinder as a function of its angular speed. Let the axes of the cylinders be parallel, but let the axis of the inner cylinder be displaced slightly from its concentric configuration.

In the second calculation, let a fluid be flowing under an assigned pressure gradient in a pipe of elliptical cross section. It is required to find the volumetric flow rate in case the cross section is nearly a circle.

The third is a heat conduction calculation for which many of the above formulas are useful. Let a hot rod lose heat by conduction to the cold wall of a pipe containing the rod. This can be worked out more or less along the lines of our example, with the result that by moving the rod off-center, the rate of heat loss is increased, but not at first order. The reader should do the calculation the hard way, by integrating the heat flow over the surface of the displaced rod. Letting Q denote the heat loss per unit length divided by the product of the thermal conductivity and the temperature difference, the reader should find

$$Q_1 = 0$$

and

$$Q_2 = \frac{1}{R_0^2} \frac{\pi}{[\kappa^2 - 1]\ln\kappa} \left[-[\kappa^2 + 1]\frac{1 - \ln\kappa}{\ln\kappa} + 6\kappa^2 \right]$$

Appendix B
The Curvature of Surfaces

B.1 Background

Let x, y and z denote the Cartesian coordinates of a point P whose position vector is denoted by $\vec{r}$. Then, let x, y and z be smooth functions of two variables u^1 and u^2. Under ordinary conditions, such a mapping of a region of the u^1, u^2- plane, given by

$$\vec{r}(u^1, \ u^2) = x(u^1, \ u^2)\vec{i} + y(u^1, \ u^2)\vec{j} + z(u^1, \ u^2)\vec{k}$$

defines a surface, denoted S, in $x, y, z-$ space.

The variables u^1 and u^2 are called surface coordinates on S. Holding one coordinate fixed and letting the other run through its range of variation defines a coordinate curve on the surface. The vectors $\vec{r}_1 = \dfrac{\partial \vec{r}}{\partial u^1}$ and $\vec{r}_2 = \dfrac{\partial \vec{r}}{\partial u^2}$ lie tangent to these coordinate curves and, at each point P of the surface, $\vec{r}_1$ and $\vec{r}_2$ must be independent. This requirement on the mapping is satisfied if and only if its Jacobian matrix, namely

$$\begin{pmatrix} \dfrac{\partial x}{\partial u^1} & \dfrac{\partial y}{\partial u^1} & \dfrac{\partial z}{\partial u^1} \\[3mm] \dfrac{\partial x}{\partial u^2} & \dfrac{\partial y}{\partial u^2} & \dfrac{\partial z}{\partial u^2} \end{pmatrix}$$

is everywhere of full rank (i.e., if and only if $\vec{r}_1 \times \vec{r}_2$ is never zero).

Then, $\vec{r}_1$ and $\vec{r}_2$ span the tangent plane to S at P and $\vec{r}_1 \times \vec{r}_2$ is perpendicular to the surface there. This is illustrated in Figure B.1. The vectors

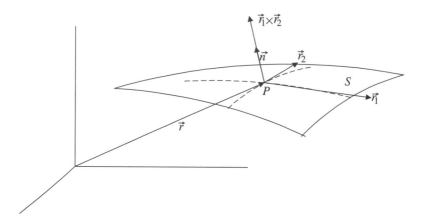

Figure B.1. *The Vectors $\vec{r}_1$, $\vec{r}_2$ and $\vec{n}$ at a Point of a Surface*

$\vec{r}_1$, $\vec{r}_2$ and $\vec{n}$, where

$$\vec{n} = \frac{\vec{r}_1 \times \vec{r}_2}{|\vec{r}_1 \times \vec{r}_2|}$$

form a right-handed set of independent vectors at every point P of S.
Introduce the notation

$$g_{\alpha\beta} = \vec{r}_\alpha \cdot \vec{r}_\beta, \quad \alpha,\ \beta = 1,\ 2$$

and

$$g = g_{11}\, g_{22} - g_{12}^2 = [\vec{r}_1 \cdot \vec{r}_1][\vec{r}_2 \cdot \vec{r}_2] - [\vec{r}_1 \cdot \vec{r}_2]^2$$

whence there obtains

$$|\vec{r}_1 \times \vec{r}_2|^2 = [\vec{r}_1 \times \vec{r}_2] \cdot [\vec{r}_1 \times \vec{r}_2] = g > 0$$

and

$$[\vec{r}_1 \times \vec{r}_2] \cdot \vec{n} = |\vec{r}_1 \times \vec{r}_2|\,\vec{n} \cdot \vec{n} = \sqrt{g}$$

The coefficients $g_{\alpha\beta}$ determine lengths and areas on the surface. Indeed, the square of the length of a displacement $d\vec{r}$, where

$$d\vec{r} = \vec{r}_1\, du^1 + \vec{r}_2\, du^2$$

is given by

$$ds^2 = d\vec{r} \cdot d\vec{r} = \sum \sum g_{\alpha\beta}\, du^\alpha\, du^\beta$$

while the vector area of the parallelogram defined by the vectors $\vec{r}_1\, du^1$ and $\vec{r}_2\, du^2$ is given by

$$d\vec{A} = \vec{r}_1 \times \vec{r}_2\, du^1\, du^2 = \vec{n}|\vec{r}_1 \times \vec{r}_2|\, du^1\, du^2 = \vec{n}\sqrt{g}\, du^1\, du^2$$

Let a function be defined at the points of the surface S. Let it be smooth. It may be defined off the surface as well, but that is not important, and it may be scalar-valued or vector-valued, and again that is not important. Denote its values $f(u^1, u^2)$. To determine its directional derivative in the direction of the tangent to a curve C lying on the surface, let the curve be specified in terms of its arc length s by $u^1 = u^1(s)$ and $u^2 = u^2(s)$. The unit tangent to this curve is denoted by $\vec{t}$, where

$$\vec{t} = \frac{d\vec{r}}{ds} = \vec{r}_1 \frac{du^1}{ds} + \vec{r}_2 \frac{du^2}{ds}$$

Let the vectors $\vec{r}^{\,1}$ and $\vec{r}^{\,2}$, lying in the tangent plane at P, be defined by

$$\vec{r}^{\,1} = \frac{\vec{r}_2 \times \vec{n}}{\vec{r}_1 \times \vec{r}_2 \cdot \vec{n}} = \frac{\vec{r}_2 \times \vec{n}}{\sqrt{g}}$$

and

$$\vec{r}^{\,2} = \frac{\vec{n} \times \vec{r}_1}{\vec{r}_1 \times \vec{r}_2 \cdot \vec{n}} = \frac{\vec{n} \times \vec{r}_1}{\sqrt{g}}$$

Then, $\{\vec{r}_1, \vec{r}_2, \vec{n}\}$ and $\{\vec{r}^{\,1}, \vec{r}^{\,2}, \vec{n}\}$ are biorthogonal sets of vectors and $\vec{r}^{\,1}$ and $\vec{r}^{\,2}$ can be used to write $\dfrac{du^1}{ds}$ and $\dfrac{du^2}{ds}$ in terms of $\vec{t}$ via

$$\vec{r}^{\,1} \cdot \vec{t} = \frac{du^1}{ds}$$

and

$$\vec{r}^{\,2} \cdot \vec{t} = \frac{du^2}{ds}$$

Now, the directional derivative of f along the curve C, denoted df/ds, can be determined by using the chain rule to carry out the differentiation. The result is

$$\frac{df}{ds} = \frac{\partial f}{\partial u^1} \frac{du^1}{ds} + \frac{\partial f}{\partial u^2} \frac{du^2}{ds} = f_1 \frac{du^1}{ds} + f_2 \frac{du^2}{ds}$$

and using $du^1/ds = \vec{t} \cdot \vec{r}^{\,1}$, etc., it becomes

$$\frac{df}{ds} = \vec{t} \cdot [\vec{r}^{\,1} f_1 + \vec{r}^{\,2} f_2]$$

The second factor on the right-hand side of the formula for df/ds (viz., $\vec{r}^{\,1} f_1 + \vec{r}^{\,2} f_2$), is defined at each point of S, and at a point P of S, it determines the directional derivative of f in any direction tangent to the surface at P. Denote it by $\nabla_s f$ and write

$$\frac{df}{ds} = \vec{t} \cdot \nabla_s f$$

where

$$\nabla_s f = \vec{r}^{\,1} \frac{\partial f}{\partial u^1} + \vec{r}^{\,2} \frac{\partial f}{\partial u^2}$$

Then, the directional derivative of f along the tangent to a curve C at P depends on the curve only via $\vec{t}$.

If, at a point P of the surface, df/ds_1 and df/ds_2 denote the directional derivatives of f in the directions $\vec{t}_1$ and $\vec{t}_2$, then

$$\nabla_s f = \vec{a}^{\,1} \frac{df}{ds_1} + \vec{a}^{\,2} \frac{df}{ds_2}$$

where $\vec{t}_1$, $\vec{t}_2$, $\vec{n}$ and $\vec{a}^{\,1}$, $\vec{a}^{\,2}$, $\vec{n}$ are biorthogonal sets of vectors. This frees $\nabla_s f$ of the surface coordinates u^1 and u^2.

If f is vector-valued, the scalar and vector invariants of $\nabla_s f$ are given by

$$\nabla_s \cdot f = \vec{r}^{\,1} \cdot f_1 + \vec{r}^{\,2} \cdot f_2$$

and

$$\nabla_s \times f = \vec{r}^{\,1} \times f_1 + \vec{r}^{\,2} \times f_2$$

The unit normal $\vec{n}$ is a vector-valued function defined over the surface and its surface gradient is given by

$$\nabla_s \vec{n} = \vec{r}^{\,1} \vec{n}_1 + \vec{r}^{\,2} \vec{n}_2$$

This tensor is important. Its vector invariant is zero and, hence, it is symmetric. To determine that this is so, write

$$\nabla_s \times \vec{n} = \vec{r}^{\,1} \times \vec{n}_1 + \vec{r}^{\,2} \times \vec{n}_2 = \frac{[\vec{r}_2 \times \vec{n}] \times \vec{n}_1}{\sqrt{g}} + \frac{[\vec{n} \times \vec{r}_1] \times \vec{n}_2}{\sqrt{g}}$$

and work out the right-hand side using $[\vec{a} \times \vec{b}] \times \vec{c} = \vec{b}[\vec{c} \cdot \vec{a}] - \vec{a}[\vec{b} \cdot \vec{c}]$. Then, use

$$\vec{n} \cdot \vec{n} = 1, \quad \vec{r}_1 \cdot \vec{n} = 0, \quad \vec{r}_2 \cdot \vec{n} = 0$$

to determine that

$$\vec{n} \cdot \vec{n}_1 = 0 = \vec{n} \cdot \vec{n}_2$$

and

$$\vec{r}_1 \cdot \vec{n}_2 = -\vec{n} \cdot \vec{r}_{12} = \vec{r}_2 \cdot \vec{n}_1$$

and, hence, that

$$\nabla_s \times \vec{n} = \vec{0}$$

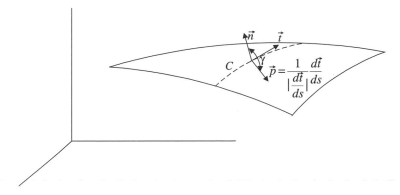

Figure B.2. *The Angle Between the Principle Normal to C and the Normal to S*

B.2 Curvature

Again, let C be a curve lying on the surface S. Let γ denote the angle between its principal normal, $\vec{p}$, and the normal to the surface, $\vec{n}$. This is illustrated in Figure B.2, where

$$\cos\gamma = \vec{p}\cdot\vec{n}$$

and where $\vec{p}$ is defined in terms of $\vec{t}$ by

$$\kappa\vec{p} = \frac{d\vec{t}}{ds}$$

This introduces κ via $\kappa = |\dfrac{d\vec{t}}{ds}|$, where κ, the rate at which the tangent is turning, is called the curvature of C.

On substituting for $\vec{p}$, the formula for $\cos\gamma$ can be written

$$\kappa\cos\gamma = \kappa\vec{p}\cdot\vec{n} = \frac{d\vec{t}}{ds}\cdot\vec{n}$$

and then, using $\vec{t}\cdot\vec{n} = 0$, it can be written in terms of $\nabla_s\vec{n}$ as

$$\kappa\cos\gamma = -\vec{t}\cdot\frac{d\vec{n}}{ds} = -\vec{t}\cdot[t\cdot\nabla_s\vec{n}] = -\vec{t}\,\vec{t}:\nabla_s\vec{n}$$

The tensor $\nabla_s\vec{n}$ is called the curvature tensor. It is defined at each point P of S. The right-hand side of this formula depends on the point P of the surface S via $\nabla_s\vec{n}$ and on the direction of the curve C lying on S, and passing through P, via $\vec{t}$.

Now, as $\cos\gamma$ is just $\vec{p}\cdot\vec{n}$, all curves lying on S passing through a point P of S and having the same tangent and principal normal must have the same curvature. This curvature then must be that of the plane curve passing through P defined by the intersection of the $\vec{t},\vec{p}-$ plane at P and the surface S. This tells us that the curvature of a surface can be described in terms of the curvature of its plane curves. The product $\kappa \cos \gamma$ is determined by the point P of the surface and the tangent direction $\vec{t}$ at that point. In terms of this product, the curvature of any plane curve passing through P in the direction $\vec{t}$ is determined only by $\cos \gamma$. The product $\kappa \cos \gamma$ is called the normal curvature and it is denoted κ_n. It is $\pm$ the curvature of the normal section of S at P in the direction $\vec{t}$. In particular, the curvature of the plane curve made by the intersection of the surface S and its normal plane at P in the direction $\vec{t}$ (i.e., the $\vec{n}, \vec{t}-$ plane), is $\pm \kappa_n$ and it depends only on whether $\vec{p} = \vec{n}$ or $\vec{p} = -\vec{n}$.

The curvature tensor, $\nabla_s \vec{n}$, at a point P then tells us κ_n at that point as it depends on $\vec{t}$, namely

$$\kappa_n = -\vec{t}\,\vec{t} : \nabla_s \vec{n}$$

where, again, κ_n is $\pm$ the curvature of the normal section at P in the direction $\vec{t}$. If $\vec{n}$ is replaced by $-\vec{n}$ over the surface, then κ_n is replaced by $-\kappa_n$.

To produce a formula by which κ_n can be determined at a point P, as it depends on the direction of $\vec{t}$ there, write

$$\kappa_n = \kappa \cos \gamma = \frac{d\vec{t}}{ds} \cdot \vec{n} = \frac{d}{ds}\frac{d\vec{r}}{ds} \cdot \vec{n}$$

$$= \frac{d}{ds} \sum \vec{r}_\alpha \frac{du^\alpha}{ds} \cdot \vec{n} = \sum \sum \vec{r}_{\alpha\beta} \frac{du^\alpha}{ds}\frac{du^\beta}{ds} \cdot \vec{n} + \sum \vec{r}_\alpha \frac{d^2 u^\alpha}{ds^2} \cdot \vec{n}$$

$$= \sum \sum \vec{r}_{\alpha\beta} \cdot \vec{n} \frac{du^\alpha}{ds}\frac{du^\beta}{ds}$$

Then, defining $b_{\alpha\beta}$ by

$$b_{\alpha\beta} = \vec{r}_{\alpha\beta} \cdot \vec{n} = b_{\beta\alpha}$$

our formula for κ_n can be written

$$\kappa_n = \sum \sum b_{\alpha\beta} \frac{du^\alpha}{ds}\frac{du^\beta}{ds}$$

where

$$\frac{du^\alpha}{ds} = \vec{r}^\alpha \cdot \vec{t}$$

and

$$b_{\alpha\beta} = \vec{r}_{\alpha\beta} \cdot \vec{n} = -\vec{r}_\alpha \cdot \vec{n}_\beta$$

$$= -\vec{r}_\alpha \vec{r}_\beta : \nabla_s \vec{n}$$

The normal curvature at a point P of S, in terms of the tangent direction, (viz., du^1, du^2) is then

$$\kappa_n = \frac{\sum \sum b_{\alpha\beta} du^\alpha du^\beta}{\sum \sum g_{\alpha\beta} du^\alpha du^\beta}$$

and the sign of κ_n is determined by the sign of $\sum b_{\alpha\beta} du^\alpha du^\beta$. If b, where $b = b_{11}b_{22} - b_{12}^2$, is positive, this sign is the same for all tangent directions. If b is equal to zero, there is one direction in which $\kappa_n = 0$, otherwise it is of one sign. If b is negative, there are two directions in which $\kappa_n = 0$ and κ_n takes both positive and negative values.

To discover the basic facts about the dependence of κ_n on $\vec{t}$, write

$$\kappa_n = -\vec{t}\,\vec{t} : \nabla_s \vec{n}$$

and let $\vec{t}$ make an angle θ with a fixed reference line in the tangent plane at P. Then, the dependence of κ_n on $\vec{t}$ is the dependence of κ_n on θ. If an arbitrary tangent $\vec{t}$ corresponds to the angle θ, let the tangent $\vec{t}'$ correspond to the angle $\theta + \frac{1}{2}\pi$ and observe that

$$\vec{t} = -\vec{n} \times \vec{t}'$$

and

$$\vec{t}' = \vec{n} \times \vec{t}$$

and, hence, that

$$\frac{d\vec{t}}{d\theta} = \vec{n} \times \vec{t} = \vec{t}'$$

and

$$\frac{d\vec{t}'}{d\theta} = \vec{n} \times \vec{t}' = -t$$

These results tell us that $[\vec{t}\,\vec{t} + \vec{t}'\vec{t}']$ does not depend on θ, namely that

$$\frac{d}{d\theta}[\vec{t}\,\vec{t} + \vec{t}'\vec{t}'] = \vec{0}$$

Now, $[\vec{t}\,\vec{t} + \vec{t}'\vec{t}']$ determines the sum of the normal curvatures in the two perpendicular directions $\vec{t}$ and $\vec{t}'$. This sum is given by

$$-[\vec{t}\,\vec{t} + \vec{t}'\vec{t}'] : \nabla_s \vec{n}$$

and, as it does not depend on θ, the sum of the normal curvatures in any two perpendicular directions must be fixed at each point of a surface. This defines what is called twice the mean curvature of the surface at that point. Denote the mean curvature by H; then,

$$2H = \kappa_n(\theta) + \kappa_n(\theta + \frac{1}{2}\pi)$$

and this is independent of θ.

The greatest, or the least, value of κ_n corresponds to a direction $\vec{t}$ such that κ_n, as it depends on θ, is stationary. This requirement is

$$\frac{d\kappa_n}{d\theta} = 0 = -[\vec{t}'\,\vec{t} + \vec{t}\,\vec{t}'] : \nabla_s \vec{n}$$

and, as $\nabla_s \vec{n}$ is symmetric, it reduces to

$$\vec{t}\,\vec{t}' : \nabla_s \vec{n} = 0$$

If κ_n is greatest or least in the direction $\vec{t}$, it must be least or greatest in the direction $\vec{t}''$.

Now, in any basis $\{\vec{t}, \vec{t}''\}$, $\nabla_s \vec{n}$ can be expanded as

$$\nabla_s \vec{n} = [\vec{t}\,\vec{t} : \nabla_s \vec{n}]\vec{t}\,\vec{t} + [\vec{t}'\,\vec{t} : \nabla_s \vec{n}]\vec{t}\,\vec{t}' + [\vec{t}\,\vec{t}' : \nabla_s \vec{n}]\vec{t}'\,\vec{t} + [\vec{t}'\,\vec{t}' : \nabla_s \vec{n}]\vec{t}'\,\vec{t}'$$

Then, if the direction $\vec{t}$ is such that κ_n is stationary as a function of θ and if $\vec{t}''$ is taken to be $\vec{n} \times \vec{t}$, this simplifies to

$$-\nabla_s \vec{n} = \kappa_n(\vec{t})\vec{t}\,\vec{t} + \kappa_n(\vec{t}')\vec{t}'\,\vec{t}'$$

whence the greatest and the least values of κ_n turn out to be the eigenvalues of $-\nabla_s \vec{n}$. Denote these eigenvalues by κ_1 and κ_2. They can be determined as the roots of

$$\det(-\nabla_s \vec{n} - \kappa I_s) = 0$$

and, hence, of

$$\det(b_{\alpha\beta} - \kappa g_{\alpha\beta}) = 0$$

or of

$$\det(b^\alpha_\beta - \kappa \delta^\alpha_\beta) = 0$$

In view of this, $2H$ can be obtained as the sum of b^1_1 and b^2_2, namely

$$2H = \kappa_1 + \kappa_2 = b^1_1 + b^2_2$$

This is a formula that can be used to calculate the mean curvature of a surface at any point P of the surface, yet sometimes it is simpler to make use of the fact that the mean curvature at a point of a surface is one-half the sum of the normal curvatures in any two perpendicular directions at that point.

B.3   Some Examples

Some mean curvature formulas needed in our essays are worked out in this section as examples of the use of the formula

$$2H = b^1_1 + b^2_2 = g^{11}b_{11} + 2g^{12}b_{12} + g^{22}b_{22}$$

Example 1

Let a surface be given by the formula

$$z = f(x, y)$$

Then, using x and y as surface coordinates (viz., $u^1 = x$, $u^2 = y$) and writing

$$\vec{r} = x\vec{i} + y\vec{j} + f(x,y)\vec{k}$$

the required calculations produce

$$\vec{r}_1 = \vec{i} + f_x\vec{k}, \quad \vec{r}_2 = \vec{j} + f_y\vec{k}$$

$$g_{11} = 1 + f_x^2, \quad g_{12} = f_x f_y, \quad g_{22} = 1 + f_y^2$$

$$g = 1 + f_x^2 + f_y^2$$

$$\vec{n} = \frac{-f_x\vec{i} - f_y\vec{j} + \vec{k}}{\sqrt{g}}$$

$$\vec{r}_{11} = f_{xx}\vec{k}, \quad \vec{r}_{12} = f_{xy}\vec{k}, \quad \vec{r}_{22} = f_{yy}\vec{k}$$

$$b_{11} = \frac{f_{xx}}{\sqrt{g}}, \quad b_{12} = \frac{f_{xy}}{\sqrt{g}}, \quad b_{22} = \frac{f_{yy}}{\sqrt{g}}$$

and

$$g^{11} = \frac{1 + f_y^2}{g}, \quad g^{12} = -\frac{f_x f_y}{g}, \quad g^{22} = \frac{1 + f_x^2}{g}$$

whence

$$2H = \frac{[1 + f_y^2]f_{xx} - 2f_x f_y f_{xy} + [1 + f_x^2]f_{yy}}{g^{\frac{3}{2}}}$$

If f_x, f_y, etc., are all small, the right-hand side of this formula can be approximated by

$$f_{xx} + f_{yy} = \nabla^2 f$$

If f is independent of y, the formula simplifies to

$$\frac{f_{xx}}{[1 + f_x^2]^{3/2}}$$

Example 2

Let a surface be given by the formula

$$r = R + f(\theta, z)$$

where R is fixed. Then, using θ and z as surface coordinates (viz., $u^1 = \theta$, $u^2 = z$) and writing

$$\vec{r} = [R + f(\theta, z)]\,\vec{i}_r(\theta) + z\vec{i}_z$$

the required calculations produce

$$\vec{r}_1 = [R + f]\, \vec{i}_\theta + f_\theta \vec{i}_r, \quad \vec{r}_2 = f_z \vec{i}_r + \vec{i}_z$$

$$g_{11} = [R + f]^2 + f_\theta^2, \quad g_{12} = f_\theta f_z, \quad g_{22} = f_z^2 + 1$$

$$g = [R + f]^2 [1 + f_z^2] + f_\theta^2$$

$$\vec{n} = \frac{[R + f]\, \vec{i}_r - f_\theta \vec{i}_\theta - [R + f] f_z \vec{i}_z}{\sqrt{g}}$$

$$\vec{r}_{11} = [-[R + f] + f_{\theta\theta}]\, \vec{i}_r + 2 f_\theta \vec{i}_\theta, \quad \vec{r}_{12} = f_{z\theta} \vec{i}_r + f_z \vec{i}_\theta, \quad \vec{r}_{22} = f_{zz} \vec{i}_r$$

$$b_{11} = \frac{-[R + f]^2 + [R + f] f_{\theta\theta} - 2 f_\theta^2}{\sqrt{g}}$$

$$b_{12} = \frac{[R + f] f_{z\theta} - f_\theta f_z}{\sqrt{g}}$$

$$b_{22} = \frac{[R + f] f_{zz}}{\sqrt{g}}$$

and

$$g^{11} = \frac{f_z^2 + 1}{g}, \quad g^{12} = -\frac{f_\theta f_z}{g}, \quad g^{22} = \frac{[R + f]^2 + f_\theta^2}{g}$$

whence

$$2H = \left[\frac{1 + f_z^2}{g}\right] \left[\frac{-[R + f]^2 - 2 f_\theta^2 + [R + f] f_{\theta\theta}}{\sqrt{g}}\right]$$

$$+ 2 \left[\frac{-f_\theta f_z}{g}\right] \left[\frac{[R + f] f_{z\theta} - f_\theta f_z}{\sqrt{g}}\right] + \left[\frac{[R + f]^2 + f_\theta^2}{g}\right] \left[\frac{[R + f] f_{zz}}{\sqrt{g}}\right]$$

If f, f_θ, f_z, etc. are all small, some approximations can be introduced. First, write

$$2H = \frac{1}{g^{\frac{3}{2}}} [-R^2 - 2Rf + Rf_{\theta\theta}] + 0 + \frac{1}{g^{\frac{3}{2}}} [R^2 + 2Rf][Rf_{zz}]$$

and

$$g = R^2 + 2Rf = R^2 \left[1 + 2\frac{f}{R}\right]$$

Then, use

$$g^{\frac{3}{2}} = R^3 \left[1 + 2\frac{f}{R}\right]^{\frac{3}{2}}$$

to get

$$2H = \left[-\frac{1}{R} + \frac{f}{R^2} + \frac{f_{\theta\theta}}{R^2} + f_{zz} \right]$$

In this formula, $-\dfrac{1}{R} + \dfrac{f}{R^2}$ is an approximation to $-\dfrac{1}{R+f}$, which is the curvature of a circle of radius $R + f$. The remaining terms account for the circumferential and the longitudinal ripples on the surface.

Example 3

Let an axisymmetric surface be given by the formula

$$z = f(r)$$

Then, using r and θ as surface coordinates (viz., $u^1 = r$, $u^2 = \theta$) and writing

$$\vec{r} = r\vec{i}_r(\theta) + f(r)\vec{i}_z$$

the required calculations lead to

$$\vec{r}_1 = \vec{i}_r + f'\vec{i}_z, \quad \vec{r}_2 = r\vec{i}_\theta$$

$$g_{11} = 1 + [f']^2, \quad g_{12} = 0, \quad g_{22} = r^2$$

$$g = r^2 \left[1 + [f']^2 \right]$$

$$\vec{n} = \frac{r\vec{i}_z - rf'\vec{i}_r}{\sqrt{g}}$$

$$\vec{r}_{11} = f''\vec{i}_z, \quad \vec{r}_{12} = \vec{i}_\theta, \quad \vec{r}_{22} = -r\vec{i}_r$$

$$b_{11} = \frac{rf''}{\sqrt{g}}, \quad b_{12} = 0, \quad b_{22} = \frac{r^2 f'}{\sqrt{g}}$$

and

$$g^{11} = \frac{1}{1 + [f']^2}, \quad g^{12} = 0, \quad g^{22} = \frac{1}{r^2}$$

Here, the surface coordinates are orthogonal and $2H$ can be determined easily by adding the normal curvatures in the r and θ directions, but continuing in the usual way, there obtains

$$2H = b_1^1 + b_2^2 = g^{11}b_{11} + 2g^{12}b_{12} + g^{22}b_{22} = \frac{f'' + \frac{1}{r}f' \left[1 + [f']^2 \right]}{\left[1 + [f']^2 \right]^{\frac{3}{2}}}$$

If f' is small, the right-hand side of this formula can be approximated by

$$f'' + \frac{1}{r}f' = \nabla^2 f$$

Now, let f depend on θ as well as on r; then, write

$$\vec{r} = r\vec{i}_r(\theta) + f(r,\ \theta)\vec{i}_z$$

Again, take $u^1 = r$ and $u^2 = \theta$, whereupon

$$\vec{r}_1 = \vec{i}_r + f_r\vec{i}_z$$

$$\vec{r}_2 = r\vec{i}_\theta + f_\theta\vec{i}_z$$

and

$$\vec{n} = \frac{-rf_r\vec{i}_r - f_\theta\vec{i}_\theta + r\vec{i}_z}{\sqrt{g}}$$

By these formulas, there obtain

$$g_{11} = 1 + f_r^2$$

$$g_{12} = f_r f_\theta$$

$$g_{22} = r^2 + f_\theta^2$$

and

$$g = r^2 + r^2 f_r^2 + f_\theta^2$$

as well as

$$\vec{r}_{11} = f_{rr}\vec{i}_z$$

$$\vec{r}_{12} = \vec{i}_\theta + f_{r\theta}\vec{i}_z$$

and

$$\vec{r}_{22} = -r\vec{i}_r + f_{\theta\theta}\vec{i}_z$$

These lead to

$$b_{11} = \frac{rf_{rr}}{\sqrt{g}}$$

$$b_{12} = \frac{-f_\theta + rf_{r\theta}}{\sqrt{g}}$$

and

$$b_{22} = \frac{r^2 f_r + rf_{\theta\theta}}{\sqrt{g}}$$

which along with

$$g^{11} = \frac{r^2 f_\theta^2}{g}$$

$$g^{12} = -\frac{f_r f_\theta}{g}$$

and

$$g^{22} = \frac{1 + f_r^2}{g}$$

produce the result

$$2H = \frac{[r^2 + f_\theta^2][r f_{rr}] - 2 f_r f_\theta[-f_\theta + r f_{r\theta}] + [1 + f_r^2][r^2 f_r + r f_{\theta\theta}]}{g^{3/2}}$$

where

$$g^{3/2} = r^3 \left[1 + \frac{1}{r^2} f_\theta^2 + f_r^2 \right]^{3/2}$$

This simplifies to the earlier result when f is independent of θ. If f_r, f_θ, etc. are all small, the right-hand side can be approximated by

$$f_{rr} + \frac{1}{r} f_r + \frac{1}{r^2} f_{\theta\theta} = \nabla^2 f$$

Appendix C
The Normal Speed of a Surface

Let a surface be denoted by

$$f(\vec{r}, \, t) = 0$$

Then, f is positive on one side of $f = 0$, negative on the other, and the normal pointing into the region where f is positive is given by

$$\vec{n} = \frac{\nabla f}{|\nabla f|}$$

Let the surface move a small distance Δs along this normal in time Δt. Then, $f(\vec{r} \pm \Delta s \, \vec{n}, \, t + \Delta t)$ is given by

$$f(\vec{r} \pm \Delta s \, \vec{n}, t + \Delta t) = f(\vec{r}, t) \pm \Delta s \, \vec{n} \cdot \nabla f(\vec{r}, t) + \Delta t \, f_t(\vec{r}, t) + \cdots$$

whence $f(\vec{r} \pm \Delta s \, \vec{n}, \, t + \Delta t) = 0 = f(\vec{r}, \, t)$ requires

$$\pm \Delta s \, \vec{n} \cdot \nabla f + \Delta t \, f_t = 0$$

The normal speed, denoted u, is then given by

$$u = \pm \frac{\Delta s}{\Delta t} = -\frac{f_t}{\vec{n} \cdot \nabla f} = -\frac{f_t}{|\nabla f|}$$

where u is positive at points of the surface $f = 0$ which are moving into the region where $f > 0$. This is illustrated in Figure C.1.

It is often useful to introduce the surface velocity. Denote this by $\vec{u}$ and take $\vec{u}$ to be $u\vec{n}$. Then, we have $\vec{n} \cdot \vec{u} = u$.

The special case where

$$r = R(\theta, \, z, \, t)$$

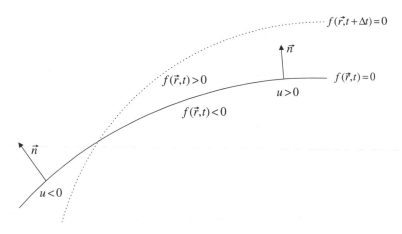

Figure C.1. *The Speed of a Moving Surface*

and, hence, where

$$f(\vec{r},\ t) = r - R(\theta,\ z,\ t)$$

is of interest in the first, third, fourth and fifth essays. There, $f(\vec{r},\ t) = 0$ denotes the surface dividing a liquid in the form of a jet from the surrounding fluid. The normal speed of this surface is given by

$$u = \frac{\dfrac{\partial R}{\partial t}}{\left[1 + \left[\dfrac{1}{R}\dfrac{\partial R}{\partial \theta}\right]^2 + \left[\dfrac{\partial R}{\partial z}\right]^2\right]^{1/2}}$$

The equation $\vec{n} \cdot \vec{v} = u$ is also of interest. It specifies no flow across a moving surface. Using $\vec{n} = \dfrac{\nabla f}{|\nabla f|}$ and $u = \dfrac{-f_t}{|\nabla f|}$, it can be written

$$\vec{v} \cdot \nabla f = -f_t$$

and, hence, if the surface is denoted by $f(x,\ y,\ z,\ t) = 0$,

$$v_x \frac{\partial f}{\partial x} + v_y \frac{\partial f}{\partial y} + v_z \frac{\partial f}{\partial z} = -f_t$$

whence the no-flow surface is seen to be material.

Index